Microorganisms for Sustainability

Volume 55

Series Editor

Naveen Kumar Arora, Environmental Microbiology
School for Environmental Sciences, Babasaheb Bhimrao Ambedkar University
Lucknow, India

Microorganisms have been in existence since the origin of life on Earth and can survive the most extreme habitats or conditions on Earth. Microorganisms are involved in regulating biogeochemical cycles, maintaining plant and animal health, and sustaining the global food chain. Moreover, they play crucial roles in addressing the challenges of climate change and achieving the targets of Sustainable Development Goals (SDGs).

This multidisciplinary book series captures the role of microbes towards building a sustainable world, while encompassing cutting-edge technologies and current needs across various fields such as agriculture sustainability, bioremediation, restoration of degraded habitats and wastelands, and food security. Additionally, this series explores microbial applications in industries, and building their utilization in clean and green energy solutions. Furthermore, themes like microbial secondary metabolites, extremophilic microbes and modern omics, including next generation sequencing and metagenomics, are also covered in this series.

With contributions from researchers across the globe, this series addresses the important call of 'One Planet-One Health-One Future'. It comprises a collection of diverse volumes that provides insights for scientists, young researchers, educators and decision-makers in the government, private sector, and non-governmental organizations, empowering their efforts to achieve the global goals.

The series invites, evaluates, and accepts book proposals to ensure a diverse, inclusive, and evolving program. The final decision regarding acceptance rests with the series editor.

Peer-review

This book series follows a stratified review process. Proposals for individual volumes are reviewed by the series editor and then the editorial board members. On a case-to-case basis, external reviewers are also invited for further evaluation of the book proposal. Review of the chapters is the responsibility of the volume editor(s). A manuscript submission platform has been recently made available to the authors, volume editors and the series editor.

Naveen Kumar Arora • Jitendra Mishra
Editors

Biotechnological Solutions for a Sustainable Future

From Soil Health to Industrial Applications

Springer

Editors
Naveen Kumar Arora
Department of Environment Science
Babasaheb Bhimrao Ambedkar University
Lucknow, Uttar Pradesh, India

Jitendra Mishra
Amity School of Biological Sciences
Amity University Punjab
Mohali, Punjab, India

ISSN 2512-1901 ISSN 2512-1898 (electronic)
Microorganisms for Sustainability
ISBN 978-981-95-3988-8 ISBN 978-981-95-3989-5 (eBook)
https://doi.org/10.1007/978-981-95-3989-5

This Springer imprint is published by the registered company Springer Nature Singapore Pte Ltd.
The registered company address is: 152 Beach Road, #21-01/04 Gateway East, Singapore 189721, Singapore

Preface

In an era of converging environmental crises—accelerating climate change, unprecedented resource depletion, and widespread pollution—humanity faces a defining moment. *Biotechnological Solutions for a Sustainable Future: From Soil Health to Industrial Applications* delivers both a roadmap and a call to action. This volume unites multidisciplinary expertise to address a singular imperative: *employing biological systems to restore planetary health while advancing human well-being.*

Part I: Soil Health and Resource Valorization

We begin at the root of sustainability—healthy soils. Chapter 1 unveils *biochar-based fertilizers* as carbon-negative tools that reverse land degradation while sequestering atmospheric CO_2. Chapter 2 transforms waste into valuable resources, converting agricultural and industrial residues into *next-generation biofuels* for circular economies. Chapter 3 pioneers the conversion of food waste into *high-performance biodegradable packaging*—simultaneously displacing petroleum plastics and valorizing organic waste streams.

Part II: Environmental Remediation and Cultural Restoration

Biological systems serve as our most precise tools for environmental detoxification. Chapter 4 deploys microbial consortia against persistent pollutants, while Chap. 5 showcases *CRISPR-engineered plastic-degrading microbes*, *mycoremediation* of oil spills, and *hyperaccumulator plants* for heavy metals—proving nature's superiority over conventional cleanup methods.

Moving beyond basic remediation, Chap. 6 harnesses *microbial biosurfactants* to emulsify hydrocarbons and chelate metals, while Chap. 7 applies *microbial nanotechnology* for sustainable wastewater treatment. Chapter 8 reimagines radioactive waste management through integrated biological and physicochemical approaches, and Chap. 9 pioneers *enzymatic restoration of cultural heritage*—safeguarding both ecological systems and human history.

Part III: Industrial Applications for Food Safety and Human Health

Life sciences transcend the research bench to transform industries. Chapter 10 engineers *nanoparticle-enhanced antimicrobial packaging* that extends food shelf-life while ensuring consumer safety. Chapter 11 exposes the carcinogenic threat of *fumonisins* and charts mitigation pathways across food production chains. Chapter 12 unlocks *microbial carotenoids* as sustainable pigments for nutraceuticals, pharmaceuticals, and industrial applications—replacing synthetic alternatives while delivering enhanced health benefits.

Scalability gaps, economic viability, and regulatory frameworks demand urgent cross-disciplinary collaboration. Yet these obstacles ignite innovation—driving optimized bioreactor designs, risk-assessed engineered microorganisms, and policy-ready life cycle assessment (LCA) models. The evidence is unequivocal: biological solutions are environmentally essential, economically viable, and industrially scalable.

We extend profound gratitude to the pioneering researchers whose work fills these pages. Together, they demonstrate how biotechnology harmonizes *ecological resilience*, *economic prosperity*, and *social equity*.

This volume represents more than a scientific compilation—it stands as a testament to biotechnology's power to build a future where healthy soils and thriving industries coexist. Within these pages lies the blueprint for a truly sustainable planet.

Lucknow, Uttar Pradesh, India Naveen Kumar Arora
Mohali, Punjab, India Jitendra Mishra

Contents

Editors and Contributors

About the Series Editor

Naveen Kumar Arora, PhD in Microbiology, Fellow of International Society of Environmental Botanists (FISEB), is a Professor in the Department of Environmental Science, Babasaheb Bhimrao Ambedkar University (A Central University), India. He is a renowned researcher in the field of environmental microbiology and biotechnology. His specific area of research is plant-microbe interactions and sustainable agriculture. He has more than 150 research articles published in premium international journals and several articles published in magazines and dailies. He is editor of 54 books, published by Springer Nature. He is the series editor of book series, Microorganisms for Sustainability, and he is the Editor-in-Chief of the journal, *Environmental Sustainability*, both published by Springer Nature.

About the Editors

Naveen Kumar Arora, PhD in Microbiology, Fellow of International Society of Environmental Botanists (FISEB); Professor in Department of Environmental Science; Former Dean, School of Earth and Environmental Sciences; Former Head, Department of Environmental Science; Former Head, Department of Environmental Microbiology; and President of Institutions Innovation Council, Babasaheb Bhimrao Ambedkar University (A Central University), Lucknow, Uttar Pradesh, India, is a renowned researcher in the field of Environmental Microbiology and Biotechnology.

His specific area of research is plant-microbe interactions and sustainable agriculture. Before joining Babasaheb Bhimrao Ambedkar University, he was Head, Department of Microbiology, Chhatrapati Shahu Ji Maharaj University, Kanpur, Uttar Pradesh, India, from 2006 to 2011. He has more than 150 research articles published in premium international journals and several articles published in magazines and dailies. He is editor of 54 books, published by Springer Nature. He is the Series Editor of book series Microorganisms for Sustainability, published by Springer Nature, and Editor of the journal *Plant Pathology*, published by Wiley. He is the Editor-in-Chief of the journal *Environmental Sustainability*, published by Springer Nature. He is also the member of several national and international societies, and in editorial board of 4 journals. As per the list released by Stanford University in collaboration with Elsevier, he is among the top 2% most cited scientists of the world for four years in a row. He has delivered lectures in conferences and seminars around the globe, some of which include Lead Lectures at University of Lisbon, Portugal; SIIRT University, Turkey; University of Tashkent, Uzbekistan; Ibn Zohr University, Agadir, Morocco; IUSS Conference at Philadelphia, USA; and conferences at Beijing (China), Hanoi (Vietnam) and several conferences in India. He has been advisor to 126 postgraduate and 22 doctoral students. He has received awards for excellence in research by the Honorable Governor of Uttar Pradesh, Asian PGPR Society, Vijnana Bharti, Samagra Vikas Welfare Society, and Research & Academic Excellence Awards in the discipline of Life Sciences by Babasaheb Bhimrao Ambedkar University (A Central University), Lucknow, Uttar Pradesh, India. He is also President of Society for Conservation of Wildlife and is Secretary, Society for Environmental Sustainability (website: www.sesindia.org).

Jitendra Mishra is Assistant Professor of Microbiology at Amity University Punjab, India. A globally recognized scientist, he ranked among the top 2% of microbiologists worldwide (Elsevier, 2024 and 2025). He previously coordinated a DST-SEED-funded project (2020–2023) at Babasaheb Bhimrao Ambedkar University, spearheading R&D in microbial technologies. With prior expertise as a DST Research Associate, he specializes in biocontrol agent development, including strain selection, carrier formulation, and quality control. His research focuses on microbial secondary metabolites, bioinoculants, and plant growth-promoting rhizobacteria (PGPR), leveraging fluorescent pseudomonads for eco-friendly phytopathogen management in saline soils. An active academic, he mentors postgraduate students, holds a patent, and has published extensively in reputed journals. He has authored three books and serves as an editor for Frontiers and Springer-Nature journals. His interdisciplinary work bridges microbial innovation and sustainable agriculture, addressing global agricultural challenges through cutting-edge biotechnological solutions.

Contributors

Neha Agarwal Department of Chemistry, Navyug Kanya Mahavidyalaya, University of Lucknow, Lucknow, Uttar Pradesh, India

Ashish Department of Bioengineering Engineering, Integral University, Lucknow, Uttar Pradesh, India

Ronakkumar Bhadaniya Department of Biotechnology and Bioengineering, Indian Institute of Advanced Research, Gandhinagar, Gujarat, India

Kanchan Bhatt Department of Biotechnology Engineering, Goel Institute of Technology and Management, Lucknow, Uttar Pradesh, India

Durdigul Botirova Institute of Fundamental and Applied Research, National Research University TIIAME, Tashkent, Uzbekistan

Shiv Kumar Dwivedi Department of Environmental Science, School of Earth and Environmental Sciences, Babasaheb Bhimrao Ambedkar University, Lucknow, India

Dilfuza Egamberdieva Institute of Fundamental and Applied Research, National Research University TIIAME, Tashkent, Uzbekistan

E. J. Ekefan Department of Crop and Environmental Protection, Joseph Sarwuan Tarka University, Makurdi, Benue, Nigeria

Reena Gupta Department of Biotechnology, Himachal Pradesh University, Summerhill, Shimla, India

Victor I. Gwa Department of Crop Protection, Faculty of Agriculture, Federal University, Dutsin-Ma, Katsina, Nigeria

Hitanshu Kapadiya Department of Biotechnology and Bioengineering, Indian Institute of Advanced Research, Gandhinagar, Gujarat, India

Gurleen Kaur Amity School of Biological Sciences, Amity University Punjab, Mohali, India

Livpreet Kaur Amity School of Biological Sciences, Amity University Punjab, Mohali, India

Manpreet Kaur Department of Biotechnology, Himachal Pradesh University, Summerhill, Shimla, India

Mamta Kodarkar Applied Environmental Biotechnology Laboratory, Department of Biological Sciences, Birla Institute of Technology and Science, Sancoale, India

Puja Kumari Department of Food and Nutrition, School of Home Science, Babasaheb Bhimrao Ambedkar University, Lucknow, Uttar Pradesh, India

Simpal Kumari Department of Microbiology, Faculty of Science & Technology, Dr. Shakuntala Misra National Rehabilitation University, Lucknow, Uttar Pradesh, India

N. Lohith Kumar Department of Agricultural Microbiology, University of Agricultural Sciences, Bengaluru, India

F. A. Lum Kings University, Odeomu, Osun, Nigeria

Aparna Mahajan Department of Food Science and Technology, School of Health Sciences, Amity University Punjab, Mohali, Punjab, India

Aharnish Maurya Department of Biotechnology Engineering, Kashi Institute of Technology, Varanasi, Uttar Pradesh, India

Sankara Rao Miditana Department of Chemistry, Government Degree College, Puttur, Tirupati, Andhra Pradesh, India

Ankita Mishra Department of Biotechnology, Mahayogi Gorakhnath University, Gorakhpur, India

Jitendra Mishra Amity School of Biological Sciences, Amity University Punjab, Mohali, Punjab, India

Shachi Mishra Department of Chemistry, Jai Prakash University, Chapra, Bihar, India

Srikanth Mutnuri Applied Environmental Biotechnology Laboratory, Department of Biological Sciences, Birla Institute of Technology and Science, Sancoale, India

Alka Nanda Department of Food and Nutrition, School of Home Science, Babasaheb Bhimrao Ambedkar University, Lucknow, Uttar Pradesh, India

Yogita Pandey Department of Biotechnology and Bioengineering, Indian Institute of Advanced Research, Gandhinagar, Gujarat, India

Sachchit Patel Department of Biotechnology and Bioengineering, Indian Institute of Advanced Research, Gandhinagar, Gujarat, India

Neela Satheesh Department of Food Science and Technology, School of Health Sciences, Amity University Punjab, Mohali, Punjab, India

Rashmi Shahi Department of Microbiology, Mahayogi Gorakhnath University, Gorakhpur, India

Arpit Shukla Department of Biotechnology and Bioengineering, Indian Institute of Advanced Research, Gandhinagar, Gujarat, India

Ayushi Singh Department of Food and Nutrition, School of Home Science, Babasaheb Bhimrao Ambedkar University, Lucknow, Uttar Pradesh, India

Neetu Singh Department of Food and Nutrition, School of Home Science, Babasaheb Bhimrao Ambedkar University, Lucknow, Uttar Pradesh, India

Tanya Singh Department of Food and Nutrition, School of Home Science, Babasaheb Bhimrao Ambedkar University, Lucknow, Uttar Pradesh, India

Muxlisa Soipova Institute of Fundamental and Applied Research, National Research University TIIAME, Tashkent, Uzbekistan

Vijendra Singh Solanki Department of Chemistry, ISR, IPS Academy, Indore, Madhya Pradesh, India

Atul Kumar Srivastava Key Laboratory of Microbiology, College of Agriculture, Guizhou University, Guiyang, People's Republic of China

Eskindir E. Tadesse Department of Animal Products Processing, University of Agriculture, Kraków, Poland

Faculty of Chemical and Food Engineering, Bahir Dar Institute of Technology, Bahir Dar University, Bahir Dar, Ethiopia

Siddraj Thakur Department of Biotechnology and Bioengineering, Indian Institute of Advanced Research, Gandhinagar, Gujarat, India

Hamza Umar Department of Bioengineering Engineering, Integral University, Lucknow, Uttar Pradesh, India

Shoxnoza Zakirova Institute of Fundamental and Applied Research, National Research University TIIAME, Tashkent, Uzbekistan

Part I
Soil Health and Resource Valorization

Chapter 1
Biochar-Based Fertilizers: A New Approach to Sustainable Soil Management

Durdigul Botirova, Shoxnoza Zakirova, Muxlisa Soipova, and Dilfuza Egamberdieva

Abstract Biochar-based fertilizers (BBFs) are an emerging type of engineered fertilizers that combine biochar with beneficial microorganisms to enhance plant nutrition and improve soil properties. BBFs are now considered to be a sustainable strategy to enhance soil health and sequester carbon, offering significant advantages over conventional mineral fertilizers. Biochar is a porous, charcoal-like material produced through pyrolysis, a process of heating organic matter in the absence of oxygen. It enhances soil health by improving chemical, physical, and biological characteristics. Its effectiveness is influenced by the feedstock used and the production conditions, both of which affect its final properties. Chemically, biochar increases nutrient retention and regulates soil pH, creating a more favorable environment for plant growth. Physically, its porous structure boosts water-holding capacity, ensuring better water availability for plants. Biologically, biochar fosters beneficial microbial activity, supporting soil biodiversity and plant health. This chapter reviews the role of biochar-based fertilizers in enhancing soil fertility and examines the key characteristics of biochar. It also examines biochar's impact on soil physicochemical properties and explores its biological effects, with an emphasis on shifts in soil microbial activity, enzyme functions, and the enhancement of soil microbial communities. By showcasing the diverse benefits of biochar-based microbial fertilizers, this chapter highlights their potential to enhance agricultural sustainability, increase crop yields, and aid in climate change mitigation through carbon sequestration. Understanding the interactions between biochar properties and soil characteristics is crucial, as the effects depend on both the type of biochar applied and the specific soil conditions.

Keywords Biochar · Fertilizers · Feedstock · Bioresources · Nutrients · Encapsulated biochar

---

D. Botirova · S. Zakirova · M. Soipova · D. Egamberdieva (✉)
International Centre for Strategic Development and Research in Food and Agriculture, Ministry of Agriculture of Uzbekistan, Tashkent, Uzbekistan

N. K. Arora, J. Mishra (eds.), *Biotechnological Solutions for a Sustainable Future*, Microorganisms for Sustainability 55,
https://doi.org/10.1007/978-981-95-3989-5_1

1.1 Introduction

The global food crisis is a pressing issue affecting numerous regions around the world. To feed the growing population is one of the biggest challenges and this has resulted in increased demand from the agricultural sector. As a result, modern agriculture plays a vital role in meeting this demand (Pang et al. 2018). Projections suggest that the global population will reach 8.6 billion by 2030 and 9.8 billion by 2050 (Da Costa et al. 2019). Significant investments have been made in the agricultural and agro-food sectors, particularly in fertilizers, to increase production. As a result, soil has been affected by high levels of agrochemical contamination (Igalavithana et al. 2015). At present, 25% of global agricultural land is classified as highly degraded, 44% as moderately degraded, and around 10% is classified as restored (El-Naggar et al. 2019).

Biochar is a carbon-rich by-product produced through the pyrolysis of various organic materials under low-oxygen conditions. It typically has a well-developed pore structure, a large surface area, effective adsorption capabilities, and a high degree of stability (Puga et al. 2020). These features enhance its ability to increase carbon storage in soil, improve water and nutrient retention, enhance soil fertility, and boost crop yields. Common feedstocks for biochar production include biomass, municipal waste, crop residues, and animal manure, sourced from both lignocellulosic and non-lignocellulosic materials. Lignocellulosic biomass, such as agricultural waste, wood by-products, energy crops, and municipal solid waste, is an excellent resource (Krishnan et al. 2021). On the other hand, non-lignocellulosic feedstocks, like sewage sludge, animal waste, and algae, pose significant management challenges due to their complex and variable compositions (Amalina et al. 2021). Biochar-based fertilizers (BBFs), which combine biochar and microbes, offer several advantages over the chemical fertilizers used rampantly at present (Joseph et al. 2021; Zhou et al. 2021; Melo et al. 2022; Rombel et al. 2022).

Many researchers consider the use of biochar, which can improve soil conditions, enhance nutrient retention, and increase nutrient use efficiency, as crucial for shifting toward sustainable agriculture (Güereña et al. 2013). The addition of biochar can also reduce soil heterotrophic respiration by enhancing the recalcitrance of soil organic carbon and reducing the activity of carbon-degrading enzymes (Li et al. 2018). Furthermore, biochar creates a physical habitat that supports microbes (Atkinson et al. 2010; Hammer et al. 2014). When used alongside chemical fertilizers, biochar can modify the composition of microbial communities and improve soil metabolism, resulting in better soil nutrient levels (Tian et al. 2016).

The selection of biomass feedstock is crucial for biochar production, as it determines the biochar's properties and potential uses (Yek et al. 2021). Biochar is made from dry feedstock (with less than 10% moisture) using various pyrolysis techniques and microwave methods (Li et al. 2020). To overcome the challenges of using wet feedstocks, hydrothermal carbonization (HTC) was developed, enabling the direct conversion of high-moisture materials into a carbon-rich solid called hydrochar. This process involves several simultaneous reactions, including hydrolysis, dehydration, decarboxylation, aromatization, and recondensation, which reduce the oxygen and hydrogen content of the feedstock (Benavente et al. 2015; Kumar et al. 2020). Although both biochar and hydrochar are derived from biomass feedstocks rich in lignin and cellulose, their properties differ significantly (Shao et al. 2019).

However, some studies suggest that the effectiveness of biochar depends on the nutrient status of the soil. In nutrient-deficient soils, biochar may have limited positive effects on plant performance due to its low nutrient content (Hussain et al. 2017). Most researchers agree that when used in conjunction with mineral fertilizers, biochar can act as a carrier that gradually releases nutrients throughout the growing season, aligning with plant nutrient requirements and minimizing nutrient leaching. This ultimately enhances fertilizer efficiency and crop yields (Puga et al. 2020). Although biochar use has several advantages in agriculture, there is a lot more to be explored regarding its role in nutrient cycling and its consistency (Kizito et al. 2019). A major reason for this is the variation in the raw material for the production of biochar. Diverse types of biomass can vary in nutrient content and may have different impacts on plants and soil. Additionally, the mechanisms by which biochar captures and releases organic or inorganic mineral nutrients are not yet fully understood. Sometimes the highly porous nature of the biochar can restrict nutrient availability, with nutrients getting immobilized after absorption, thus making them inaccessible (Bruun et al. 2012). To improve fertilizer efficiency, BBFs are produced by combining biochar with inorganic fertilizers using various physical or chemical methods, such as impregnation, pelletizing, and encapsulation (Barbosa et al. 2022; Das and Ghosh 2021). These biochar-based slow-release fertilizers (SRFs) can provide nutrients for an entire growing season with a single application, reducing stress and toxicity caused by excessive nutrients in the root zones. The use of biochar-based microbial fertilizers enhances soil quality, crop yields, nutrient retention, and carbon sequestration (Ahmad et al. 2014). Jain et al. (2020) examined the effects of biochar on the yield of palmarosa in the soil contaminated with mine waste, including heavy metals. The high palmarosa yield was linked to enhanced pH, cation-exchange capacity, water and nutrient retention, and the reduction of heavy metals in the mine waste/soil due to biochar application.

The effects of BBFs on soil microorganisms and their role in improving plant yield and quality are not yet well understood. This is especially true in highly acidic soils, such as those found in tea orchards, where fungi play a vital role in the breakdown of soil organic matter (Sun et al. 2016; Ji et al. 2020). A deeper understanding of how these fertilizers affect tea growth and microbial dynamics could provide valuable insights for enhancing tea production and soil management practices in tea gardens (Yang et al. 2021). Given biochar's unique properties, including its porous structure, large surface area, and numerous surface functional groups, it holds potential as a fertilizer carrier.

1.2 Biochar Characterization

Biochar's large surface area and negative surface charge contribute to its strong adsorption capabilities, helping to reduce the leaching of heavy metals, pesticides, herbicides, and other contaminants (Tomczyk et al. 2020; Kapoor et al. 2022). However, biochar's properties and effectiveness vary depending on the feedstock and pyrolysis conditions. Studies have found that biochar produced from materials high in hemicellulose, cellulose, and lignin is particularly effective at reducing soil pollutants (Susilawati et al. 2020). Additionally, its stability and resistance to decomposition enhance its potential for long-term carbon sequestration in soil (Barbhuiya et al. 2024). Three main methods are used to produce BBFs: the impregnation technique, the mixed granulation technique, and the co-pyrolysis technique. In the impregnation method, biochar is soaked in a liquid fertilizer solution to incorporate nutrients, followed by drying to create nutrient-enriched biochar (Sim et al. 2021). Similarly, Lateef et al. (2019) described a process in which biochar is immersed in a nutrient solution for several hours to maximize absorption, then dried. The mixed granulation technique involves grinding biochar with one or more types of solid fertilizers into a uniform powder, which is then pelletized (Shin and Park 2018). A comparative study of lignocellulose pyrolysis and its key components explored the characteristics and distribution of the resulting biochar and volatiles (Biswas et al. 2017). These processes typically generate three products: biochar, bio-oil, and gases. Conditions like moderate temperatures, high heating rates, low pressure, and smaller biomass particle sizes tend to produce a higher yield of biochar with significant carbon content, increased porosity, and a larger surface area (Rangabhashiyam and Balasubramanian 2019). Specifically, cow bone biochar shows exceptional adsorptive capabilities, making it effective for remediating various contaminants in water and soil (Bayartsengel et al. 2021). Studies by Lateef et al. (2019) demonstrated strong nutrient adsorption capabilities of biochar produced from corn cobs at temperatures between 300 and 400 °C. Recently, a novel approach has emerged that combines encapsulation technology with nutrient-loaded biochar to enhance the sustained release of nutrients, aiming to minimize leaching and improve nitrogen use efficiency (Sim et al. 2021). Biochar derived from wood-based feedstocks exhibits improved physical properties, including greater porosity and specific surface area. In contrast, biochar sourced from straw-based feedstocks is recognized for its superior elemental composition, characterized by higher levels of oxygen, nitrogen, and hydrogen. Additionally, biochar produced from straw at lower pyrolysis temperatures contains more functional groups that enhance charge transfer potential and adsorption stability by increasing surface charge density, charge distribution, and bonding orbitals (He et al. 2024a). The selection of feedstock has a significant impact on the physicochemical properties and elemental composition of biochar. Different feedstocks, such as woody and herbaceous plants, have distinct structures and compositions. Typically, woody plants exhibit a denser structure and higher levels of lignin and cellulose than herbaceous plants (Zhang et al. 2022). Although biochars offer numerous benefits, they typically have low nutrient content and limited nutrient-release capabilities when applied to soil (Ding et al. 2016). To boost their nutrient content, biochars can be enhanced by adsorbing ions (Lateef et al. 2019).

1.3 The Effect of Biochar on Soil Physicochemical Properties

Soil physico chemical properties are essential for promoting plant growth and productivity. Biochar contains high concentrations of nutrients and organic C in various forms. Biochar also has diverse functional groups on its surface, which play several important roles (Ndoung et al. 2021). The biochar enhances soil porosity, decreases bulk density, improves soil aggregation, and boosts moisture retention. In comparison to standard biochar, biochar composites possess distinctive colloidal properties, including a wider particle size distribution, increased surface roughness, a larger specific surface area, enhanced surface charge heterogeneity, and varying elemental and functional group content (He et al. 2024b). Biochar typically has a basic pH due to the inorganic materials it retains from the original biomass. Adding biochar to acidic soils is helpful in neutralizing and stabilizing soil pH, which helps in the thriving of diverse microbial communities that would have been impacted by acidic environments (Halmi and Simarani 2021). Research by Khan et al. (2024) showed that applying biochar at a rate of 60 t ha^{-1} improved soil physical properties compared to a control group over 3 years of crop cultivation. Some biochars are alkaline, which helps improve nutrient availability in acidic soils. However, their addition to alkaline soils may lead to increased salinity (Premalatha et al. 2023). Biochar generally reduces sodium ion concentrations in soil colloids through strong adsorption or by exchanging sodium ions for calcium (Ca) or magnesium (Mg) ions on its surface, which aids in salt leaching during irrigation. Due to its alkaline properties, biochar is often used to enhance acidic soils (Gao et al. 2024). Due to its larger surface area and negative charge, biochar is more effective at absorbing and retaining cations compared to other forms (Clough et al. 2013). Furthermore, biochar can act as an adsorbent to remove both organic and inorganic contaminants from the soil. Its incorporation enhances overall soil quality, boosts microbial and enzymatic activity, increases soil organic carbon content, and improves nutrient retention and availability (Elkhlifi et al. 2023). Adding biochar can also help reduce ammonia volatilization and N_2O emissions. It promotes the solubilization of available phosphorus while decreasing phosphorus fixation in soil minerals. Additionally, the high levels of extractable potassium in biochar positively impact the potassium cycle in soil (Jindo et al. 2020).

1.4 The Effect of Biochar on Soil Biological Properties

The application of biochar as a soil amendment has been widely studied for its potential to improve soil properties, enhance enzyme activities, and support soil health. Biochar, derived from various feedstocks and produced at different temperatures, has demonstrated beneficial effects on soil fertility, microbial communities, and enzymatic processes that are essential for nutrient cycling. Incorporating biochar into the soil can modify microbial abundance, activity, and community structure. However, the long-term consequences of these alterations remain unclear, which is significant given that biochar can persist in the soil for centuries (Hardy et al. 2019). According to Xu et al. (2023), biochar primarily influences only a few microbial phyla, notably *Acidobacteria* and *Gemmatimonadetes*, while having a limited effect on most others. These changes are associated with the properties of biochar and its impact on soil conditions, suggesting that the influence on soil microbial diversity can be managed by modifying biochar characteristics. Additionally, smaller bacteria and fungi can inhabit the pores of specific biochars, shielding them from soil predators. Soluble compounds such as sugars, alcohols, acids, ketones, and water stored in the mesopores and micropores of biochar can enhance microbial activity and alter microbial composition (Jaafar et al. 2015; Adnan et al. 2020). For example, biochar applied to pepper cultivation increased the activity and populations of *Pseudomonas* spp., filamentous fungi, and *Bacillus* species (Gao et al. 2017). A study by Zhou et al. (2019) found that adding biochar fertilizer increased microbial biomass and bacterial diversity in karst calcareous soils, highlighting its important role in rehabilitating degraded areas. This beneficial effect was attributed to improved soil physical quality, enhanced soil water retention capacity, increased nutrient availability, and greater microbial biodiversity involved in nutrient cycling (Egamberdieva et al. 2017, 2023, Ma et al. 2019a, b). Several studies have reported improved plant-microbe interactions with biochar application (Egamberdieva et al. 2019, 2022). Chew et al. (2020) found that applying biochar developed from wheat straw resulted in improved microbial activity in the rice rhizosphere. The biochar improved the population and activity of plant growth-promoting rhizobacteria and fungi in the rhizosphere, helping the rice plants to sequester nutrients. Additionally, Joseph et al. (2015) observed that application of biochar along with iron nanoparticles to the wheat crop resulted in better mycorrhizal colonization, resulting in improved nutrient uptake and yields. Nielsen et al. (2014) reported that the application of biochar-mineral complex (BMC) resulted in improved microbial activity as compared to the addition of chemical fertilizers only. Biochar is now universally known to enhance the activities of microbial communities, particularly in the rhizosphere region, which helps in the better establishment of plant-microbe symbiosis. Changes in microbial communities are associated with the high porosity of biochar, which provides microorganisms with habitats within its pores, protecting them from predators and promoting their growth and diversity (Lehmann et al. 2011; Nguyen et al. 2018). As a result, BBFs serve as shelters for soil microorganisms. Additionally, the pyrolysis process can transform specific nutrients into forms that are accessible for microbial metabolic activities (Ameloot et al. 2013; Ma et al. 2022a, b). Furthermore, biochar can immobilize harmful soil contaminants, reducing their negative impacts on microbial growth and function (Bolan et al. 2023).

Enzymes produced by both plant roots and microorganisms are essential for nutrient availability in the rhizosphere, as they are involved in the hydrolysis of carbon substrates and organic nutrients, such as nitrogen, phosphorus, and sulfur (Gianfreda 2015). The addition of biochar affects soil enzyme activity. The interplay between biochar, the substrate, and enzymes has a significant impact on soil enzymatic functions (Lammirato et al. 2011). Bailey et al. (2011) found that the effects of biochar on soil enzyme activities differ depending on the specific soil type and the enzyme in question. Studies have demonstrated that the activity of urease and sucrase in the soil significantly increases with higher doses and concentrations of biochar. This enhancement occurs because biochar improves the organic matter and nutrient content in the soil, creating favorable conditions for soil microorganisms and thereby increasing both their activity and enzymatic functions (Gao et al. 2020; Wu et al. 2020). The response of soil enzymes to biochar varies greatly depending on the amount of biochar applied, its type, and the soil characteristics (Xu et al. 2019). Additionally, biochar applications have been shown to enhance urease activity, which is responsible for breaking down urea in the soil, due to improved levels of organic matter and nutrients (Wu et al. 2020; Jiang et al. 2021). Biochar can affect soil enzyme activity through multiple mechanisms. First, its influence on soil microorganisms can alter the production and release of enzymes. Second, changes in soil pH can either enhance or suppress the activity of existing enzymes. Lastly, biochar particles may absorb enzymes directly, impacting their overall function (Foster et al. 2018). The synergistic effect of biochar on the activity and colonization of PGPR in the rhizosphere is now very well-known and documented (Egamberdiyeva and Höflich 2003a, b). PGPR can help plants in various direct and indirect ways (Egamberdieva et al. 2020; Arora and Bouizgarne 2024). Soil biodiversity is crucial to ensure food security (Arora 2023), and biochar can play a very important role in the maintenance of soil diversity and quality. Figure 1.1 showcases the various positive aspects of using biochar in agriculture.

The following table summarizes key findings from recent studies that explore the impact of different biochar types, application rates, and soil conditions on various soil properties, such as enzyme activity, soil porosity, nutrient content, and microbial health (Table 1.1).

1.5 Future Prospects and Conclusion

Biochar-based fertilizers are a promising strategy for sustainable agriculture that combines biochar with mineral fertilizers or beneficial microorganisms to enhance soil properties and improve plant nutrition. This chapter examines the diverse advantages of BBFs, including their capacity to increase nitrogen use efficiency,

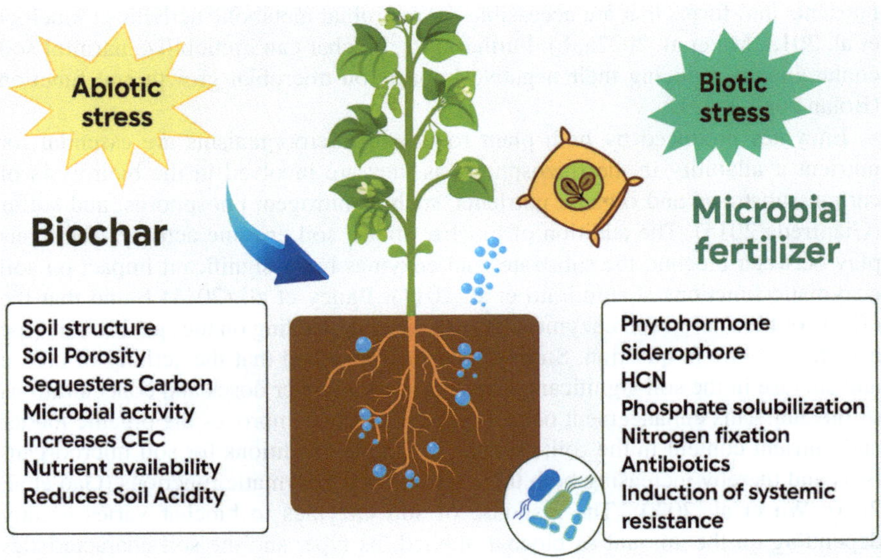

Fig. 1.1 Positive effects of biochar on soil and agro-ecosystem

boost soil fertility, and aid in carbon sequestration. By influencing soil microbial abundance and enzyme activity, biochar enhances nutrient availability and promotes overall soil health. Biochar's long-term stability in soil makes it an effective tool for carbon sequestration, contributing to climate change mitigation. Its porous structure enhances soil aeration, water retention, and nutrient availability, enabling climate-resilient agriculture. The impact of biochar on soil characteristics depends on the type of feedstock and production conditions, highlighting the importance of customizing biochar applications for specific soil environments to achieve optimal results. When combined with microbial inoculants, biochar provides a synergistic effect, offering a protective habitat for microbes, which enhances their role in promoting plant growth, increasing nutrient cycling, and improving resistance to plant diseases. The use of biochar-based microbial fertilizers supports sustainable waste management by converting agricultural residues into valuable resources. This contributes to a circular bioeconomy, reducing reliance on chemical fertilizers while recycling organic waste. The formulations can be tailored to specific crop or soil requirements, allowing them to address particular deficiencies or environmental conditions. Furthermore, these fertilizers show promise in rehabilitating degraded, arid, saline, and nutrient-poor soils, making them valuable for land restoration efforts. Digital agriculture further enhances the potential of biochar-based microbial fertilizers. Precision farming technologies, such as sensors and artificial intelligence, can optimize their application and monitor their effects on soil health and crop yield. Digital platforms also provide opportunities for research on biochar's interaction with soil microbiomes and its impact on plant development. Biochar is also suitable for integration into agroforestry systems, where they can support soil

Table 1.1 Impact of different biochar types and application rates on soil properties and soil enzyme activities

Biochar type	Soil types	Temperature (°C)	Application rates	Positive effect	References
Swine-manure	Laterite	350 °C and 500 °C	1–2%	Sucrase, phosphatase, catalase, urease	Jiang et al. (2021)
Corn stalks + Fe3Mn	As-contaminated paddy soil	300 °C and 600 °C	5–20 g/kg^{-1}	Abundance of microorganisms and soil enzyme activity	Zhang et al. (2020)
Peanut shell	Solonchak	350 °C–550 °C	34–101 t/ha	Catalase, alkaline phosphatase	Yao et al. (2021)
Rice husks	Sandy clay	350 °C	3 t–12 t/ha	Invertase, urease, alkaline phosphatase, catalase	Oladele (2019)
Maize, wheat, peanut shells	Coastal saline	600 °C	4–6 kg/m^2	Alkaline phosphatase	Zhang et al. (2024)
Cassava straw	Ultisol paddy soil	400 °C	30 t/ha	Urease, glucosidase, catalase	Ullah et al. (2020)
Wheat straw	Clay	350 °C–550 °C	20–40 t/ha	Soil aggregation, microbial community	Lu et al. (2020)
Sewage sludge	Sandy loam	300 °C	2%	Total organic carbon, nitrogen	Velli et al. (2021)
Wheat straw	Loamy sand	300 °C	2–4%	Water content	Głąb et al. (2016)
Forest biomass waste	Desert sandy	450 °C	10 t/ha	Soil porosity	Baiamonte et al. (2019)
Corn straw	Typical soil	500 °C	40–120 t/ha	Soil porosity	Fu et al. (2019)
Pistachio residues	Sandy loam	300 °C	1–2%	Soil structure	Daneshvar et al. (2023)
Corn cob	Silty clay	500 °C	1–2%	Soil carbon	Majeed et al. (2022)

quality and tree-crop interactions. Commercialization opportunities include the development of region-specific and crop-specific formulations, catering to the growing global demand for organic and sustainable farming practices. However, challenges remain, such as the cost of production, the need for standardization to ensure product quality and efficacy, and the importance of farmer education to encourage adoption.

References

Adnan M, Fahad S, Zamin M, Shah S, Mian IA, Danish S, Zafar-Ul-Hye M, Battaglia ML, Naz RMM, Saeed B, Saud S, Ahmad I, Yue Z, Brtnicky M, Holatko J, Datta R (2020) Coupling phosphate-solubilizing bacteria with phosphorus supplements improve maize phosphorus acquisition and growth under lime induced salinity stress. Plants 9(7):900. https://doi.org/10.3390/plants9070900

Ahmad M, Rajapaksha AU, Lim JE, Zhang M, Bolan N, Mohan D, Vithanage M, Lee SS, Ok YS (2014) Biochar as a sorbent for contaminant management in soil and water: a review. Chemosphere 99:19–33. https://doi.org/10.1016/j.chemosphere.2013.10.071

Amalina IF, Haziq JM, Syukor ARA, Ridwan AFA, Rashid AHM (2021) Study of palm acid oil (PAO) from sludge palm oil mill effluent (POME) as goat's feed. Mater Today 41:96–101. https://doi.org/10.1016/j.matpr.2020.11.1013

Ameloot N, De Neve S, Jegajeevagan K, Yildiz G, Buchan D, Funkuin YN, Prins W, Bouckaert L, Sleutel S (2013) Short-term CO2 and N2O emissions and microbial properties of biochar amended sandy loam soils. Soil Biol Biochem 57:401–410. https://doi.org/10.1016/j.soilbio.2012.10.025

Arora P (2023) Life below land: the need for a new sustainable development goal. Front Environ Sci 11:1215282. https://doi.org/10.3389/fenvs.2023.1215282

Arora NK, Bouizgarne B (2024) Microbial biotechnology for sustainable agriculture volume 2. In: Arora NK (ed) Microorganisms for sustainability, vol 34. Springer, Singapore, p 384

Atkinson CJ, Fitzgerald JD, Hipps NA (2010) Potential mechanisms for achieving agricultural benefits from biochar application to temperate soils: a review. Plant Soil 337(1–2):1–18. https://doi.org/10.1007/s11104-010-0464-5

Baiamonte G, Crescimanno G, Parrino F, De Pasquale C (2019) Effect of biochar on the physical and structural properties of a sandy soil. Catena 175:294–303. https://doi.org/10.1016/j.catena.2018.12.019

Bailey VL, Fansler SJ, Smith JL, Bolton H Jr (2011) Reconciling apparent variability in effects of biochar amendment on soil enzyme activities by assay optimization. Soil Biol Biochem 43(2):296–301. https://doi.org/10.1016/j.soilbio.2010.10.014

Barbhuiya S, Bhusan Das B, Kanavaris F (2024) Biochar-concrete: a comprehensive review of properties, production and sustainability. Case Stud Constr Mater 20:e02859. https://doi.org/10.1016/j.cscm.2024.e02859

Barbosa CF, Correa DA, Carneiro JS d S, Melo LCA (2022) Biochar phosphate fertilizer loaded with urea preserves available nitrogen longer than conventional urea. Sustainability 14(2):686. https://doi.org/10.3390/su14020686

Bayartsengel B, Erdene-Ochir N, Battulga S, Tyeliubek S, Jantsanpurev E, Chuluun B (2021) Characterization of biochars produced from various biowastes. In: Proceedings of the 5th international conference on chemical investigation and utilization of natural resource (ICCIUNR-2021). Atlantis Press, Paris

Benavente V, Calabuig E, Fullana A (2015) Upgrading of moist agro-industrial wastes by hydrothermal carbonization. J Anal Appl Pyrolysis 113:89–98. https://doi.org/10.1016/j.jaap.2014.11.004

Biswas B, Pandey N, Bisht Y, Singh R, Kumar J, Bhaskar T (2017) Pyrolysis of agricultural biomass residues: comparative study of corn cob, wheat straw, rice straw and rice husk. Bioresour Technol 237:57–63. https://doi.org/10.1016/j.biortech.2017.02.046

Bolan S, Hou D, Wang L, Hale L, Egamberdieva D, Tammeorg P, Li R, Wang B, Xu J, Wang T, Sun H, Padhye LP, Wang H, Siddique KHM, Rinklebe J, Kirkham MB, Bolan N (2023) The potential of biochar as a microbial carrier for agricultural and environmental applications. Sci Total Environ:163968. https://doi.org/10.1016/j.scitotenv.2023.163968

Bruun EW, Ambus P, Egsgaard H, Hauggaard-Nielsen H (2012) Effects of slow and fast pyrolysis biochar on soil C and N turnover dynamics. Soil Biol Biochem 46:73–79. https://doi.org/10.1016/j.soilbio.2011.11.019

Chew J, Zhu L, Nielsen S, Graber E, Mitchell DRG, Horvat J, Mohammed M, Liu M, van Zwieten L, Donne S, Munroe P, Taherymoosavi S, Pace B, Rawal A, Hook J, Marjo C, Thomas DS, Pan G, Li L, Bian R, McBeath A, Bird M, Thomas T, Husson O, Solaiman Z, Joseph S, Fan X (2020) Biochar-based fertilizer: supercharging root membrane potential and biomass yield of rice. Sci Total Environ 713:136431. https://doi.org/10.1016/j.scitotenv.2019.136431

Clough T, Condron L, Kammann C, Müller C (2013) A review of biochar and soil nitrogen dynamics. Agronomy 3(2):275–293. https://doi.org/10.3390/agronomy3020275

Da Costa TP, Westphalen G, Nora FBD, de Zorzi Silva B, da Rosa GS (2019) Technical and environmental assessment of coated urea production with a natural polymeric suspension in spouted bed to reduce nitrogen losses. J Clean Prod 222:324–334. https://doi.org/10.1016/j.jclepro.2019.03.069

Daneshvar S, Mosaddeghi MR, Afyuni M (2023) Effect of biochar and hydrochar of pistachio residues on physical quality indicators of a sandy loam soil. Geoderma Reg 36:e00740. https://doi.org/10.1016/j.geodrs.2023.e00740

Das SK, Ghosh GK (2021) Development and evaluation of biochar-based secondary and micronutrient enriched slow release nano-fertilizer for reduced nutrient losses. Biomass Convers Biorefin. https://doi.org/10.1007/s13399-021-01880-5

Ding Y, Liu Y, Liu S, Li Z, Tan X, Huang X, Zeng G, Zhou L, Zheng B (2016) Biochar to improve soil fertility. A review. Agron Sustain Dev 36(2). https://doi.org/10.1007/s13593-016-0372-z

Egamberdieva D, Reckling M, Wirth S (2017) Biochar-based inoculum of *Bradyrhizobium* sp. improves plant growth and yield of lupin (*Lupinus albus* L.) under drought stress. Eur J Soil Biol 78:38–42. https://doi.org/10.1016/j.ejsobi.2016.11.007

Egamberdieva D, Li L, Ma H, Wirth S, Bellingrath-Kimura SD (2019) Soil amendment with different maize biochars improves chickpea growth under different moisture levels by improving symbiotic performance with *Mesorhizobium ciceri* and soil biochemical properties to varying degrees. Front Microbiol 10:2423. https://doi.org/10.3389/fmicb.2019.02423

Egamberdieva D, Shurigin V, Alaylar B, Ma H, Müller MEH, Wirth S, Reckling M, Bellingrath-Kimura SD (2020) The effect of biochars and endophytic bacteria on growth and root rot disease incidence of *Fusarium* infested narrow-leafed Lupin (*Lupinus angustifolius* L.). Microorganisms 8(4):496. https://doi.org/10.3390/microorganisms8040496

Egamberdieva D, Ma H, Shurigin V, Alimov J, Wirth S, Bellingrath-Kimura SD (2022) Biochar additions alter the abundance of p-cycling-related bacteria in the rhizosphere soil of *Portulaca oleracea* L. under salt stress. Soil Syst 6(3):64. https://doi.org/10.3390/soilsystems6030064

Egamberdieva D, Alaylar B, Alimov J, Jabborov Z, Bellingrath-Kimura SD (2023) Combined effects of biochar and plant growth promoting bacteria *Pseudomonas putida* TSAU1 on plant growth, nutrient uptake of wheat, and soil enzyme activities. Turk J Agric For 47. https://doi.org/10.55730/tar-2108-79

Egamberdiyeva D, Höflich G (2003a) The effect of associative bacteria from different climates on plant growth of pea at different soils and temperatures. Arch Agron Soil Sci 49(2):203–213. https://doi.org/10.1080/0365034031000101355a

Egamberdiyeva D, Höflich G (2003b) Influence of growth promoting bacteria on the growth of wheat at different soils and temperatures. Soil Biol Biochem 35:973–978. https://doi.org/10.1016/S0038-0717(03)00158-5

Elkhlifi Z, Iftikhar J, Sarraf M, Ali B, Saleem MH, Ibranshahib I, Bispo MD, Meili L, Ercisli S, Torun Kayabasi E et al (2023) Potential role of biochar on capturing soil nutrients, carbon sequestration and managing environmental challenges: a review. Sustainability 15(3):2527. https://doi.org/10.3390/su15032527

El-Naggar A, Lee SS, Rinklebe J, Farooq M, Song H, Sarmah AK, Zimmerman AR, Ahmad M, Shaheen SM, Ok YS (2019) Biochar application to low fertility soils: a review of current status, and future prospects. Geoderma 337:536–554. https://doi.org/10.1016/j.geoderma.2018.09.034

Foster E, Fogle E, Cotrufo M (2018) Sorption to biochar impacts β-glucosidase and phosphatase enzyme activities. Agriculture (Basel) 8(10):158. https://doi.org/10.3390/agriculture8100158

Fu Q, Zhao H, Li T, Hou R, Liu D, Ji Y, Zhou Z, Yang L (2019) Effects of biochar addition on soil hydraulic properties before and after freezing-thawing. Catena 176:112–124. https://doi.org/10.1016/j.catena.2019.01.008

Gao L, Wang R, Shen G, Zhang J, Meng G, Zhang J (2017) Effects of biochar on nutrients and the microbial community structure of tobacco-planting soils. J Soil Sci Plant Nutr 17(4):884–896. https://doi.org/10.4067/s0718-95162017000400004

Gao WH, Guo ZH, Xue C, Yin HF, Chang MY, Zhou QX, Wang GL, Li F, Liu Y (2020) Effect of biochar and biochar compound fertiliser on soil enzyme activity of soybean. J Huaibei Norm Univ (Nat Sci) 41:48–53

Gao G, Yan L, Tong K, Yu H, Lu M, Wang L, Niu Y (2024) The potential and prospects of modified biochar for comprehensive management of salt-affected soils and plants: a critical review. Sci Total Environ 912:169618. https://doi.org/10.1016/j.scitotenv.2023.169618

Gianfreda L (2015) Enzymes of importance to rhizosphere processes. J Soil Sci Plant Nutr. https://doi.org/10.4067/s0718-95162015005000022

Głąb T, Palmowska J, Zaleski T, Gondek K (2016) Effect of biochar application on soil hydrological properties and physical quality of sandy soil. Geoderma 281:11–20. https://doi.org/10.1016/j.geoderma.2016.06.028

Güereña D, Lehmann J, Hanley K, Enders A, Hyland C, Riha S (2013) Nitrogen dynamics following field application of biochar in a temperate North American maize-based production system. Plant Soil 365(1–2):239–254. https://doi.org/10.1007/s11104-012-1383-4

Halmi MFA, Simarani K (2021) Effect of two contrasting biochars on soil microbiota in the humid tropics of Peninsular Malaysia. Geoderma 395:115088. https://doi.org/10.1016/j.geoderma.2021.115088

Hammer EC, Balogh-Brunstad Z, Jakobsen I, Olsson PA, Stipp SLS, Rillig MC (2014) A mycorrhizal fungus grows on biochar and captures phosphorus from its surfaces. Soil Biol Biochem 77:252–260. https://doi.org/10.1016/j.soilbio.2014.06.012

Hardy B, Sleutel S, Dufey JE, Cornelis J-T (2019) The long-term effect of biochar on soil microbial abundance, activity and community structure is overwritten by land management. Front Environ Sci. https://doi.org/10.3389/fenvs.2019.00110

He D, Luo Y, Zhu B (2024a) Feedstock and pyrolysis temperature influence biochar properties and its interactions with soil substances: insights from a DFT calculation. Sci Total Environ 922:171259. https://doi.org/10.1016/j.scitotenv.2024.171259

He X, Wang Q, Jin Y, Chen Y, Huang L (2024b) Properties of biochar colloids and behaviors in the soil environment: influencing the migration of heavy metals. Environ Res 247:118340. https://doi.org/10.1016/j.envres.2024.118340

Hussain M, Farooq M, Nawaz A, Al-Sadi AM, Solaiman ZM, Alghamdi SS, Ammara U, Ok YS, Siddique KHM (2017) Biochar for crop production: potential benefits and risks. J Soils Sediments 17(3):685–716. https://doi.org/10.1007/s11368-016-1360-2

Igalavithana AD, Ok YS, Usman ARA, Al-Wabel MI, Oleszczuk P, Lee SS (2015) The effects of biochar amendment on soil fertility, SSSA special publications. American Society of Agronomy and Soil Science Society of America, Madison

Jaafar NM, Clode PL, Abbott LK (2015) Biochar-soil interactions in four agricultural soils. Pedosphere 25(5):729–736. https://doi.org/10.1016/s1002-0160(15)30054-0

Jain S, Khare P, Mishra D, Shanker K, Singh P, Singh RP, Das P, Yadav R, Saikia BK, Baruah BP (2020) Biochar aided aromatic grass [*Cymbopogon martini* (Roxb.) Wats.] vegetation: a sustainable method for stabilization of highly acidic mine waste. J Hazard Mater 390:121799. https://doi.org/10.1016/j.jhazmat.2019.121799

Ji L, Ni K, Wu Z, Zhang J, Yi X, Yang X, Ling N, You Z, Guo S, Ruan J (2020) Effect of organic substitution rates on soil quality and fungal community composition in a tea plantation with long-term fertilization. Biol Fertil Soils 56(5):633–646. https://doi.org/10.1007/s00374-020-01439-y

Jiang Y, Wang X, Zhao Y, Zhang C, Jin Z, Shan S, Ping L (2021) Effects of biochar application on enzyme activities in tea garden soil. Front Bioeng Biotechnol 9:728530. https://doi.org/10.3389/fbioe.2021.728530

Jindo K, Audette Y, Higashikawa FS, Silva CA, Akashi K, Mastrolonardo G, Sánchez-Monedero MA, Mondini C (2020) Role of biochar in promoting circular economy in the agriculture sector. Part 1: a review of the biochar roles in soil N, P and K cycles. Chem Biol Technol Agric 7(1). https://doi.org/10.1186/s40538-020-00182-8

Joseph S, Anawar HM, Storer P, Blackwell P, Chia C, Lin Y, Munroe P, Donne S, Horvat J, Wang J, Solaiman ZM (2015) Effects of enriched biochars containing magnetic iron nanoparticles on mycorrhizal colonisation, plant growth, nutrient uptake and soil quality improvement. Pedosphere 25(5):749–760. https://doi.org/10.1016/s1002-0160(15)30056-4

Joseph S, Cowie AL, Van Zwieten L, Bolan N, Budai A, Buss W, Cayuela ML, Graber ER, Ippolito JA, Kuzyakov Y, Luo Y, Ok YS, Palansooriya KN, Shepherd J, Stephens S, Weng Z, Lehmann J (2021) How biochar works, and when it doesn't: a review of mechanisms controlling soil and plant responses to biochar. Glob Change Biol Bioenergy 13:1731–1764. https://doi.org/10.1111/gcbb.12885

Kapoor A, Sharma R, Kumar A, Sepehya S (2022) Biochar as a means to improve soil fertility and crop productivity: a review. J Plant Nutr 45(15):2380–2388. https://doi.org/10.1080/01904167.2022.2027980

Khan Z, Zhang K, Khan MN, Zhu K, Hu L (2024) Effects of biochar persistence on soil physiochemical properties, enzymatic activities, nutrient utilization, and crop yield in a three-year rice-rapeseed crop rotation. Eur J Agron 154:127096. https://doi.org/10.1016/j.eja.2024.127096

Kizito S, Luo H, Lu J, Bah H, Dong R, Wu S (2019) Role of nutrient-enriched biochar as a soil amendment during maize growth: exploring practical alternatives to recycle agricultural residuals and to reduce chemical fertilizer demand. Sustainability 11(11):3211. https://doi.org/10.3390/su11113211

Krishnan S, Zulkapli NS, Kamyab H, Taib SM, Din MFBM, Majid ZA, Chaiprapat S, Kenzo I, Ichikawa Y, Nasrullah M, Chelliapan S, Othman N (2021) Current technologies for recovery of metals from industrial wastes: an overview. Environ Technol Innov 22:101525. https://doi.org/10.1016/j.eti.2021.101525

Kumar A, Saini K, Bhaskar T (2020) Hydochar and biochar: production, physicochemical properties and techno-economic analysis. Bioresour Technol 310:123442. https://doi.org/10.1016/j.biortech.2020.123442

Lammirato C, Miltner A, Kaestner M (2011) Effects of wood char and activated carbon on the hydrolysis of cellobiose by β-glucosidase from *Aspergillus niger*. Soil Biol Biochem 43(9):1936–1942. https://doi.org/10.1016/j.soilbio.2011.05.021

Lateef A, Nazir R, Jamil N, Alam S, Shah R, Khan MN, Saleem M, Rehman S (2019) Synthesis and characterization of environmental friendly corncob biochar based nano-composite – a potential slow release nano-fertilizer for sustainable agriculture. Environ Nanotechnol Monit Manag 11:100212. https://doi.org/10.1016/j.enmm.2019.100212

Lehmann J, Rillig MC, Thies J, Masiello CA, Hockaday WC, Crowley D (2011) Biochar effects on soil biota – a review. Soil Biol Biochem 43(9):1812–1836. https://doi.org/10.1016/j.soilbio.2011.04.022

Li Y, Li Y, Chang SX, Yang Y, Fu S, Jiang P, Luo Y, Yang M, Chen Z, Hu S (2018) Biochar reduces soil heterotrophic respiration in a subtropical plantation through increasing soil organic carbon recalcitrancy and decreasing carbon-degrading microbial activity. Soil Biol Biochem 122:173–185. https://doi.org/10.1016/j.soilbio.2018.04.019

Li C, Zhang C, Gao G, Gholizadeh M, Zhang S, Xu L, Zhang L, Li Q, Hu X (2020) Interaction of the volatiles from co-pyrolysis of pig manure with cellulose/glucose and their effects on char properties. J Environ Chem Eng 8(6):104583. https://doi.org/10.1016/j.jece.2020.104583

Lu H, Bian R, Xia X, Cheng K, Liu X, Liu Y, Wang P, Li Z, Zheng J, Zhang X et al (2020) Legacy of soil health improvement with carbon increase following one time amendment of biochar in a paddy soil – a rice farm trial. Geoderma 376:114567

Ma H, Egamberdieva D, Wirth S, Bellingrath-Kimura SD (2019a) Effect of biochar and irrigation on soybean-rhizobium symbiotic performance and soil enzymatic activity in field rhizosphere. Agronomy 9(10):626. https://doi.org/10.3390/agronomy9100626

Ma H, Egamberdieva D, Wirth S, Li Q, Omari RA, Hou M, Bellingrath-Kimura SD (2019b) Effect of biochar and irrigation on the interrelationships among soybean growth, root nodulation, plant p uptake, and soil nutrients in a sandy field. Sustainability 11(23):6542. https://doi.org/10.3390/su11236542

Ma H, Li Q, Egamberdieva D, Bellingrath-Kimura SD (2022a) A case study in desertified area: soybean growth responses to soil structure and biochar addition integrating ridge regression models. Agronomy 12(6):1341. https://doi.org/10.3390/agronomy12061341

Ma H, Shurigin V, Jabborova D, dela Cruz JA, dela Cruz TE, Wirth S, Bellingrath-Kimura SD, Egamberdieva D (2022b) The integrated effect of microbial inoculants and biochar types on soil biological properties, and plant growth of lettuce (*Lactuca sativa* L.). Plants 11(3):423. https://www.mdpi.com/2223-7747/11/3/423

Majeed M, Ali JS, Li Y, Günal E, Ali El Enshasy H, Salmen SH, Sürücü A (2022) The impact of corncob biochar and poultry litter on pepper (*Capsicum annuum* L.) growth and chemical properties of a silty-clay soil. Saudi J Biol Sci 29(4):2998–3005. https://doi.org/10.1016/j.sjbs.2022.01.037

Melo LCA, Lehmann J, Carneiro JS d S, Camps-Arbestain M (2022) Biochar-based fertilizer effects on crop productivity: a meta-analysis. Plant Soil. https://doi.org/10.1007/s11104-021-05276-2

Ndoung OCN, de Figueiredo CC, Ramos MLG (2021) A scoping review on biochar-based fertilizers: enrichment techniques and agro-environmental application. Heliyon 7(12):e08473. https://doi.org/10.1016/j.heliyon.2021.e08473

Nguyen TTN, Wallace HM, Xu C-Y, Zwieten LV, Weng ZH, Xu Z, Che R, Tahmasbian I, Hu H-W, Bai SH (2018) The effects of short term, long term and reapplication of biochar on soil bacteria. Sci Total Environ 636:142–151. https://doi.org/10.1016/j.scitotenv.2018.04.278

Nielsen S, Minchin T, Kimber S, van Zwieten L, Gilbert J, Munroe P, Joseph S, Thomas T (2014) Comparative analysis of the microbial communities in agricultural soil amended with enhanced biochars or traditional fertilisers. Agric Ecosyst Environ 191:73–82. https://doi.org/10.1016/j.agee.2014.04.006

Oladele SO (2019) Effect of biochar amendment on soil enzymatic activities, carboxylate secretions and upland rice performance in a sandy clay loam Alfisol of Southwest Nigeria. Sci Afr 4:e00107. https://doi.org/10.1016/j.sciaf.2019.e00107

Pang W, Hou D, Wang H, Sai S, Wang B, Ke J, Wu G, Li Q, Holtzapple M (2018) Preparation of microcapsules of slow-release NPK compound fertilizer and the release characteristics. J Braz Chem Soc. https://doi.org/10.21577/0103-5053.20180117

Premalatha RP, Poorna Bindu J, Nivetha E, Malarvizhi P, Manorama K, Parameswari E, Davamani V (2023) A review on biochar's effect on soil properties and crop growth. Front Energy Res. https://doi.org/10.3389/fenrg.2023.1092637

Puga AP, Grutzmacher P, Cerri CEP, Ribeirinho VS, de Andrade CA (2020) Biochar-based nitrogen fertilizers: greenhouse gas emissions, use efficiency, and maize yield in tropical soils. Sci Total Environ 704:135375. https://doi.org/10.1016/j.scitotenv.2019.135375

Rangabhashiyam S, Balasubramanian PJ (2019) The potential of lignocellulosic biomass precursors for biochar production: performance, mechanism and wastewater application—a review. Ind Crop Prod 128:405–423. https://doi.org/10.1016/j.indcrop.2018.11.041

Rombel A, Krasucka P, Oleszczuk P (2022) Sustainable biochar-based soil fertilizers and amendments as a new trend in biochar research. Sci Total Environ 816:151588. https://doi.org/10.1016/j.scitotenv.2021.151588

Shao Y, Long Y, Wang H, Liu D, Shen D, Chen T (2019) Hydrochar derived from green waste by microwave hydrothermal carbonization. Renew Energy 135:1327–1334. https://doi.org/10.1016/j.renene.2018.09.041

Shin J, Park S (2018) Optimization of blended biochar pellet by the use of nutrient releasing model. Appl Sci (Basel) 8(11):2274. https://doi.org/10.3390/app8112274

Sim DHH, Tan IAW, Lim LLP, Hameed BH (2021) Encapsulated biochar-based sustained release fertilizer for precision agriculture: a review. J Clean Prod 303:127018. https://doi.org/10.1016/j.jclepro.2021.127018

Sun R, Dsouza M, Gilbert JA, Guo X, Wang D, Guo Z, Ni Y, Chu H (2016) Fungal community composition in soils subjected to long-term chemical fertilization is most influenced by the type of organic matter. Environ Microbiol 18(12):5137–5150. https://doi.org/10.1111/1462-2920.13512

Susilawati A, Maftuah E, Fahmi A (2020) The utilization of agricultural waste as biochar for optimizing swampland: a review. IOP Conf Ser Mater Sci Eng 980(1):012065. https://doi.org/10.1088/1757-899x/980/1/012065

Tian J, Wang J, Dippold M, Gao Y, Blagodatskaya E, Kuzyakov Y (2016) Biochar affects soil organic matter cycling and microbial functions but does not alter microbial community structure in a paddy soil. Sci Total Environ 556:89–97. https://doi.org/10.1016/j.scitotenv.2016.03.010

Tomczyk A, Sokołowska Z, Boguta P (2020) Biochar physicochemical properties: pyrolysis temperature and feedstock kind effects. Rev Environ Sci Biotechnol 19(1):191–215. https://doi.org/10.1007/s11157-020-09523-3

Ullah S, Liang H, Ali I, Zhao Q, Iqbal A, Wei S, Shah T, Yan B, Jiang L (2020) Biochar coupled with contrasting nitrogen sources mediated changes in carbon and nitrogen pools, microbial and enzymatic activity in paddy soil. J Saudi Chem Soc 24(11):835–849. https://doi.org/10.1016/j.jscs.2020.08.008

Velli P, Manolikaki I, Diamadopoulos E (2021) Effect of biochar produced from sewage sludge on tomato (*Solanum lycopersicum* L.) growth, soil chemical properties and heavy metal concentrations. J Environ Manag 297:113325. https://doi.org/10.1016/j.jenvman.2021.113325

Wu S, Zhang Y, Tan Q, Sun X, Wei W, Hu C (2020) Biochar is superior to lime in improving acidic soil properties and fruit quality of *Satsuma mandarin*. Sci Total Environ 714:136722. https://doi.org/10.1016/j.scitotenv.2020.136722

Xu YX, He LL, Liu YX, Lv HH, Wang YY, Chen JY, Yang SM (2019) Effects of biochar addition on enzyme activity and fertility in paddy soil after six years. Chin J Appl Ecol 30:1110–1118

Xu W, Xu H, Delgado-Baquerizo M, Gundale MJ, Zou X, Ruan H (2023) Global meta-analysis reveals positive effects of biochar on soil microbial diversity. Geoderma 436:116528. https://doi.org/10.1016/j.geoderma.2023.116528

Yang W, Li C, Wang S, Zhou B, Mao Y, Rensing C, Xing S (2021) Influence of biochar and biochar-based fertilizer on yield, quality of tea and microbial community in an acid tea orchard soil. Appl Soil Ecol 166:104005. https://doi.org/10.1016/j.apsoil.2021.104005

Yao T, Zhang W, Gulaqa A, Cui Y, Zhou Y, Weng W, Wang X, Liu Q, Jin F (2021) Effects of peanut shell biochar on soil nutrients, soil enzyme activity, and rice yield in heavily saline-sodic paddy field. J Soil Sci Plant Nutr. https://doi.org/10.1007/s42729-020-00390-z

Yek PNY, Cheng YW, Liew RK, Wan Mahari WA, Ong HC, Chen W-H, Peng W, Park Y-K, Sonne C, Kong SH et al (2021) Progress in the torrefaction technology for upgrading oil palm wastes to energy-dense biochar: a review. Renew Sust Energ Rev 151:111645. https://doi.org/10.1016/j.rser.2021.111645

Zhang Q, Song Y, Wu Z, Yan X, Gunina A, Kuzyakov Y, Xiong Z (2020) Effects of six-year biochar amendment on soil aggregation, crop growth, and nitrogen and phosphorus use efficiencies in a rice-wheat rotation. J Clean Prod 242:118435. https://doi.org/10.1016/j.jclepro.2019.118435

Zhang L, Larsson A, Moldin A, Edlund U (2022) Comparison of lignin distribution, structure, and morphology in wheat straw and wood. Ind Crop Prod 187:115432. https://doi.org/10.1016/j.indcrop.2022.115432

Zhang S, Wang L, Gao J, Zhou B, Hao W, Feng D, Sun X (2024) Effect of biochar on biochemical properties of saline soil and growth of rice. Heliyon 10(1):e23859. https://doi.org/10.1016/j.heliyon.2023.e23859

Zhou Z, Gao T, Van Zwieten L, Zhu Q, Yan T, Xue J, Wu Y (2019) Soil microbial community structure shifts induced by biochar and biochar-based fertilizer amendment to Karst calcareous soil. Soil Sci Soc Am J 83(2):398–408. https://doi.org/10.2136/sssaj2018.08.0297

Zhou J, Qu T, Li Y, Van Zwieten L, Wang H, Chen J, Song X, Lin Z, Zhang X, Luo Y, Cai Y, Zhong Z (2021) Biochar-based fertilizer decreased while chemical fertilizer increased soil N_2O emissions in a subtropical Moso bamboo plantation. Catena 202:105257. https://doi.org/10.1016/j.catena.2021.105257

Chapter 2
Bioconversion of Agricultural/Industrial Waste into Biofuels

Hamza Umar, Kanchan Bhatt, Aharnish Maurya, and Ashish

Abstract Rising demand for renewable energy has increased interest in biofuels as sustainable alternatives to fossil fuels. This chapter examines the bioconversion of agricultural and industrial waste into first- and second-generation biofuels, addressing waste valorization, energy security, and environmental sustainability. The study explores lignocellulosic biomass and agricultural residues as feedstocks, utilizing both biochemical and thermochemical conversion pathways, including fermentation, anaerobic digestion, and pyrolysis. Key focus areas include microbial consortia optimization, engineered strains, and advanced bioreactor design to enhance biofuel yields. The chapter evaluates environmental benefits, economic feasibility, and policy frameworks for widespread adoption. Life cycle assessment (LCA) demonstrates long-term sustainability through carbon footprint reduction and integration with existing industrial systems. This waste-to-biofuel approach promotes a circular bioeconomy, reduces waste disposal challenges, and aligns global climate action goals through multidisciplinary strategies.

Keywords Biofuels · Bioconversion · Lignocellulosic biomass · Anaerobic digestion · Microbial consortia · Circular bioeconomy · Waste valorization · Sustainable energy

H. Umar · Ashish (✉)
Department of Bioengineering Engineering, Integral University,
Lucknow, Uttar Pradesh, India
e-mail: ashish@iul.ac.in

K. Bhatt
Department of Biotechnology Engineering, Goel Institute of Technology and Management,
Lucknow, Uttar Pradesh, India

A. Maurya
Department of Biotechnology Engineering, Kashi Institute of Technology,
Varanasi, Uttar Pradesh, India

N. K. Arora, J. Mishra (eds.), *Biotechnological Solutions for a Sustainable Future*, Microorganisms for Sustainability 55,
https://doi.org/10.1007/978-981-95-3989-5_2

2.1 Introduction

In response to the world's rising energy consumption, growing environmental sustainability concerns, and the utilization of natural sources of oil and gas reserves, scientists are looking into alternative and renewable energy sources. Biofuels derived from biomass provide an excellent alternative as they exhibit carbon neutrality and have the capability to mitigate greenhouse gas emissions. Hence, bioconversion is the process of converting agricultural and industrial waste into biofuels, and biomass utilization stands out as a viable and sustainable approach that promotes the concepts of the circular economy. Industrial by-products such as food processing waste, lignocellulosic effluents, and glycerol from the production of biodiesel, as well as agricultural residues including straw, husks, bagasse, and manure, are adequate and unexplored sources of organic carbon.

Bioconversion methods, such as anaerobic digestion, fermentation, enzymatic hydrolysis and microbial transformation, are all capable of transforming waste or biomass into bioethanol, biobutanol, biogas, biodiesel, and other advanced biofuels. The growing generation of waste is a significant environmental issue resulting from the expansion of the farming and industrial sectors. The agriculture industry generates five million tons of garbage annually, according to Ravindran et al. (2018). These waste streams contain a lot of lipids, proteins, and carbohydrates that biological processes can transform into products with additional value.

These wastes include household, garden, and timber waste as well as agricultural and industrial waste, and their main constituents are starch and lignocellulose. Lignocellulose is the most abundantly available raw material on Earth, it is composed of two kinds of carbohydrate polymers, cellulose and hemicellulose, and an aromatic-rich polymer called lignin, it is composed of three main polymers: 35–50% cellulose, 20–35% hemicellulose, and 15–25% lignin, where these polymers are interconnected through a combination of covalent and non-covalent bonds (Bharathiraja et al. 2017; Kohli et al. 2019). In addition, lignocellulose contains other organic substances in minute quantities, including fats, proteins, dietary fibers, vitamins and some key nutrients. It also contains some inorganic substances like carbon, water, sulphates, nitrates, and silicates (Mehmood et al. 2019).

According to (Yazid et al. 2017), agro-industrial waste such as bagasse, grass stems, cobs, fruit and husk may be used as a feedstock for a value-added product. Utilizing waste materials to create useful products that boost financial benefits in a range of enterprises is currently a global trend. Lignocellulosic biomass is an abundant, readily available bio-renewable natural resource that is used as a precursor for a value-added product formulation in the modern industrial sector (Mehmood et al. 2019). Ngbede et al. (2019) also reported that a vast amount of agro-industrial waste is globally generated by food and agricultural industries. Such waste poses a menace to the ecosystem due to its negative effect on the environment.

Agro-industrial waste that contains a high amount of nutrients shouldn't be considered waste. To enhance the industrial output of value-added products such as enzymes, biofuels, biofertilizers, and a variety of other industrial products, these materials can alternatively serve as substrates for microbial cultures. By using the appropriate microbes to transform waste into value-added products, these technologies aim to alleviate disposal-related issues and lessen environmental contamination (Ezejiofor 2014). Nowadays, it is still essential to lessen the negative environmental consequences of these pollutants and efficiently and economically transform these wastes into important industrial and profitable products (Mehmood et al. 2019).

2.2 Overview of Global Energy Demand and Environmental Concerns

The global energy demand during the twenty-first century has been encountering new and serious problems due to more energy demands, and the environment is damaged due to the consequences of energy usage. The world population is expected to reach 9.7 billion by 2050; however, this population rise has led to a rise in global energy consumption (Awogbemi and Von Kallon 2022). Predictions indicate that the world's primary energy consumption will rise from 543 exajoules (EJ) in 2015 to 575 EJ in 2018 and 725 EJ in 2050 (Awogbemi and Von Kallon 2022). Climate change is the negative impact of global warming, which is caused by the rise in energy consumption of burning fossil fuels, which also contributes to over 36 billion metric tons of CO_2 emissions annually (Awogbemi and Von Kallon 2022; Leong et al. 2021).

Due to high demand and usage of fossil fuels, which causes global warming and havoc to the ecosystem by depleting the ozone layer, there is a need for a global transition to cleaner and greener energy sources. Biofuels, derived from biomass and organic waste, are becoming a competitive alternative to fossil fuels. The transition from fossil fuel to renewable energy provides some environmental and socioeconomic benefits, including greenhouse gas emissions reduction, less dependency on fossil fuels and the potential for rural economic growth (Fatih Demirbas et al. 2011; Mekunye and Makinde 2024).

Agro-industrial waste generated globally is estimated at 998 million tons of organic waste annually, and most of this waste is left to decompose or is burned by an open fire, thereby causing harm to the ecosystem (Awogbemi and Von Kallon 2022). Bioconversion of waste into a value-added product like bioethanol, biodiesel, and biogas can help reduce the harmful impact these wastes have on the environment. This idea matches the circular bioeconomy approach, where waste is reprocessed into raw material to make new products and energy. This helps reduce greenhouse gas emissions and supports long-term environmental sustainability Leong et al. (2021).

Additionally, Biofuels, during combustion, the CO_2 released when they are burned is balanced by the CO_2 the plants absorbed while growing, creating a cycle that doesn't add extra CO_2 to the air, which makes them environmentally desirable. Thus, biofuels effectively reduce the effect of greenhouse gases on the environment, and also contribute to the advancement of global climate goals (Fatih Demirbas et al. 2011). Transition of waste to biofuels helps the United Nations in achieving its agenda of Sustainable Development Goals (SDGs), with an interest towards achieving affordable clean energy, fighting climate change, and using resources responsibly. Therefore, converting agro-industrial waste into biofuels is helpful in two ways: it addresses pressing environmental challenges and contributes to global energy security. This study also looks at the progress, problems, and future opportunities in this important area.

2.3 Significance of Biofuels in the Circular Economy

Concern over resource depletion, waste accumulation, and environmental degradation has grown globally, leading to a paradigm change in economic systems toward more sustainable ones. A great solution to these problems is the circular economy, which focuses on reducing waste, resource optimization, and the regeneration of natural systems. In this climate, biofuels derived from the bioconversion of industrial and agricultural waste are essential for fostering circularity since they transform waste streams into beneficial bioproducts and sustainable energy sources.

Biofuels are an eco-friendly alternative to fossil fuels, with the advantage of being biodegradable, low-carbon footprint and being accommodated by the existing systems. Unlike traditional linear models that rely on extract-produce-dispose cycles, the circular economy emphasizes reuse, recycling, and value-adding. This concept is demonstrated by converting organic waste, such as lignocellulosic residues, food processing wastes, and agro-industrial effluents, into biofuels, which reduces environmental burdens and transforms liabilities into resources (Leong et al. 2021).

The idea of a biorefinery is an important part of the circular bioeconomy. It helps connect waste-to-biofuel processes with bigger industrial setups, making everything work together more efficiently. Biotechnological and thermochemical processes like fermentation, anaerobic digestion, pyrolysis, and gasification can be used to transform waste into value-added products such as bioethanol, biodiesel, biogas, and biohydrogen (Mekunye and Makinde 2024; Awogbemi and Von Kallon 2022). This appraisal includes helping to generate clean energy as well as waste management, pollution reduction, and soil nutrient recycling, all crucial pillars of sustainable development.

Circular bioeconomy approaches attempt to reduce dependency on finite fossil resources by shortening material loops and increasing resource efficiency. The Sustainable Development Goals (SDGs) agenda is to provide an eco-friendly environment, where biofuels generate clean and sustainable energy, fighting climate change, and use resources responsibly. Leong et al. (2021) state that transforming waste into energy through a circular economy has numerous advantages, which are reduced greenhouse gas emissions, decreased reliance on petrochemicals, and enhanced economic resilience through the development of regional bioenergy systems.

Additionally, biofuels created from trash reduce competition with food resources, in contrast to first-generation biofuels that rely on edible crops. Using non-edible and leftover biomass, the technique increases food security while protecting arable land (Fatih Demirbas et al. 2011). With the correct technology and accompanying regulations, bioconversion of industrial and agricultural waste into biofuels has the potential to drive systemic change toward a low-carbon, resource-efficient, and circular global economy with the right technology and supportive policies.

2.4 Rationale for Using Agricultural and Industrial Waste

Managing the growing amounts of organic waste and supplying the growing need for renewable energy are two pressing issues facing the modern world. Waste from industry and agriculture, which was formerly seen to be a burden on the environment, is now a viable feedstock for bioconversion processes that produce biofuels. This strategy's justification is based on imperatives pertaining to the environment, the economy, and energy.

Annually, more than 998 million tons of agricultural waste are produced worldwide; the majority of this waste is left to decompose or is burned by an open fire, thereby seriously degrading the environment through greenhouse gas emissions and air pollution (Awogbemi and Von Kallon 2022). Similarly, Industries like food processing, pulp and paper, and fermentation industries generate large amounts of solid wastes and rich organic effluents. Because of their biochemical oxygen demand (BOD), microbial contamination, and role in landfill overpopulation, these waste streams represent major risks to the environment and public health if left untreated (Ngbede et al. 2019).

Nevertheless, lignocellulosic material, carbohydrates, sugars, and protein components that may be efficiently transformed into biofuels through biological or thermochemical processes are plentiful in these wastes. For instance, cellulose and hemicellulose found in rice husks, cassava peels, sugarcane bagasse, and food industry waste may be hydrolyzed and fermented to produce bioethanol or biogas (Mekunye and Makinde 2024; Fatih Demirbas et al. 2011). In addition to keeping garbage out of landfills, the capacity to turn waste into energy lowers the cost of producing biofuel by substituting inexpensive, locally accessible biomass for pricey synthetic feedstocks.

Furthermore, the "food vs. fuel" controversy surrounding first-generation biofuels, which depend on edible crops like maize and sugarcane, is lessened by using waste-based feedstocks. In poor nations where food shortages and energy poverty frequently occur, the bioconversion strategy's transition to non-food, waste-derived biomass protects food security while improving energy security (Awogbemi and Von Kallon 2022; Leong et al. 2021).

This idea focuses on recycling resources and reducing waste, which fits well with the main goals of a circular economy. Transformation of waste to energy is an example of a closed-loop system, where leftover biomass is not just thrown away but is instead valorized into a value-added product such as bioethanol, biodiesel, and biogas (Leong et al. 2021). Renewable energy's contribution to society is numerous, including lowering greenhouse gas emissions, protecting natural resources, and generating jobs in rural areas via the collecting and processing of biomass, which promotes sustainable development.

In conclusion, using industrial and agricultural waste as feedstocks for biofuels provides a comprehensive approach to environmental preservation, energy production, and waste management. Waste-derived biofuels will play an ever-more-important part in the worldwide shift to sustainable energy systems as technology and legislative frameworks develop.

2.5 Agricultural Waste/Feedstocks for Biofuel Production

After crops are harvested, agro-waste such as stalks, husks and other plant waste is often not used much, but they can be transformed into valuable products instead of being burned or composted, making them an eco-friendlier option. Waste conversion into biofuels such as ethanol and biodiesel significantly reduces the effect of greenhouse gas emissions in the environment, particularly NOx and SOx gases, which are potent contributors to climate change (de Freitas et al. 2021). Bioethanol production technology maximizes the utilization of cellulose and hemicellulose by transforming them through a series of processes into an economically valuable product via a biological approach called fermentation (Shukla et al. 2023). However, proper utilization of agricultural waste or residue addresses the detrimental effects of them within the society, also optimizing the biofuel supply chain helps address the seasonal and geographic variability of residue availability, enhancing efficiency and supporting cleaner energy generation from agricultural biomass (Saleem 2022).

2.5.1 Crop Waste/Residues

Agro-residues such as straw, husks, stalks, and leaves are all underutilized by discarding them in the field after harvest, either for animals to graze on them or to be decomposed or degraded by microorganisms or termites. Crop residues are divided into two categories: plant residues, which include stems, leaves, and roots, and field residues, which include stalks, stubble, and seed pods (Grzyb et al. 2020). Some of these crop residues are considered valuable as fuel materials, and some have the potential to be good precursors for biofuel production. When processed, crop residues can be transformed into solid biofuels or refined into liquid fuels such as ethanol and biodiesel. Although biofuels generally have a smaller carbon footprint compared to fossil fuels during consumption, their production can demand significant water and oil resources, raising environmental concerns (Jeswani et al. 2020). Despite being viewed as carbon neutral, biofuels can still have environmental impacts similar to fossil fuels if unsustainably managed. Key energy crops like corn, millet, and white sweet clover are widely used in biofuel generation, particularly in sustainable transport and thermal energy applications (Yang et al. 2023b). The United States is the highest producer of biofuel in the world. In the U.S., among all the precursors of biodiesel, corn is the most utilized due to its potential and availability, while Brazil utilizes sugarcane. Biomass refers to any organic matter used to create energy, while biofuels are specific energy products derived from biomass. Advanced biofuels are liquid fuels produced from non-food biomass sources, such as agricultural residues, forestry residues, algae, and waste materials, using advanced technologies. Advanced biofuels also aim to reduce greenhouse gas emissions compared to fossil fuels and promote energy sustainability (Yap et al. 2021).

2.5.2 *Animal Manure and Slurry*

Animal waste can be transformed into a renewable source of energy through an effective and environmentally friendly means. Animal manure and slurry are significant renewable resources that can be effectively converted into biofuels through biological processes (Rout et al. 2023). Conversion of animal waste into value-added products contributes to the realization of the UNDP agenda of promoting low-emission, climate-friendly, and fair development by protecting natural resources, reducing pollution, and supporting long-term environmental health (Raihan and Tuspekova 2022). Anaerobic digestion, a biological process in which animal manure is transformed into biofuel by the action of microorganisms to break down organic matter without oxygen to produce biogas, primarily consisting of methane (Rasapoor et al. 2020). In contrast to conventional cow dung cakes, cow dung can be converted into biogas, a cleaner and more effective fuel. The by-product of the biogas production is an organic fertilizer which is used in farms and gardens for efficient yield (De Sousa et al. 2022), shown in Fig. 2.1. Moreover, dried animal feces can be directly used as fuel in some rural communities. Animal fats and used cooking oils can also undergo transesterification to produce biodiesel and glycerin. Utilizing waste into a value-added product and encouraging eco-friendly waste management solutions. In the United States, most biofuel produced and used is ethanol; however, biogas from animal waste still has potential worldwide. Cow dung and slurry are useful bioresources that can be used for clean energy generation (Shaibur et al. 2021).

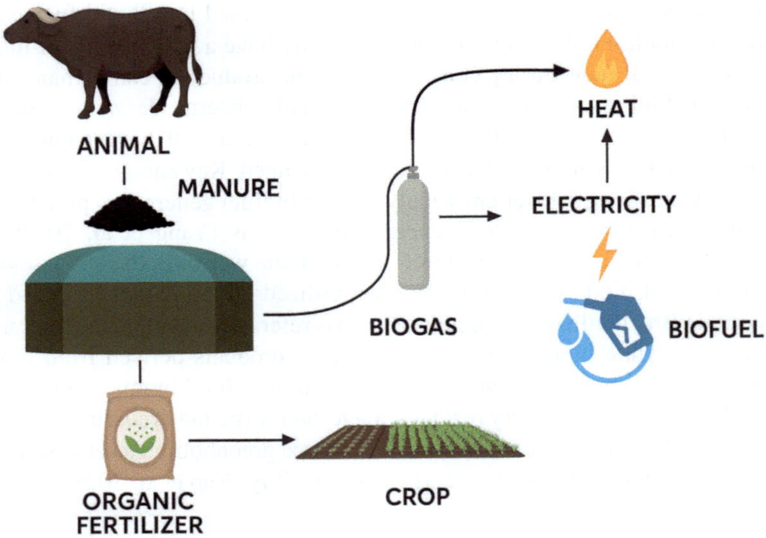

Fig. 2.1 Animal waste utilization into a value-added product

2.5.3 Agro-Industrial By-Products

Bagasse and molasses are valuable renewable resources with much potential in biofuel production. Bagasse, a by-product of sugar processing, is a precursor for biofuel production, which serves as an alternative to fossil fuels, and also complies with the UNDP agenda for an environmentally sustainable alternative fuel by reducing toxic gases that deplete the ozone layer (Ajala et al. 2021). Bagasse, a cellulose material that is the leftover material after extracting juice from sugarcane or sorghum, is utilized as a raw material for biofuel production. It also finds applications in paper production and the manufacture of building materials, enhancing its economic value (Crini et al. 2020). Molasses is a by-product of the sugar processing industry, which is obtained after the crystallization of sugar. Molasses is a thick, dark syrup produced as a by-product of sugar refining, typically from sugar cane or sugar beets. It's a viscous liquid that remains after sugar crystals are removed from the juice during the sugar-making process (Jeremiah and Kabeyi 2023). Molasses is used in numerous ways; it can be used as a sweetener, it can be used as a flavor in food, it can also be used as a key component in certain commercial products, and molasses can also be used as a feedstock for ethanol production. Utilizing agro-industrial waste and industrial by-products into a value-added product contributes to sustainable energy systems and circular economy practices. The transformation of agro-industrial waste into a useful product promotes environmental pollution, and it also enhances the overall efficiency and sustainability of industries.

2.5.4 Industrial Waste

Industrial waste is regularly viewed as an environmental nuisance; meanwhile, it has great potential to be a feedstock or precursor for biofuel production. These wastes are obtained from various sectors such as food processing, paper mills, breweries, and chemical manufacturing. However, waste rich in sugars, fats, and proteins can be converted into biogas, biodiesel and bioethanol via processes like anaerobic digestion, transesterification and fermentation, respectively (Dhanya et al. 2020). This method is a sustainable and eco-friendly alternative source of renewable energy, which encourages the transformation of waste into a value-added product and also protects the environment from pollution and reduces the effects of greenhouse gas emissions (Mozhiarasi and Natarajan 2022). Moreover, industrial waste conversion into a value-added product supports the circular economy by transforming waste residues into useful products.

2.5.5 Food Processing Waste

Food waste processing is an important opportunity for biofuel production, as it provides both environmental and economic benefits. Waste food rich in sugars, fats, and proteins can be transformed into value-added products since it contains rich organic matter (Yukesh Kannah et al. 2020). Also, waste food can be valorized into biofuels such as bioethanol, biodiesel and biogas using different conversion technologies. Energy can be recovered from food waste using several key methods, including anaerobic digestion, fermentation, transesterification, and pyrolysis (Srivastava et al. 2018). However, the breaking down of organic materials in an inert environment under the influence of biological enzymes to produce biogas, a mixture primarily of methane and carbon dioxide, is referred to as Anaerobic digestion. However, during the process, theoretically, a kilogram of food waste can produce up to 421 liters of biogas, through pyrolysis by decomposition of food residues, a combustible gas containing hydrogen, methane, and carbon monoxide, which is used for power generation or as feedstock for further fuel synthesis (Blasi et al. 2023). Waste cooking oils can be transformed into biodiesel. Utilizing food waste into a value-added product for biofuel contributes to reducing environmental pollution and reduces the emission of toxic gases into the environment (Yukesh Kannah et al. 2020). This approach supports sustainable waste management while addressing growing global energy demands.

2.5.6 Waste from the Pulp and Paper Industry

Waste from the Pulp and Paper industries contains a high content of lignocellulose, composed of cellulose, hemicellulose, and lignin, which is considered a valuable source for biofuel production (Haile et al. 2021). By-products such as black liquor and paper sludge can also be transformed into biofuels through gasification, anaerobic digestion, and fermentation. Black liquor is suitable for producing syngas or heat, while paper sludge can generate biogas or ethanol (Jian et al. 2021). These methods significantly benefit the environment and play a crucial role in energy recovery and supporting sustainability. Utilizing the potential of these waste streams encourages a circular economy and offers a renewable alternative to fossil fuels within industrial processes (Yang et al. 2023a).

2.5.7 Textile and Brewery Waste

Textile and brewery industries discharge waste that has significant importance, specifically to the energy sector. Textile industry effluent contains organic-rich material and natural fiber residues like cotton and wool, which can be processed into a valuable product through anaerobic digestion or fermentation to produce biogas or bioethanol (Iragavarapu et al. 2023). However, brewery waste, like spent grains, yeast, and wastewater, contains a lot of carbohydrates and proteins, making it a good material for producing bioethanol or biogas (Mitri et al. 2022). These residues can effectively be used through microbial conversion processes, reducing both environmental pollution and the waste disposal dilemma. Conversion of energy from this waste supports renewable energy development and also promotes circular economy principles. By converting textile and brewery byproducts into an economically valuable resource, industries can now operate on a zero-waste production, whereby a by-product of one production process will be the feedstock of another production and curtailing the problem or burden of waste management and also having a cleaner, eco-friendly, green energy atmosphere (Kabeyi and Olanrewaju 2022).

2.5.8 Municipal Solid Waste

This is the type of waste generated by humans due to the day-to-day domestic activities such municipal solid waste include food scraps, yard trimmings, woods, and biodegradable plastics, keeping these waste within the community serves as a threat to health and a comfortable environment, and atmosphere, some of these waste are non-degradable which if left it will stay in the landfill for a long period without degrading, however, these wastes have the potentials to be transformed into a valuable feedstock for biofuel production (Goel et al. 2021). Moreover, this waste can

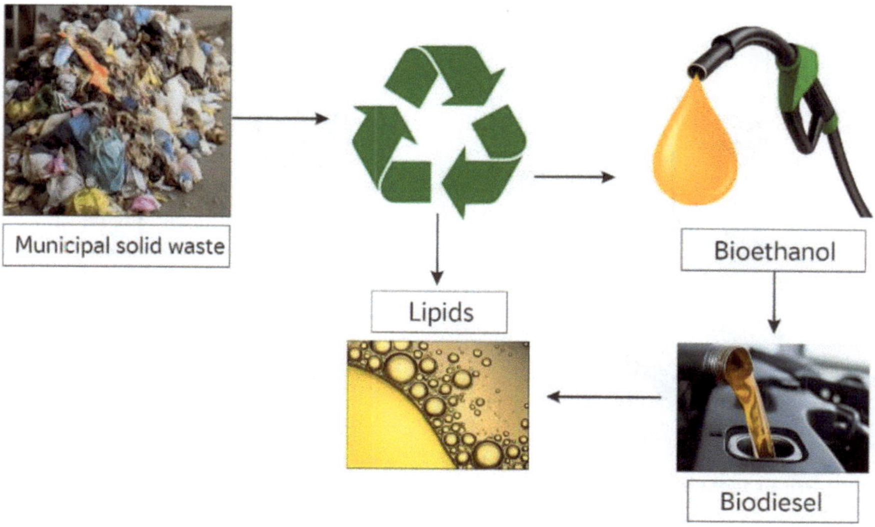

Fig. 2.2 Conversion of waste into a valuable product

be valorized into a value-added product of biogas, bioethanol, or biodiesel through biochemical and thermochemical conversion processes via anaerobic digestion, fermentation, and pyrolysis (Jha et al. 2022). The organic fraction of MSW contains the potential of a renewable energy source and also contributes to waste management, which also contributes to lower greenhouse gas emissions, and promotes sustainable urban waste management strategies, as shown in Fig. 2.2.

2.6 Types of Biofuels Derived from Waste

2.6.1 Bioethanol

Bioethanol is a renewable biofuel produced mainly by fermenting sugars found in materials like sugarcane, corn, and lignocellulosic agricultural residues. Ethanol is classified as a first-generation biofuel because it is obtained from food crops, and the second-generation category is obtained from non-food plant materials (e.g., crop residues, wood, and lignocellulosic materials) (Jha et al. 2022). The first generation of ethanol was sidelined because it is affecting the food chain in such a way that food intended for humans is being diverted to be used by machines (Rika Widianita 2023). Bioethanol, a renewable fuel, is primarily produced from sugar and starch-containing materials through a process involving fermentation and distillation. Lignocellulosic biomass, like agricultural residues, is being converted into a valuable renewable source of energy through enzymatic hydrolysis to break down complex carbohydrates under the influence of an enzyme into simple sugars via

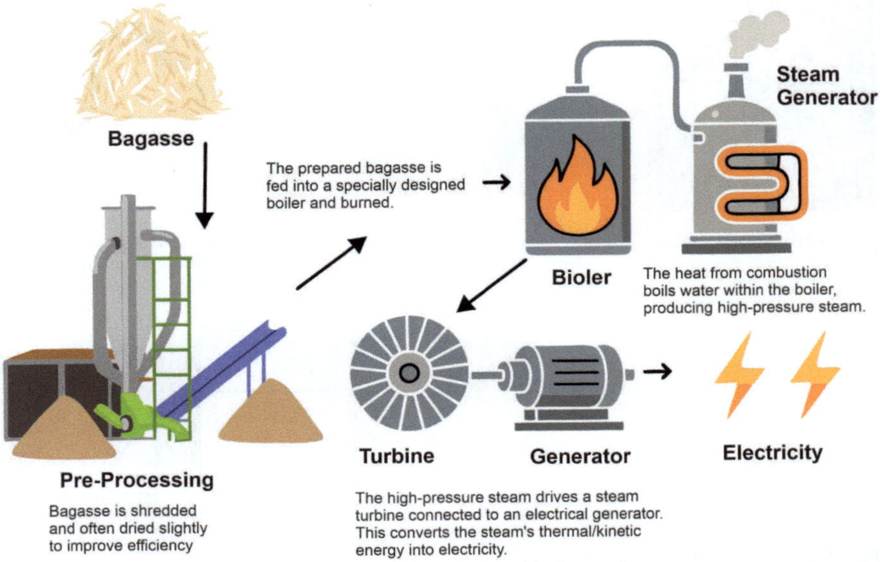

Fig. 2.3 Bagasse transformation into Bioenergy

fermentation. Bioethanol is an alternative to fossil fuels. Bioethanol, unlike fossil fuels, emits several gases when burned, just like fossil fuels, but generally in lower amounts, which contributes to reducing greenhouse gas emissions and improving air quality. It is widely used in blends such as E10 or E85 (Jhang et al. 2020) (Fig. 2.3).

2.6.2 Biobutanol

Biobutanol is another form of renewable energy fuel source, where sugars and starches are transformed into biofuel through fermentation under the influence of a bacterium called *Clostridium* species using the acetone-butanol-ethanol (ABE) process (Riaz et al. 2022). Biobutanol, which is also regarded as a next-generation biofuel, has some differences with bioethanol. Table 2.1 shows the Comparative analysis between Bioethanol and Biobutanol. Biobutanol can be used without modification or blending in an internal combustion engine (Dessie et al. 2020). Feedstocks for the production of biobutanol are almost similar to bioethanol, which include agricultural residues, lignocellulosic biomass, and food waste (Toor et al. 2020). Biobutanol is becoming popular as a clean, renewable fuel that can help reduce the use of fossil fuels.

Table 2.1 Comparative analysis between bioethanol and biobutanol

Feature	Bioethanol	Biobutanol
Chemical formula	C_2H_5OH	C_4H_9OH
Energy content	~21 MJ/L (lower than gasoline)	~29 MJ/L (closer to gasoline)
Production	Fermentation of sugars/starches using yeast (*Saccharomyces cerevisiae*)	ABE (Acetone-Butanol-Ethanol) fermentation using *Clostridium* species
Volatility	More volatile	Less volatile
Water solubility	Highly soluble (hygroscopic)	Lower solubility—better in moist environments
Compatibility	Corrosive to pipelines and engines; often requires blending (e.g., E10)	More compatible with existing gasoline infrastructure
Octane rating	High (~108)—good anti-knock	Moderate (~96)—still good anti-knock properties
Boiling point	78 °C	117 °C

2.6.3 *Biodiesel*

Biodiesel is a renewable fuel made by a process called transesterification. Fats from vegetable oils, animal fats, or used cooking oil can be turned into fuel by reacting them with alcohol like methanol, using a catalyst (for example, sodium or potassium hydroxide) to speed up the process (Salaheldeen et al. 2021). This reaction produces fatty acid methyl esters (FAME) and glycerol as by-products. Biodiesel is biodegradable, non-toxic, and compatible with conventional diesel engines in its pure form (B100) or mixes such as B20 or B5 (Randive et al. 2024). It helps the environment by releasing fewer greenhouse gases and reducing harmful air pollutants like carbon monoxide. Biodiesel costs in India vary by region, with rates quoted as $ 0.835/liter in New Delhi and Noida and Rs. 0.80/liter elsewhere (Randive et al. 2024).

2.6.4 *Biogas*

Biogas is a renewable fuel with a composition of methane (CH_4, 50–70%) and carbon dioxide (CO_2, 25–50%) and some trace elements. Biogas is a product of the fermentation of food scraps, animal dung and agro-waste in a reactor or vessel under the influence of an enzyme (Kabeyi and Olanrewaju 2022). Biogas is produced via a process called anaerobic digestion; this process involves the decomposition of organic material by an enzyme in a reactor in the absence of air. After the production the produced biogas is being purified, the purified biomethane can be used as a source of heat in the household or industries, or it can be used for the production of electricity (Dalpaz et al. 2020). It is a sustainable substitute for fossil fuels because of its great energy potential due to its methane content. However, after the conversion of the precursor into biogas, the biomass or the leftover organic waste, which is a nutrient-rich digestate, is used for soil improvement (Bhatia and Singh 2024). Biogas production also encourages waste management, lowers the effect of greenhouse gas emissions on the environment, and supports the ideas of the circular economy.

2.6.5 Biohydrogen

Biohydrogen is considered a clean, carbon-free biofuel produced by microorganisms through a process called dark fermentation. These methods utilize organic substrates to generate hydrogen gas, providing an alternative to conventional hydrogen production from fossil fuels (Megia et al. 2021). Biohydrogen can be produced through various methods, such as using light-based fermentation, anaerobic fermentation, specialized enzymes or microbes, or a combination of these methods. The theoretical biohydrogen yield is around 12 mol H_2/mol glucose, with practical microbiological yields of approximately 3.8 mol/mol. Biohydrogen plays a significant role in clean energy transitions due to its high energy density and zero carbon emissions when combusted. It is relevant in transportation, electricity generation, and fuel cells, aiding efforts to combat climate change and decrease dependence on fossil fuels.

2.6.6 Biochar

Biochar is a carbon-rich material produced through a thermal process called pyrolysis by heating biomass in low-oxygen conditions. Biochar is a renewable fuel that has a 27–32 MJ/kg calorific value, which may be the same or greater than coal (Lee et al. 2020). However, biochar is generated after the pyrolysis reaction. The precursors for the production of biochar involve feedstocks like agro-waste, forestry waste, animal manure, and plastic waste. Biochar has minimal greenhouse gas emissions. Biochar has a lot of environmental benefits, and its economic value is on the rise, with a market value of about $600 million by 2023 (Ahmed et al. 2024).

2.7 Bioconversion Technologies and Pathways

2.7.1 Thermochemical Conversion

Thermochemical conversion is the method of transforming biomass into value-added products and fuels using a thermochemical process. The methods applied for such processes or approaches include combustion, pyrolysis, gasification, hydrothermal liquefaction (HTL), hydrothermal carbonization (HTC), etc. (Teoh et al. 2024). This thermochemical process involves different methods or approaches: (a) combustion, the process of burning biomass with oxygen to generate heat and electricity, (b) pyrolysis, a method whereby biomass decomposes due to the influence of heat in an inert environment to produce bio-oil, biochar, and syngas, (c) gasification, a thermochemical process that converts carbonaceous materials like biomass or fossil fuels into a gaseous product called syngas, primarily composed of carbon monoxide and hydrogen, (d) HTC and HTL, the processes where water is used at high temperature and pressure to produce solid or liquid fuels. These thermochemical processes provide a scalable and flexible innovation for converting different biomass into sustainable and renewable energy (Wang and Wu 2023).

2.7.1.1 Pyrolysis

This is the thermal breakdown of materials at elevated temperatures in the absence of oxygen. During the process, complex organic molecules are broken down into simpler compounds, which frequently results in a mixture of liquids, gases, and solid-like char. Pyrolysis is operated at a temperature range of 400–600 °C (Fahmy et al. 2020). Pyrolysis is the mechanism that breaks down complex biopolymers like cellulose, hemicellulose, and lignin into simpler compounds. Unlike combustion, which uses oxygen and gives off heat and emissions, pyrolysis works without oxygen, producing useful products like biochar, bio-oil, and syngas (Jackowski et al. 2020). Pyrolysis is divided into three main types based on how quickly the material is heated: slow pyrolysis (which mostly produces biochar), fast pyrolysis (which gives more bio-oil), and flash pyrolysis (which uses extremely fast heating). Further classification includes the mode of heat transfer—using solid, gas, or no external carrier. Pyrolysis supports energy recovery, waste valorization, and greenhouse gas reduction.

2.7.1.2 Gasification

Gasification is a thermochemical process that converts solid organic materials like biomass or fossil fuels into a gaseous product known as synthesis gas (syngas), with a composition of hydrogen (H_2), carbon monoxide (CO), and methane. Gasification occurs under limited oxygen and high temperature (800–1000 °C) to obtain efficiency and cleaner emissions through partial oxidation phenomena. However, several types of waste can be processed as feedstocks so as to obtain the desired result; such types of waste include municipal solid waste, plastics, and tires. However, environmental concerns persist, particularly the formation of tar and complex waste management, necessitating efficient gas cleanup systems.

2.7.1.3 Combustion

This is a process where biomass and organic matter, like wood, agricultural residues, or plastics, are burned in the presence of oxygen to produce heat, carbon dioxide, and water vapor. This process effectively releases the stored chemical energy in biomass, making it a dependable source of renewable energy. The heat generated through combustion is utilized for heat generation in homes and industries, and also for electricity generation through boilers, furnaces, and engines (Ononogbo et al. 2023). Biomass combustion systems are widely used for low-temperature applications (up to 100 °C), including individual heating and centralized district heating networks. However, the combustion process does emit greenhouse gases and particulates, necessitating the use of a proper emission control strategy to minimize the effect of the SOx and NOx gases, so as not to deplete the ozone layer and ensure environmental compliance and sustainability (Kalair et al. 2021).

2.7.2 Biochemical Conversion

Biochemical conversion is the bioconversion and breakdown of organic biomass by using microorganisms into high-value-added products such as biofuels (Dessie et al. 2020). The techniques used to achieve such a task are anaerobic digestion, fermentation, and composting. This method exploits the metabolic flexibility of bacteria, yeasts, and microalgae to break down complex organic matter under mild conditions. However, the transformation of waste to a high-value product via biochemical means is classified as anaerobic digestion, where biogas is produced as the final product, while fermentation yields bioethanol, and all of these are achieved under the influence of microbial metabolites. Conversion of biomass via the utilization of microorganisms is environmentally friendly and energy efficient. New trends have shown the potential of using metabolic engineering in a lot of ways. This makes biochemical conversion an important part of managing waste in an eco-friendly way, producing renewable energy, and supporting biorefinery processes (Intasian et al. 2021).

2.7.2.1 Anaerobic Digestion

Biogas is a renewable source of energy produced through the decomposition of organic biomass under the influence of microorganisms in the absence of oxygen. The feedstocks that are used to achieve this process are agro-waste, municipal solid waste, wastewater sludge, and food residues in a sealed, oxygen-free digester. The produced biogas can be directly combusted to heat, converted to electricity via combined heat and power (CHP) units, and upgraded to biomethane, which can then be used as vehicle fuel or injected into natural gas pipelines. Following the decomposition process, a nutrient-rich residue remains, which can be used as an organic fertilizer to improve soil quality (Gunina and Kuzyakov 2022). Beyond energy production, anaerobic digestion reduces greenhouse gas emissions, odor pollution and provides a sustainable environment.

2.7.2.2 Fermentation

Biochemical valorization turns otherwise problematic agricultural residues, food scraps, animal manures, and industrial by-products into clean energy carriers through the concerted action of microbes and their enzymes. Under oxygen-limited conditions, fermentation remains the major pathway where various heterotrophic microbes depolymerize carbohydrates into fermentable sugars and redirect them toward ethanol, butanol, or methane-rich biogas (Mengqi et al. 2023; Singh and Singh 2022). Among these, ethanol produced through microbial metabolism from agri-feedstock is a cheaper and more sustainable method for biofuel production (Fig. 2.4).

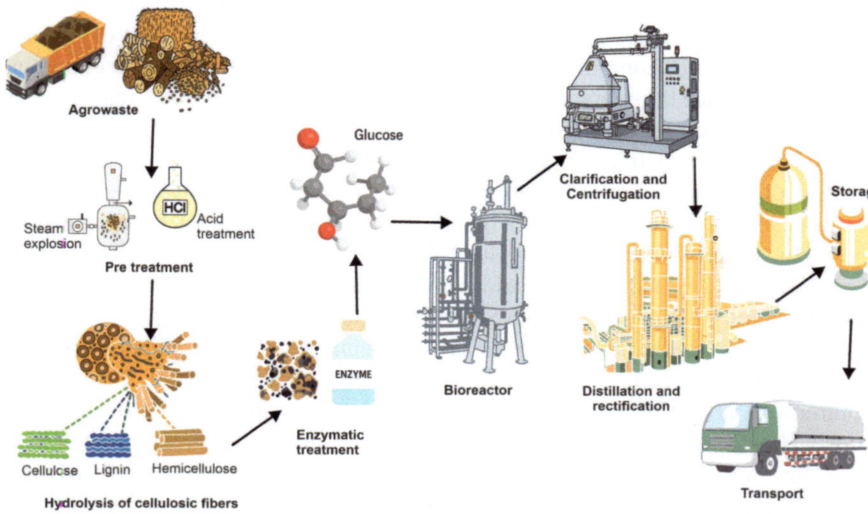

Fig. 2.4 Biochemical conversion route for producing liquid biofuel from lignocellulosic agrowaste. Collected agricultural residues are delivered to the plant (left) and subjected to thermochemical pretreatment, typically steam explosion and dilute acid hydrolysis, which disrupts the lignocellulosic matrix (cellulose, hemicellulose, lignin) and increases substrate accessibility. Enzymatic saccharification then converts the liberated polysaccharides to fermentable glucose. The sugar stream enters a stirred-tank bioreactor, where ethanologenic microbes ferment it to ethanol. The fermentation broth is subsequently clarified and centrifuged to remove biomass and insoluble, after which multistage distillation and rectification yield fuel-grade ethanol. The purified product is stored in transport for commercial utility

The biochemical conversion of waste to biofuels employs two primary fermentation modes, wet fermentation for low-solids slurries and dry fermentation for high-solids stacks, enabling reactor designs to be tailored to feedstock moisture content. Energy recovery is optimized by integrating the four key anaerobic digestion stages (hydrolysis, acidogenesis, acetogenesis, and methanogenesis) in specialized biogas digesters. These processes are powered by a carefully selected consortium of microbial specialists. *Clostridium* species (*C. acetobutylicum*, *C. beijerinckii*) excel in acetone-butanol-ethanol (ABE) fermentation, efficiently metabolizing mixed hexose-pentose streams derived from lignocellulosic waste (Birgen et al. 2019). The industrial workhorse *Saccharomyces cerevisiae* has been genetically engineered to expand its substrate range, enabling the fermentation of hemicellulosic sugars like xylose and arabinose (Nijland and Driessen 2020). Meanwhile, *Escherichia coli* is a versatile organism in synthetic biology, capable of producing isobutanol, fatty acid ethyl esters (FAEEs), and biohydrogen, thanks to its fast growth and ease of genetic modification (Wang et al. 2017).

Methanogenic archaea (*Methanosarcina*, *Methanobacterium*) complete the anaerobic digestion cycle by converting volatile fatty acids and H_2/CO_2 into methane, while lipase-producing bacteria (*Bacillus*, *Pseudomonas*) facilitate enzymatic biodiesel production. Additionally, oleaginous microalgae (*Chlorella*, *Scenedesmus*) contribute to biofuel production through their high lipid content, and engineered synthetic consortia, combining cellulolytic, fermentative, and lipid-accumulating microbes, enhance the degradation of complex, heterogeneous waste streams, maximizing carbon-to-fuel conversion efficiency (Ravindran et al. 2018; Hossain et al. 2024). The efficiency of each biochemical process depends on using the right combination of enzymes or microorganisms that work together effectively.

Thermostable cellulases, a synergistic trio of endoglucanases, exoglucanases, and β-glucosidases, cleave cellulose into glucose monomers (Chaudhari et al. 2023). Hemicellulases (xylanases, β-xylosidases) dismantle hemicellulose, while ligninolytic enzymes (laccases, peroxidases) from white-rot fungi (*Trametes versicolor*, *Phanerochaete chrysosporium*) open up the lignin barrier that shields structural polysaccharides, greatly increasing hydrolysis rates (Goldbeck et al. 2014; Pérez-Contreras et al. 2025). For lipid-rich waste materials, immobilized lipases from fungus (*Rhizomucor miehei*) and a yeast species (*Candida antarctica*) are used to carry out transesterification under mild, solvent-free conditions. This process produces high-purity fatty acid methyl esters (biodiesel) and reduces the need for extensive purification afterwards (Salaheldeen et al. 2021).

Modern biorefineries therefore design enzyme cocktails, often optimized through protein engineering and synthetic biology, to align with feedstock composition and target fuel. Integrating cellulolytic, hemicellulolytic, ligninolytic, and lipolytic activities in a single step slashes residence times, raises sugar or ester titers, and improves the techno-economic outlook of waste-to-biofuel schemes (Ufitikirezi et al. 2024). By amalgamating tailored microbial consortia with next-generation enzymes, biochemical conversion platforms now offer a scalable, circular pathway for transforming diffuse organic wastes into liquid and gaseous fuels, advancing both energy security and sustainable waste management.

2.7.2.3 Metabolic Engineering and Synthetic Biology Approaches

Development in converting industrial and agricultural waste into biofuels using microbes has focused mainly on improving their genetic and metabolic pathways through engineering. As discussed in the earlier section, specialized microorganisms like *S. cerevisiae* and *E. coli* have been adapted to different varieties of sugar substrates from lignocellulosic biomass by altering important metabolic pathways. This has led to increased yields of ethanol, butanol, and other biofuels. For instance, a significant drawback of conventional ethanol fermentation has been addressed by engineering *S. cerevisiae* to ferment cellulose-derived sugars efficiently (Oh and Jin 2020). Furthermore, by rerouting carbon flux and enhancing tolerance to ethanol stress, targeted gene deletions utilizing CRISPR-Cas9, such as the ablation of GPD2, FPS1, and ADH2, have greatly increased ethanol productivity (Bian et al. 2022). Simultaneously, metabolic flow analysis (MFA) is an essential instrument for comprehending intracellular energy and carbon distribution quantitatively. MFA facilitates the detection and resolution of pathway bottlenecks, improving total metabolic efficiency by directing rational strain engineering (Kim et al. 2008). Together, these integrated approaches, which include pathway balance, promoter engineering, and genome editing, improve microbial robustness and biofuel production. Additionally, they make it easier to use complicated waste feedstocks, which increases the viability and cost sustainability of bioconversion on an industrial scale.

Synthetic biology has revolutionized the bioconversion of agricultural and industrial waste into biofuels by enabling the engineering of microbial cell factories with tailored metabolic capabilities (Jian et al. 2021). Complex waste materials such as lignocellulose and lipids can be effectively broken down into biofuels by genetically modifying microorganisms like *E. coli*, *S. cerevisiae*, and microalgae with engineered pathways. This enables the production of biofuels such as ethanol, butanol, biodiesel, and biogas (Keasling 2012). Synthetic biology overcomes the limitations of native microbial strains by equipping robust microbial hosts with optimized artificial pathways, such as those for isobutanol or fatty acid ethyl esters (FAEEs) (Liao et al. 2016). Furthermore, by leveraging complementary enzymatic activities, it facilitates the design of synthetic microbial consortia capable of cooperatively degrading diverse polymers, including cellulose, starch, and lignin (Shong et al. 2012). Dynamic regulatory circuits and biosensors enable real-time metabolic control, enhancing fuel yields while minimizing by-products (Shong et al. 2012).

By using CRISPR-Cas9 genome editing and minimal genome strategies, scientists can simplify and optimize microbial processes, leading to more energy-efficient biofuel production (Jakočinas et al. 2015), while substrate-specific pathway engineering allows microorganisms to valorize unconventional waste streams like glycerol, whey, and municipal solid waste (Clomburg et al. 2017). In the near future computational tools, including metabolic modeling, machine learning, and automated genome synthesis, may improve the efficiency, scalability, and economic viability of waste-to-biofuel conversion.

2.8 Challenges and Limitations

2.8.1 *Technical Barriers*

Large-scale implementation of bioconversion of industrial and agricultural waste into biofuels is hampered by several technical obstacles. One of the main challenges is the complex and resistant nature of lignocellulosic biomass, which is made up of cellulose, hemicellulose, and lignin. Because of their covalent and hydrogen bonding, these constituents produce a stiff matrix that is resistant to enzymatic hydrolysis. It takes efficient pretreatment techniques, like ammonia fiber expansion, steam explosion, or acid hydrolysis, to break this structure and reveal fermentable sugars. Nevertheless, harmful by-products like furfural and hydroxymethyl furfural are frequently produced by these pretreatment procedures, which prevent microbial fermentation and lower output (Vasileiadou 2024).

Another major technological challenge is the inefficiency of microbial fermentation where traditional yeast strains like *Saccharomyces cerevisiae* cannot ferment pentose sugars derived from hemicellulose. Although promising, genetically modified microorganisms still have problems with process optimization, yield consistency, and stability. An in situ method, which combines fermentation, biomass hydrolysis, and enzyme production in one step, has the potential to lower costs and boost efficiency, but it is currently constrained by issues with scalability and strain development (Mehmood et al. 2019).

Techniques like thermochemical processes, gasification and pyrolysis are further hampered by catalyst deactivation and low conversion efficiency. Because of partial conversions and adverse responses, biofuel yields are frequently low. Furthermore, it is still difficult to recover and repurpose biocatalysts like enzymes and microbial cells. These obstacles will keep limiting the economic viability of producing ethanol from waste biomass unless bioreactor design, process integration, and catalyst engineering advances are made (Quevedo-Amador et al. 2024).

2.8.2 *Economic Constraints*

The costs of producing biofuel from industrial and agricultural waste pose a significant barrier to its widespread use. The costs of collecting, transportation, storage, and preparation can be high, even though waste biomass is frequently free or inexpensive. In rural and remote regions, the cost-benefit of waste-based biofuels is reduced where transporting and handling feedstock can account for up to 40% of the total production cost (Thakur et al. 2024). Energy-intensive and capital-expensive pretreatment procedures are necessary to make lignocellulosic biomass vulnerable to enzymatic assault. Another significant cost factor is the manufacture of enzymes, specifically cellulases and hemicellulases. Even with attempts to lower enzyme loads by genetic engineering and better fermentation techniques, their large-scale commercial production is still not profitable (Borin et al. 2019).

Additionally, the biofuels market is extremely erratic and susceptible to changes in the price of oil globally. Fossil fuels are strongly subsidized in many areas, which makes biofuels less viable. Although tax breaks, renewable energy requirements, and carbon credits are important governmental incentives for economic viability, their uneven application and frequent changes make the investment climate insecure. Private sector involvement is further discouraged by the significant capital investment needed to establish integrated biorefineries, particularly in developing nations with little access to green finance (Srivastava et al. 2020). The commercialization of waste-derived biofuels is further limited by the lack of developed supply chains and established market connections. The economic viability of bioconversion technologies will continue to be limited until notable cost reductions are attained through technological innovation and encouraging legislative frameworks.

2.8.3 Feedstock Variability

When it comes to turning industrial and agricultural waste into biofuels, feedstock variability presents a significant obstacle. The physical and chemical properties of agricultural wastes like sugarcane bagasse, rice husk, maize stover, and wheat straw vary greatly due to their build-up composition. According to Mehmood et al. (2019), these variations call for customized pretreatment techniques, which make process standardization more difficult and result in inefficient biofuel yields. For example, the amount of lignin varies significantly amongst biomass types and has a direct impact on enzymatic digestibility. Enzyme access to cellulose and hemicellulose is decreased by a high lignin content, which functions as a physical barrier. Similarly, in thermochemical processes like pyrolysis and gasification, changes in ash content and particle size affect the efficiency of thermal conversion and ash-related issues. Because of this heterogeneity, systems that are designed for one feedstock may not work well with another, making equipment design and reactor operation more difficult (Brar et al. 2014).

Supply networks for biomass are made more complex by regional distribution and seasonal availability. Seasonality in the generation of agricultural leftovers causes feedstock shortages during off-seasons unless expensive storage facilities are used. Supply networks that are inconsistent have the potential to interfere with plant operations and impact total financial results. There is an additional layer of uncertainty because industrial waste streams differ in terms of both amount and quality based on production schedules (Ahmed et al. 2024). These issues are made worse by the absence of approved quality standards for biomass feedstock. Predicting conversion efficiencies and optimizing process parameters are challenging without uniform characterization and quality control procedures. Reducing feedstock-related constraints might be possible with the development of reliable feedstock evaluation instruments and the establishment of decentralized biomass preprocessing centers.

2.8.4 Environmental and Regulatory Issues

Despite its potential as a sustainable energy source, waste-to-biofuel conversion faces various environmental and regulatory concerns. Although the procedure promotes waste minimization and greenhouse gas mitigation, improper management may have unforeseen negative effects. According to Carrillo-Nieves et al. (2022), some pretreatment techniques, including chemical hydrolysis, produce hazardous by-products and wastewater streams that need additional treatment to prevent environmental pollution.

Without proper emission controls, thermochemical processes like gasification and pyrolysis can release harmful pollutants such as tar, volatile organic compounds (VOCs), and other hazardous substances. These pollutants can provide major health and environmental hazards in areas with lax environmental controls. Another major issue is the disposal of process leftovers, including fermentation stillage and lignin-rich slurries. The overall sustainability of the production of biofuel may be jeopardized in the absence of suitable waste management techniques (Quevedo-Amador et al. 2024).

Market integration is hampered on the regulatory front by the lack of defined regulations and sustainability criteria for biofuels sourced from trash. In many regions of the world, regulations pertaining to carbon accounting, environmental compliance, and fuel quality are still developing. Investor confidence is impacted by this uncertainty, which also restricts private sector involvement. Global trade and knowledge transfer are further complicated by the uneven application of renewable energy targets, biofuel blending requirements, and subsidy programs in various geographical areas (Borin et al. 2019) To address these issues, an environmental impact analysis regulatory body or life cycle assessment (LCA) is needed to evaluate the environmental impact of biofuels across the entire supply chain, along with globally standardized regulations. It will be crucial to advance environmental compliance, regulatory clarity, and the concepts of the circular economy to make sure that bio-conversion technologies fulfil their sustainability promises.

2.9 Conclusion and Future Outlook

Technology is evolving on a day-to-day basis, but still, the world is facing setbacks due to some pressing global challenges, such as waste management and energy security. These challenges can only be tackled by the bioconversion of agricultural and industrial waste into high-value products for a promising and sustainable environment. This study explores the potential of utilizing different biomass, like agro-industrial waste, lignocellulosic and food processing residues, into a value-added product by transforming them via a biological, mechanical and chemical process to produce biofuels such as bioethanol, biobutanol, biodiesel, biogas, and biohydrogen. Advanced biotechnological interventions, such as pretreatment methods, microbial strain engineering, enzymatic hydrolysis, and consolidated bioprocessing, have shown significant improvements in conversion efficiencies and process economics. Moreover, integrating waste valorization with circular economy principles ensures minimal environmental impact and promotes sustainable industrial practices.

Despite considerable progress, several bottlenecks still hinder the large-scale deployment of bioconversion technologies. These include the high cost of enzymes, limited availability of robust microbial strains, challenges in scale-up, and regulatory hurdles. Additionally, feedstock variability and logistics pose serious challenges to the consistent operation of biorefineries. Addressing these limitations will require interdisciplinary research, policy support, and public-private partnerships.

In the future, emerging tools such as synthetic biology, metabolic engineering, CRISPR-based genome editing, and artificial intelligence-driven process optimization are expected to revolutionize biofuel production from waste. Building integrated biorefineries that produce both biofuels and high-value bioproducts can greatly improve economic sustainability. Furthermore, government incentives, carbon credit schemes, and international collaborations will play a vital role in accelerating commercialization. In summary, transitioning from a fossil-fuel-reliant energy system to a bio-based circular economy is both achievable and critically necessary, and also essential for a sustainable future. The bioconversion of waste into high-value biofuels represents a promising and sustainable development, reducing greenhouse gas emissions and enhancing energy self-sufficiency, which aligns with the United Nations Sustainable Development Goals (SDGs) for climate change mitigation efforts.

References

Ahmed SF, Mehejabin F, Chowdhury AA, Almomani F, Khan NA, Badruddin IA, Kamangar S (2024) Biochar produced from waste-based feedstocks: mechanisms, affecting factors, economy, utilization, challenges, and prospects. GCB Bioenergy 16(8):1–37. https://doi.org/10.1111/gcbb.13175

Ajala EO, Ighalo JO, Ajala MA, Adeniyi AG, Ayanshola AM (2021) Sugarcane bagasse: a biomass sufficiently applied for improving global energy, environment and economic sustainability. Bioresour Bioprocess 8(1). https://doi.org/10.1186/s40643-021-00440-z

Awogbemi O, Von Kallon DV (2022) Valorization of agricultural wastes for biofuel applications. Heliyon 8(10):e11117. https://doi.org/10.1016/j.heliyon.2022.e11117

Bharathiraja S, Suriya J, Krishnan M, Manivasagan P, Kim SK (2017) Production of enzymes from agricultural wastes and their potential industrial applications. In: Advances in food and nutrition research, vol 80, 1st edn. Elsevier Inc. https://doi.org/10.1016/bs.afnr.2016.11.003

Bhatia T, Singh S (2024) Sustainable management of organic agricultural wastes: contributions in nutrients availability, pollution mitigation and crop production. Discov Agric 2:130. Springer International Publishing. https://doi.org/10.1007/s44279-024-00147-7

Bian H, Yang Y, Tu P, Chen JY (2022) Value-added utilization of wheat straw: from cellulose and cellulose nanofiber to all-cellulose nanocomposite film. Membranes 12(5). https://doi.org/10.3390/membranes12050475

Birgen C, Dürre P, Preisig HA, Wentzel A (2019) Butanol production from lignocellulosic biomass: revisiting fermentation performance indicators with exploratory data analysis. Biotechnol Biofuels 12(1):1–15. https://doi.org/10.1186/s13068-019-1508-6

Blasi A, Verardi A, Lopresto CG, Siciliano S, Sangiorgio P (2023) Lignocellulosic agricultural waste valorization to obtain valuable products: an overview. Recycling 8(4):1–46. https://doi.org/10.3390/recycling8040061

Borin GP, Alves RF, Ferraz Júnior ADN (2019) Current status of biotechnological processes in the biofuel industries. In: Bioprocessing for biomolecules production, pp 47–69. https://doi.org/10.1002/9781119434436.ch3

Brar SK, Dhillon GS, Soccol CR (2014) Biotransformation of waste biomass into high value biochemicals. In: Biotransformation of waste biomass into high value biochemicals, vol 9781461480051. https://doi.org/10.1007/978-1-4614-8005-1

Carrillo-Nieves D, Saldarriaga-Hernandez S, Gutiérrez-Soto G, Rostro-Alanis M, Hernández-Luna C, Alvarez AJ, Iqbal HMN, Parra-Saldívar R (2022) Biotransformation of agro-industrial waste to produce lignocellulolytic enzymes and bioethanol with a zero waste. Biomass Convers Biorefinery 12(2):253–264. https://doi.org/10.1007/s13399-020-00738-6

Chaudhari YB, Várnai A, Sørlie M, Horn SJ, Eijsink VGH (2023) Engineering cellulases for conversion of lignocellulosic biomass. Protein Eng Des Sel 36(March):1–12. https://doi.org/10.1093/protein/gzad002

Clomburg JM, Crumbley AM, Gonzalez R (2017) Industrial biomanufacturing: the future of chemical production. Science 355(6320). https://doi.org/10.1126/science.aag0804

Crini G, Lichtfouse E, Chanet G, Morin-Crini N (2020) Applications of hemp in textiles, paper industry, insulation and building materials, horticulture, animal nutrition, food and beverages, nutraceuticals, cosmetics and hygiene, medicine, agrochemistry, energy production and environment: a review. Environ Chem Lett 18(5):1451–1476. https://doi.org/10.1007/s10311-020-01029-2

Dalpaz R, Konrad O, da Silva C, Cyrne C, Panis Barzotto H, Hasan C, Guerini Filho M (2020) Using biogas for energy cogeneration: an analysis of electric and thermal energy generation from agro-industrial waste. Sustain Energy Technol Assess 40(67):100774. https://doi.org/10.1016/j.seta.2020.100774

de Freitas EN, Salgado JCS, Alnoch RC, Contato AG, Habermann E, Michelin M, Martínez CA, Polizeli M d LTM (2021) Challenges of biomass utilization for bioenergy in a climate change scenario. Biology 10(12). https://doi.org/10.3390/biology10121277

De Sousa MH, Sabrina A, Correia RC, Leite NP, Eduardo C, Bueno G, Luís R, De Santana JS, Luna J, Sales AT, De Souza CC, Aparecida K, De Souza RB, Pinheiro IO, Henríquez JR, Ricardo A, Schuler P, Valadares E, Barretto DS et al (2022) Valorizing municipal organic waste to produce biodiesel, biogas, organic fertilizer, and value-added chemicals: an integrated biorefinery approach Brazilian National Agency of Petroleum Brazilian National Policy on Solid Wastes. Biomass Convers Biorefin 12:827–841. https://doi.org/10.1007/s13399-020-01252-5

Dessie W, Luo X, Wang M, Feng L, Liao Y, Wang Z, Yong Z, Qin Z (2020) Current advances on waste biomass transformation into value-added products. Appl Microbiol Biotechnol 104(11):4757–4770. https://doi.org/10.1007/s00253-020-10567-2

Dhanya BS, Mishra A, Chandel AK, Verma ML (2020) Development of sustainable approaches for converting the organic waste to bioenergy. Sci Total Environ 723:138109. https://doi.org/10.1016/j.scitotenv.2020.138109

Ezejiofor T (2014) Waste to wealth- value recovery from agro-food processing wastes using biotechnology: a review. Br Biotechnol J 4(4):418–481. https://doi.org/10.9734/bbj/2014/7017

Fahmy TYA, Fahmy Y, Mobarak F, El-Sakhawy M, Abou-Zeid RE (2020) Biomass pyrolysis: past, present, and future. Environ Dev Sustain 22(1):17–32. https://doi.org/10.1007/s10668-018-0200-5

Fatih Demirbas M, Balat M, Balat H (2011) Biowastes-to-biofuels. Energy Convers Manag 52(4):1815–1828. https://doi.org/10.1016/j.enconman.2010.10.041

Goel V, Luthra P, Kapur GS, Ramakumar SSV (2021) Biodegradable/bio-plastics: myths and realities. J Polym Environ 29(10):3079–3104. https://doi.org/10.1007/s10924-021-02099-1

Goldbeck R, Damásio ARL, Gonçalves TA, Machado CB, Paixão DAA, Wolf LD, Mandelli F, Rocha GJM, Ruller R, Squina FM (2014) Development of hemicellulolytic enzyme mixtures for plant biomass deconstruction on target biotechnological applications. Appl Microbiol Biotechnol 98(20):8513–8525. https://doi.org/10.1007/s00253-014-5946-6

Grzyb A, Wolna-Maruwka A, Niewiadomska A (2020) Environmental factors affecting the mineralization of crop residues. Agronomy 10(12). https://doi.org/10.3390/agronomy10121951

Gunina A, Kuzyakov Y (2022) From energy to (soil organic) matter. Glob Chang Biol 28(7):2169–2182. https://doi.org/10.1111/gcb.16071

Haile A, Gelebo GG, Tesfaye T, Mengie W, Mebrate MA, Abuhay A, Limeneh DY (2021) Pulp and paper mill wastes: utilizations and prospects for high value-added biomaterials. Bioresour Bioprocess 8(1). https://doi.org/10.1186/s40643-021-00385-3

Hossain GS, Liang Y, Foo JL, Chang MW (2024) Engineered microbial consortia for next-generation feedstocks. Biotechnol Notes 5:23–26. https://doi.org/10.1016/j.biotno.2024.01.002

Intasian P, Prakinee K, Phintha A, Trisrivirat D, Weeranoppanant N, Wongnate T, Chaiyen P (2021) Enzymes, in vivo biocatalysis, and metabolic engineering for enabling a circular economy and sustainability. Chem Rev 121(17):10367–10451. https://doi.org/10.1021/acs.chemrev.1c00121

Iragavarapu GP, Imam SS, Sarkar O, Mohan SV, Chang YC, Reddy MV, Kim SH, Amradi NK (2023) Bioprocessing of waste for renewable chemicals and fuels to promote bioeconomy. Energies 16(9). https://doi.org/10.3390/en16093873

Jackowski M, Niedźwiecki Ł, Jagiełło K, Uchańska O, Trusek A (2020) Brewer's spent grains—valuable beer industry by-product. Biomolecules 10(12):1–18. https://doi.org/10.3390/biom10121669

Jakočinas T, Bonde I, Herrgård M, Harrison SJ, Kristensen M, Pedersen LE, Jensen MK, Keasling JD (2015) Multiplex metabolic pathway engineering using CRISPR/Cas9 in saccharomyces cerevisiae. Metab Eng 28:213–222. https://doi.org/10.1016/j.ymben.2015.01.008

Jeremiah M, Kabeyi B (2023) Sugarcane molasses to energy conversion for sustainable production and energy Transition, August, 2299–2311. https://doi.org/10.46254/an12.20220405

Jeswani HK, Chilvers A, Azapagic A (2020) Environmental sustainability of biofuels: a review: environmental sustainability of biofuels. Proc R Soc A: Math Phys Eng Sci 476(2243):20200351. https://doi.org/10.1098/rspa.2020.0351

Jha S, Nanda S, Acharya B, Dalai AK (2022) A review of thermochemical conversion of waste biomass to biofuels. Energies 15(17):1–23. https://doi.org/10.3390/en15176352

Jhang SR, Lin YC, Chen KS, Lin SL, Batterman S (2020) Evaluation of fuel consumption, pollutant emissions and well-to-wheel GHGs assessment from a vehicle operation fueled with bioethanol, gasoline and hydrogen. Energy 209:118436. https://doi.org/10.1016/j.energy.2020.118436

Jian Z, Yuan-Fang P, Wan-Li W, Qin W, Gong-Nan X, Hong-Fei L, Tian X, Shuang-Fei W (2021) Black liquor increases methane production from excess pulp and paper industry sludge. Chemosphere 280(March):130665. https://doi.org/10.1016/j.chemosphere.2021.130665

Kabeyi MJB, Olanrewaju OA (2022) Biogas production and applications in the sustainable energy transition. J Energy 2022:1–43. https://doi.org/10.1155/2022/8750221

Kalair AR, Seyedmahmoudian M, Stojcevski A, Abas N, Khan N (2021) Waste to energy conversion for a sustainable future. Heliyon 7(10):e08155. https://doi.org/10.1016/j.heliyon.2021.e08155

Keasling JD (2012) Synthetic biology and the development of tools for metabolic engineering. Metab Eng 14(3):189–195. https://doi.org/10.1016/j.ymben.2012.01.004

Kim HU, Kim TY, Lee SY (2008) Metabolic flux analysis and metabolic engineering of microorganisms. Mol BioSyst 4(2):113–120. https://doi.org/10.1039/b712395g

Kohli K, Prajapati R, Sharma BK (2019) Bio-based chemicals from renewable biomass for integrated biorefineries. Energies 12(2). https://doi.org/10.3390/en12020233

Lee M, Lin YL, Te Chiueh P, Den W (2020) Environmental and energy assessment of biomass residues to biochar as fuel: a brief review with recommendations for future bioenergy systems. J Clean Prod 251:119714. https://doi.org/10.1016/j.jclepro.2019.119714

Leong HY, Chang CK, Khoo KS, Chew KW, Chia SR, Lim JW, Chang JS, Show PL (2021) Waste biorefinery towards a sustainable circular bioeconomy: a solution to global issues. Biotechnol Biofuels 14(1):1–15. https://doi.org/10.1186/s13068-021-01939-5

Liao JC, Mi L, Pontrelli S, Luo S (2016) Fuelling the future: microbial engineering for the production of sustainable biofuels. Nat Rev Microbiol 14(5):288–304. https://doi.org/10.1038/nrmicro.2016.32

Megia PJ, Vizcaino AJ, Calles JA, Carrero A (2021) Hydrogen production technologies: from fossil fuels toward renewable sources. A mini review. Energy Fuels 35(20):16403–16415. https://doi.org/10.1021/acs.energyfuels.1c02501

Mehmood T, Saman T, Asgher M, Irfan M, Anwar Z, Nadeem F, Siddiqa A (2019) Optimization of cultural parameters for pectin methylestrase and polygalacturonase production from Schizophyllum commune in solid state fermentation. Bangladesh J Bot 48(1):65–74. https://doi.org/10.3329/bjb.v48i1.47417

Mekunye F, Makinde P (2024) Production of biofuels from agricultural waste. Asian J Agric Hortic Res 11(3):37–49. https://doi.org/10.9734/ajahr/2024/v11i3328

Mengqi Z, Shi A, Ajmal M, Ye L, Awais M (2023) Comprehensive review on agricultural waste utilization and high-temperature fermentation and composting. Biomass Convers Biorefinery 13(7):5445–5468. https://doi.org/10.1007/s13399-021-01438-5

Mitri S, Salameh SJ, Khelfa A, Leonard E, Maroun RG, Louka N, Koubaa M (2022) Valorization of brewers' spent grains: pretreatments and fermentation, a review. Fermentation 8(2):50. https://doi.org/10.3390/fermentation8020050

Mozhiarasi V, Natarajan TS (2022) Slaughterhouse and poultry wastes: management practices, feedstocks for renewable energy production, and recovery of value added products. Biomass Convers Biorefin:1705–1728. https://doi.org/10.1007/s13399-022-02352-0

Ngbede DE, Ahmadu U, Gana M, Attah F (2019) Bioconversion of agricultural and industrialised waste to generate valuable products. Braz J Biol Sci 6(14):e415. https://doi.org/10.21472/bjbs.061411

Nijland JG, Driessen AJM (2020) Engineering of pentose transport in Saccharomyces cerevisiae for biotechnological applications. Front Bioeng Biotechnol 7(January):1–13. https://doi.org/10.3389/fbioe.2019.00464

Oh EJ, Jin YS (2020) Engineering of Saccharomyces cerevisiae for efficient fermentation of cellulose. FEMS Yeast Res 20(1):1–11. https://doi.org/10.1093/femsyr/foz089

Ononogbo C, Nwosu EC, Nwakuba NR, Nwaji GN, Nwufo OC, Chukwuezie OC, Chukwu MM, Anyanwu EE (2023) Opportunities of waste heat recovery from various sources: review of technologies and implementation. Heliyon 9(2):e13590. https://doi.org/10.1016/j.heliyon.2023.e13590

Pérez-Contreras S, Avalos-de la Cruz DA, Lizardi-Jiménez MA, Herrera-Corredor JA, Baltazar-Bernal O, Hernández-Martínez R (2025) Production of ligninolytic and cellulolytic fungal enzymes for agro-industrial waste valorization: trends and applicability. Catalysts 15(1):1–17. https://doi.org/10.3390/catal15010030

Quevedo-Amador RA, Escalera-Velasco BP, Arias AMR, Reynel-Ávila HE, Moreno-Piraján JC, Giraldo L, Bonilla-Petriciolet A (2024) Application of waste biomass for the production of biofuels and catalysts: a review. Clean Technol Environ Policy 26(4):943–997. Springer Berlin Heidelberg. https://doi.org/10.1007/s10098-023-02728-4

Raihan A, Tuspekova A (2022) Nexus between economic growth, energy use, agricultural productivity, and carbon dioxide emissions: new evidence from Nepal. Energy Nexus 7(July):100113. https://doi.org/10.1016/j.nexus.2022.100113

Randive AB, SVERI's College of Engineering (Poly), Pandharpur, India, Khadake SB, Mallad HM (2024) Biodiesel: a renewable source of fuel. Int J Adv Res Sci Commun Technol 3:225–240. https://doi.org/10.48175/ijarsct-22836

Rasapoor M, Young B, Brar R, Sarmah A, Zhuang WQ, Baroutian S (2020) Recognizing the challenges of anaerobic digestion: critical steps toward improving biogas generation. Fuel 261:116497. https://doi.org/10.1016/j.fuel.2019.116497

Ravindran R, Hassan SS, Williams GA, Jaiswal AK (2018) A review on bioconversion of agro-industrial wastes to industrially important enzymes. Bioengineering 5(4):1–20. https://doi.org/10.3390/bioengineering5040093

Riaz S, Mazhar S, Abidi SH, Syed Q, Abbas N, Saleem Y, Nadeem AA, Maryam M, Essa R, Ashfaq S (2022) Biobutanol production from sustainable biomass process of anaerobic ABE fermentation for industrial applications. Arch Microbiol 204(11):1–9. https://doi.org/10.1007/s00203-022-03284-z

Rout PR, Pandey DS, Haynes-Parry M, Briggs C, Manuel HLC, Umapathi R, Mukherjee S, Panigrahi S, Goel M (2023) Sustainable valorisation of animal manures via thermochemical conversion technologies: an inclusive review on recent trends. Waste Biomass Valorization 14(2):553–582. https://doi.org/10.1007/s12649-022-01916-5

Salaheldeen M, Mariod AA, Aroua MK, Rahman SMA, Soudagar MEM, Fattah IMR (2021) Current state and perspectives on transesterification of triglycerides for biodiesel production. Catalysts 11(9):1–37. https://doi.org/10.3390/catal11091121

Saleem M (2022) Possibility of utilizing agriculture biomass as a renewable and sustainable future energy source. Heliyon 8(2):e08905. https://doi.org/10.1016/j.heliyon.2022.e08905

Shaibur MR, Husain H, Arpon SH (2021) Utilization of cow dung residues of biogas plant for sustainable development of a rural community. Curr Res Environ Sustain 3:100026. https://doi.org/10.1016/j.crsust.2021.100026

Shong J, Jimenez Diaz MR, Collins CH (2012) Towards synthetic microbial consortia for bioprocessing. Curr Opin Biotechnol 23(5):798–802. https://doi.org/10.1016/j.copbio.2012.02.001

Shukla A, Kumar D, Girdhar M, Kumar A, Goyal A, Malik T, Mohan A (2023) Strategies of pretreatment of feedstocks for optimized bioethanol production: distinct and integrated approaches. Biotechnology for Biofuels and Bioproducts, 16:44. https://doi.org/10.1186/s13068-023-02295-2

Singh A, Singh A (2022) Microbial degradation and value addition to food and agriculture waste. Curr Microbiol 79(4):1–17. https://doi.org/10.1007/s00284-022-02809-5

Srivastava N, Srivastava M, Mishra PK, Gupta VK, Molina G, Rodriguez-Couto S, Manikanta A, Ramteke PW (2018) Applications of fungal cellulases in biofuel production: advances and limitations. Renew Sust Energ Rev 82:2379–2386. https://doi.org/10.1016/j.rser.2017.08.074

Srivastava RK, Shetti NP, Reddy KR, Aminabhavi TM (2020) Sustainable energy from waste organic matters via efficient microbial processes. Sci Total Environ 722:137927. https://doi.org/10.1016/j.scitotenv.2020.137927

Teoh RH, Mahajan AS, Moharir SR, Abdul Manaf N, Shi S, Thangalazhy-Gopakumar S (2024) A review on hydrothermal treatments for solid, liquid and gaseous fuel production from biomass. Energy Nexus 14(April):100301. https://doi.org/10.1016/j.nexus.2024.100301

Thakur V, Baghmare P, Verma A, Verma JS, Geed SR (2024) Recent progress in microbial biosurfactants production strategies: applications, technological bottlenecks, and future outlook. Bioresour Technol 408(July):131211. https://doi.org/10.1016/j.biortech.2024.131211

Toor M, Kumar SS, Malyan SK, Bishnoi NR, Mathimani T, Rajendran K, Pugazhendhi A (2020) An overview on bioethanol production from lignocellulosic feedstocks. Chemosphere 242. https://doi.org/10.1016/j.chemosphere.2019.125080

Ufitikirezi J d DM, Filip M, Ghorbani M, Zoubek T, Olšan P, Bumbálek R, Strob M, Bartoš P, Umurungi SN, Murindangabo YT, Heřmánek A, Tupý O, Havelka Z, Stehlík R, Černý P, Smutný L (2024) Agricultural waste valorization: exploring environmentally friendly approaches to bioenergy conversion. Sustainability (Switzerland) 16(9):1–24. https://doi.org/10.3390/su16093617

Vasileiadou A (2024) From organic wastes to bioenergy, biofuels, and value-added products for urban sustainability and circular economy: a review. Urban Sci 8(3). https://doi.org/10.3390/urbansci8030121

Wang C, Pfleger BF, Kim SW (2017) Reassessing Escherichia coli as a cell factory for biofuel production. Curr Opin Biotechnol 45:92–103. https://doi.org/10.1016/j.copbio.2017.02.010

Yang M, Chen L, Wang J, Msigwa G, Osman AI, Fawzy S, Rooney DW, Yap PS (2023a) Circular economy strategies for combating climate change and other environmental issues. Environ Chem Lett 21(1):55–80. https://doi.org/10.1007/s10311-022-01499-6

Yang L, Lamont LD, Liu S, Guo C, Stoner S (2023b) A review on potential biofuel yields from cover crops. Fermentation 9(10):1–15. https://doi.org/10.3390/fermentation9100912

Yap JK, Sankaran R, Chew KW, Halimatul Munawaroh HS, Ho SH, Rajesh Banu J, Show PL (2021) A review manuscript submitted to chemosphere advancement of green technologies: a comprehensive review on the potential application of microalgae biomass. Chemosphere 281(May):130886. https://doi.org/10.1016/j.chemosphere.2021.130886

Yazid NA, Barrena R, Komilis D, Sánchez A (2017) Solid-state fermentation as a novel paradigm for organic waste valorization: a review. Sustainability (Switzerland) 9(2):1–28. https://doi.org/10.3390/su9020224

Yukesh Kannah R, Merrylin J, Poornima Devi T, Kavitha S, Sivashanmugam P, Kumar G, Rajesh Banu J (2020) Food waste valorization: biofuels and value added product recovery. Bioresour Technol Rep 11(August):100524. https://doi.org/10.1016/j.biteb.2020.100524

Chapter 3
Sustainable Biodegradable Packaging Materials from Food Waste

Neela Satheesh, Aparna Mahajan, and Eskindir E. Tadesse

Abstract The growing environmental concern posed by petroleum-based plastic packaging has fueled worldwide action to find more sustainable alternatives. One of the most viable solutions is the conversion of food waste like fruit peel, crop wastes, and agro-industrial by-products into a raw material for biodegradable packaging. Food waste rich in biopolymers like starch, cellulose, pectin, proteins, and chitosan is a source for producing eco-friendly materials. Innovations in material science, green chemistry, and nanotechnology have enabled the conversion of these waste-derived biopolymers into functional packaging films with enhanced mechanical strength, barrier properties, and bioactive functions. This chapter comprehensively reviews the potential of biodegradable packaging materials derived from plant-based, seafood, dairy, animal, and microbial-derived polymers from food wastes, offering insights into extraction techniques, material modifications, and applications. This approach aligns with circular economy principles and presents an integrated strategy for reducing both plastic pollution and food waste.

Keywords Agro-industrial by-products · Circular economy · Food waste valorization, Functional packaging films · Microbial polymers · Eco-friendly packaging

N. Satheesh (✉) · A. Mahajan
Department of Food Science and Technology, School of Health Sciences, Amity University Punjab, Mohali, Punjab, India

E. E. Tadesse (✉)
Department of Animal Products Processing, University of Agriculture, Kraków, Poland

Faculty of Chemical and Food Engineering, Bahir Dar Institute of Technology, Bahir Dar University, Bahir Dar, Ethiopia

N. K. Arora, J. Mishra (eds.), *Biotechnological Solutions for a Sustainable Future*, Microorganisms for Sustainability 55,
https://doi.org/10.1007/978-981-95-3989-5_3

3.1 Introduction

The environmental crisis driven by the overuse of synthetic plastics has drawn global attention to the urgent need for sustainable packaging solutions. As per the safe food Advocacy (2025), in 2024, around 220 million tons of plastic was used around the globe. Around 69.5 million tons is mismanaged and ends up in the natural environment. Traditional plastic packaging, particularly in the food sector, has contributed significantly to the generation of non-biodegradable waste in the food packaging sector. This alarming trend, combined with the environmental degradation caused by fossil fuel extraction and plastic pollution, has prompted researchers, industries, and policymakers to explore eco-friendly alternatives that align with the principles of sustainability and circular economy (Rajeshkumar 2021).

The Food and Agriculture Organization (FAO) estimates that 1.3 billion tons of food valued at $750 billion is discarded globally each year (Di Bartolo et al. 2021). Food waste poses an environmental burden due to its perishability. One of the promising solutions for both the problems is the development of biodegradable packaging materials derived from food waste, an abundant, renewable, and underutilized resource. Food waste, including fruit peels, crop residues, and agro-industrial by-products, is rich in biopolymers such as starch, cellulose, pectin, and proteins, which can be harnessed to produce bio-based packaging materials (Wadhwa et al. 2016). These materials possess inherent advantages such as biodegradability, compostability, and reduced environmental footprint, offering a viable substitute to conventional plastics in food packaging applications.

Innovations in green chemistry, biotechnology, bio process engineering and nanotechnology have enabled the transformation of food waste into functional bioplastics with improved mechanical strength, barrier properties, and active functionalities like antimicrobial or antioxidant activity (Fig. 3.1). Such developments not only support waste valorization and reduce dependency on finite fossil resources but also enhance food preservation, safety, and shelf life. However, challenges such as high production costs, performance limitations, and scalability remain critical hurdles for widespread adoption (Ortiz et al. 2018; Zhou et al. 2024).

This chapter explores the potential of utilizing food waste-derived polymers for sustainable biodegradable packaging, reviewing recent advances, processing technologies, and the environmental benefits of this approach.

Fig. 3.1 (continued) production. The process begins with diverse biological sources (Plant starch, Seafood/Chitosan, Dairy/Animal collagen/gelatin, Microbial PLA/PHA). Polymers are extracted using mechanical (e.g., High-Pressure Homogenization, Milling), chemical (e.g., Acid/Alkali, Microwave), or green-tech methods (e.g., Deep Eutectic Solvents, Enzyme Cocktail), yielding purified polymers. Reinforcement mechanisms (Microfibril formation, Entanglement, Percolation, Interfacial Adhesion) enhance polymer quality. The superior quality polymers are then processed via molding methods (Extrusion, Injection molding, Blow-molding) to form final sustainable packaging materials such as films, coatings, and composites

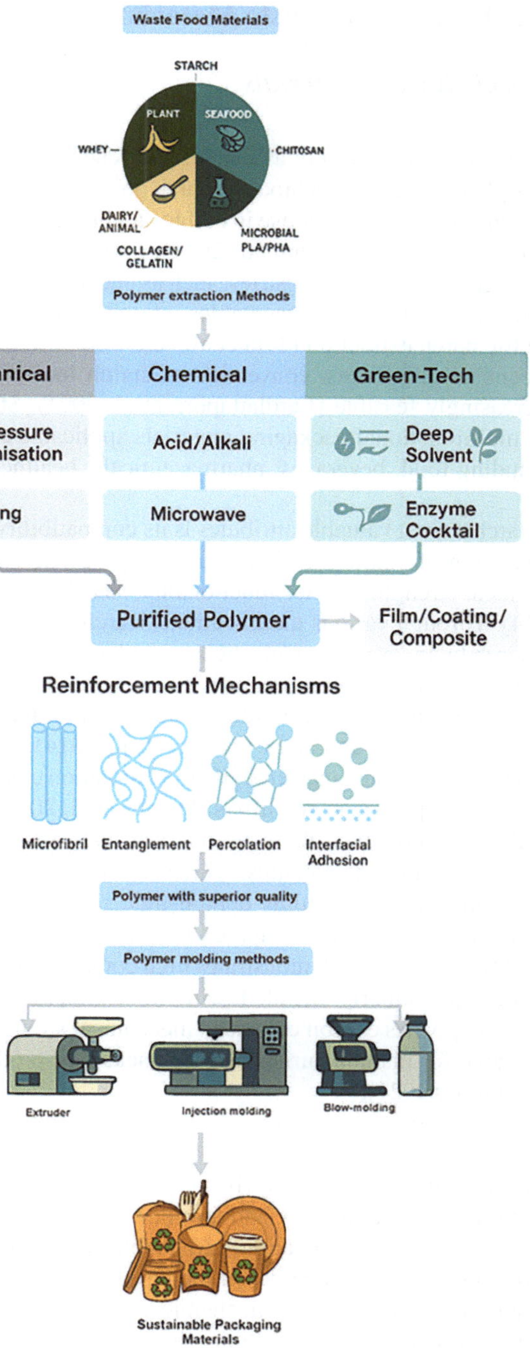

Fig. 3.1 Schematic overview of the valorization of waste food materials for sustainable packaging

3.2 Plant-Based Food Waste Materials

3.2.1 Starch-Based Materials

Starch is a cost-effective, widely available, and renewable polysaccharide derived from various plant sources. As a biodegradable polymer, it exhibits versatile properties, making it highly suitable for use in edible and biodegradable packaging materials (Behrooznia and Nourmohammadi 2024). Significantly, food waste generated from various agro-processing industries, such as peels, seeds (Table 3.1), trimmings, skins, inferior-quality fruits and vegetables, grains, and tubers, serves as a sustainable source for starch extraction (Tafa et al. 2023).

With recent advancements, conversion of starch into biodegradable films has become increasingly feasible (Lerdlattaporn et al. 2021). Starch-based films have emerged as multifunctional packaging materials applicable across a broad range of sectors, including food, beverages, pharmaceuticals, healthcare, and personal care products.

One of starch's most valuable attributes is its compatibility and flexibility, allowing the production of edible films with tailored properties, such as variable thickness, mechanical strength, barrier functionality, saleability, and printability (Dutta and Sit 2024). Through various modifications, starch can be transformed into thermoplastic starch by incorporating plasticizers under heat and shear conditions. This thermoplastic starch form is compatible with conventional plastic processing methods, such as extrusion, injection molding, and blow molding, enabling the production of films, containers, and bags (Zhang et al. 2014). Although thermoplastic starch exhibits lower mechanical strength, this limitation can be addressed by incorporating reinforcing polymers and fillers (Diyana et al. 2021).

Among the most effective fillers are nanoparticles, including nanoclay, nanofibers, nanocellulose, and nanocrystals, which significantly enhance film properties. These nano-reinforced starch films demonstrate improved strength, rigidity, and thermal degradation resistance compared to unmodified starch films (Bangar and Whiteside 2021). Starch also demonstrates high compatibility in the development of bio-composites, particularly blended with biodegradable polymers and natural fibers. These composites exhibit enhanced mechanical strength and flexibility, making them suitable for manufacturing durable packaging products such as trays and films (Rajeshkumar 2021).

Furthermore, starch can be effectively blended with other biodegradable polyesters such as polycaprolactone (PCL) or polyhydroxyalkanoates (PHAs), either individually or in combination. These blended films exhibited higher mechanical strength and processing efficiency and are thus suitable for high-performance and industrial-scale packaging applications (Ali Akbari Ghavimi et al. 2015).

Recent research has also focused on the development of modified starch-based packaging materials through physical, chemical, and enzymatic modifications. Such modifications improve the structural integrity, water resistance, and environmental responsiveness of starch, making it more adaptable for diverse packaging needs.

Table 3.1 Starch from different fruit waste and its properties

Fruit waste type	Properties of the starch	References
Mango seed kernel	High amylopectin, smooth granules, good swelling power, low gelatinization temperature	Akhter et al. (2024)
Jackfruit seed	Moderate amylose, white color, good pasting, suitable for edible films	Mukprasirt and Sajjaanantakul (2004)
Durian seed	Smooth granules, moderate swelling, high clarity paste	Baraheng and Karrila (2019)
Avocado seed	Oval granules, high peak viscosity, low retrogradation, biodegradable potential	Rivera–González et al. (2019)
Banana peel	Low amylose, high swelling index, film-forming, moderate viscosity	Chandrasekar et al. (2023)
Mango peel	Moderate viscosity, biodegradable film-forming potential	Chen et al. (2019)
Sapodilla seed	White, pure starch with small granules, industrially viable	Punia Bangar et al. (2022)
Papaya seed	Small-sized granules, moderate amylose, suitable for edible packaging	Sethulakshmi and Saravanakumar (2024)
Longan seed	High swelling power, spherical granules, water-binding capacity	Xiao et al. (2025)
Lychee seed	Irregular granules, fast gelatinization, suitable as a thickener	Pareek (2015)
Rambutan seed	Good swelling capacity, high paste clarity, sticky consistency	Jahurul et al. (2020)
Tamarind seed	High viscosity, excellent retrogradation resistance, good for noodles	Xie et al. (2020)
Passion fruit peel	Mixed fiber-starch blend, good for edible films	Moro et al. (2017)
Pineapple peel	Sticky pastes, moderate viscosity, thickening potential	Kumar et al. (2021)

Modified starch is widely utilized in coatings, films, and laminates, offering customized performance characteristics that align with specific environmental conditions and product requirements (Shah et al. 2016).

3.2.2 Cellulosic Materials

Cellulose is a linear, fibrous polysaccharide and the most abundant bio-renewable macromolecule on earth, primarily sourced from terrestrial and aquatic plant biomass from the food and agriculture source (Trache et al. 2017). Structurally, cellulose is a hydrophilic, biodegradable polymer characterized by a high degree of crystallinity and elevated molecular weight. In its native form, cellulose exhibits a semicrystalline architecture comprising tightly packed linear chains of β-1,4-linked D-glucopyranose units. These chains are stabilized by an extensive intra- and intermolecular hydrogen bonding network, giving rise to supramolecular structures (Heinze 2015).

Cellulose can be fractionated into several discrete morphologies with tailored properties by various extraction and processing techniques. These techniques ranged from controlled acid hydrolysis to mechanical treatments such as high-pressure homogenization and ultrasonication. Microcrystalline cellulose (MCC) produced via diluted acid hydrolysis (Lupidi et al. 2023), nanocrystalline cellulose (NCC) is obtained via aggressive acid hydrolysis, nanofibrillated cellulose (NFC) from intensive mechanical treatment. The NFC offers tunable mechanical, optical, and barrier properties, making them ideal for advanced applications in nanocomposites and packaging (Lindström 2017).

A wide array of cellulose derivatives have been synthesized to expand their chemical reactivity and functional performance. These include esters and ethers such as cellulose acetate (CA), cellulose nitrate (CN), cellulose sulfate (CS), ethyl cellulose (EC), methylcellulose (MC), carboxymethyl cellulose (CMC), hydroxyethyl cellulose (HEC), hydroxypropyl cellulose (HPC), and hydroxypropyl methylcellulose (HPMC). These derivatives enhance cellulose's solubility, film-forming ability, and biocompatibility, thereby extending its application to food packaging, pharmaceuticals, textiles, and environmental technologies (Doelker 2005).

The CA forms films with exceptional UV-blocking, mechanical integrity, and antimicrobial activity, particularly in blends with PLA (Cindradewi et al. 2021). The CS is water-soluble and possesses antimicrobial, antiviral, and anticoagulant activities. Its low cost and excellent biocompatibility make it suitable for the preparation of edible films for extending the shelf life of fresh produce (Cindradewi et al. 2021). Despite its reactivity and combustibility, CN films incorporating natural extracts like grape pomace exhibit antioxidant potential in food packaging, albeit with safety limitations. The EC is hydrophobic, water-insoluble, and stable in organic solvents. It serves as active food packaging by incorporating bioactives (e.g., capsaicin, polyphenols), enhancing mechanical properties and microbial resistance (Zhou et al. 2024). The MC is water-soluble, approved as GRAS by the FDA, and it exhibits thermogelling behavior, emulsification, and pH stability. Composite films of MC by plant extracts or chitosan enhance antioxidant activity, UV shielding, and mechanical performance (Cindradewi et al. 2021).

The HPMC films can be tailored through the addition of essential oils, nanoparticles, or polyphenols to improve water vapor barrier, antimicrobial, and antioxidant properties (Klangmuang and Sothornvit 2016). The CMC is an anionic, water-soluble cellulose derivative synthesized via etherification with monochloroacetic acid. It acts as a stabilizer, thickener, and film-former, with applications in pH-sensitive intelligent packaging, antimicrobial films (Yildirim-Yalcin et al. 2022). The HEC offers biodegradability, rheological control, and compatibility with active agents (Yildirim-Yalcin et al. 2022). The HPC is a thermoplastic cellulose soluble at low temperatures; its composites (cellulose nanocrystals) exhibit enhanced mechanical and barrier performance (Walters et al. 2020).

Despite their versatility, cellulose-based films often exhibit limitations in water vapor resistance. For instance, MC and HPMC films effectively block gases and lipids but allow moisture permeability, while CA suffers from moderate mechanical and thermal stability (Grzybek et al. 2024). To overcome these drawbacks, functionalization through blending of bioactive agents (e.g., antioxidants, antimicrobials) has been shown to significantly improve barrier integrity, extend shelf life, and enhance food safety (Doelker 2005).

Cellulose-based intelligent packaging represents a transformative innovation in food quality monitoring. These systems utilize natural, pH-responsive pigments, particularly anthocyanin, to signal food freshness through visible color changes. For example, red cabbage anthocyanins exhibit reliable pH-induced chromatic transitions, while blackcurrant anthocyanins offer complementary spectral responses (Grzybek et al. 2024). Several prototypes, such as CMC-gelatin films with curcumin and chitosan (for pork freshness) (Yildirim-Yalcin et al. 2022), HPMC/MCC films with butterfly pea extract (for fish spoilage), and CMC-ovalbumin films with blueberry anthocyanins (for mushrooms), demonstrate real-time food spoilage detection (Boonsiriwit et al. 2021). These smart films provide sustainable, non-invasive tools for enhancing consumer safety and reducing food waste.

3.2.3 Pectin-Based Materials

Pectin is a functionally significant polysaccharide with both nutritional and economic importance. It is found in substantial quantities within plant-derived residues and by-products (Mohnen 2008). Chemically, pectin is a complex heteropolysaccharide predominantly constituted by D-galacturonic acid units that occur in homogalacturonan (HG), rhamnogalacturonan-I (RG-I), and rhamnogalacturonan-II (RG-II). It is primarily localized in the middle lamella of plant cell walls, featuring a linear backbone of homogalacturonic acid adorned with diverse side chains comprising approximately 17 distinct monosaccharides (Caffall and Mohnen 2009).

Pectin can be extracted from the various fruit waste by the different traditional and advanced methods. The conventional pectin extraction is a multi-stage physicochemical process, in which temperature, pH, solvent properties, source characteristics and duration of extraction influence solubilization of pectin from plant-tissues, its hydrolysis and extraction (Kumar et al. 2023). The conventional methods include acid extraction at pH 1–3 at higher temperatures from 80 to 100 °C for 30 min. In case of the alkali extraction at a pH range of 9–13 and temperature of 32 and 80 °C. The chelating agents-assisted extraction, a traditional chelating agent like cyclohexanediaminetetraacetic acid, ethylenediamine tetraacetic acid, sodium hexametaphosphate, ammonium oxalate and sodium citrate (Caffall and Mohnen 2009; Kumar et al. 2023).

However, traditional extraction of pectin has several drawbacks since it requires an excessive amount of solvents, increased duration of extraction, degradation of the desired pectin, and even possible effects on the environment. Various advanced methods like microwave-assisted extraction (MAE), ultrasound-assisted extraction (UAE), supercritical water extraction (SCWE), ohmic heating-assisted extraction

(OHAE), cold plasma-assisted extraction (CPAE), electromagnetic induction heating (EIH)-assisted extraction, high-pressure extraction (HPE), pulsed electric field extraction (PEFE), radio frequency-assisted extraction (RFAE), Ionic liquid-assisted extraction (ILAE), and deep eutectic solvent extraction (DESE) were reported. Advanced extraction methods reduce solvent usage, increase efficiency by speeding up extraction, and mostly conserve pectin's functional properties (Adetunji et al. 2017).

The peels and pomace from various citrus fruits such as orange (*Citrus sinensis*), lemon (*Citrus limon*), lime (*Citrus aurantifolia*), grapefruit (*Citrus paradisi*), mandarin (*Citrus reticulata*), pomelo (*Citrus maxima*), kumquat (*Citrus japonica*), and citron (*Citrus medica*) can yield approximately 25–30% pectin (Asfaw et al. 2023). Similarly, the mango peel contains 20–35% pectin (Bekele and Satheesh 2020). Melon peel (*Cucumis melo* L.) contains a substantial amount of pectin (approximately 29.48%), and recent studies have demonstrated successful pectin or pectic polysaccharide recovery. The outer, fibrous peel of ripe jackfruit contains a high pectin content ranging from 14.8% to 18.6% (w/w) (Raji et al. 2017). Likewise, banana peel contains significant levels of pectin (5.89–24.08%) depending on the type and variety (Khamsucharit et al. 2018). Furthermore, pomegranate peel is another notable source, with a pectin content of approximately 20–25% (Yang et al. 2018).

Production of pectin edible films can be achieved using different approaches, including casting, extrusion, spraying, and knife-coating (Asfaw et al. 2023). The plasticizer has an important role to play in the formation of the pectin film. The plasticizers modify the three-dimensional organization and decrease attractive intermolecular forces. These molecular organizational changes enhance extensibility, dispensability, and film flexibility that are formed, but simultaneously, cohesion and rigidity of the films are reduced. Food-grade plasticizers, including glycerol, are the most popular plasticizers used due to stability and compatibility with hydrophilic biopolymeric packaging (Pasini Cabello et al. 2015).

Food-grade plasticizers include polyols (e.g., glycerol, sorbitol, and polyethylene glycol), sugars (e.g., glucose and honey), and lipids (e.g., monoglycerides, phospholipids, and surfactants). Acetylated monoglycerides, polyethylene glycol, and sucrose are commonly incorporated into film formulations to modify mechanical properties, particularly to reduce brittleness (Espitia et al. 2014). While the addition of plasticizers enhances film flexibility and mechanical strength, it can also significantly alter the barrier properties of the pectin films. Sothornvit and Krochta (2000) reported that plasticizers are hygroscopic and attract water molecules, which can lead to increased water vapor permeability. Although plasticizers improve handling and film integrity, they typically reduce the film's effectiveness as a barrier against moisture, gases, and aroma compounds (Janjarasskul and Krochta 2010).

3.2.4 Hemicellulose-Based Materials

Hemicelluloses are the second most common polysaccharides found in plants that are typically located in the primary and secondary cell walls. Heteropolysaccharides of glucomannans, mannans, xyloglucans, xylans, and β-(1, 3 and 1, 4)-glucans. The variability in the hemicelluloses is attributed to the composition of the various monomeric units like pentoses (arabinose and xylose) and hexoses (galactose, glucose, and mannose), and other minor components like uronic acid and sugars (rhamnose and fucose) (Spiridon and Popa 2008). Based on the sugar units present, hemicelluloses are classified into four primary categories, i.e., xylans, mannans, xyloglucans, and mixed-linkage β-glucans (Ebringerová et al. 2005).

Hemicellulose extraction from the plant waste requires the breakdown of the complex chemical linkages between cellulose, hemicellulose, and lignin. Researchers reported various approaches to enhance hemicellulose recovery such as treatments with organic acids, organic bases, deep eutectic solvents (DES), and enzyme-assisted methods. Deep eutectic solvent (choline chloride and urea) offers a new and sustainable method for dissolving xylan-type hemicellulose (Lu et al. 2021). Organic acids (Toluene sulfonic acid, salicylic acid and 1-amino-2-naphthol 4-sulfonic acid, Formic acid) are commonly utilized in the extraction of hemicellulose from lignocellulosic biomass due to their ability to cleave intermolecular bonds in the complex biomass structure, thereby facilitating the release of hemicellulose (Liu et al. 2022). In the case of the alkali (such as sodium hydroxide, calcium hydroxide, potassium hydroxide, and ammonium hydroxide) extraction methods, both organic and inorganic bases are used (Soares Rodrigues et al. 2016). Inorganic alkali use may increase environmental issues. Hence, researchers reported the usage of the imidazole, tetra-n-butylphosphonium hydroxide (TBPH), and tetraethylammonium hydroxide (TEAH) (He et al. 2024).

Enzyme-assisted extraction of the hemicellulose presents a number of benefits compared to conventional chemical techniques, such as reduced energy requirements, shorter extraction times, improved yields, and lower solvent utilization (Costa et al. 2020). For example, enzymes such as xylanases selectively degrade and hydrolyze the complex hemicellulose polymer into soluble hemicellulose fragments. Later chemical treatments efficiently separate and purify hemicellulose from the remaining biomass matrix with high yield, low use of harsh chemicals and facilitating sustainable biomass processing practices (Streimikyte et al. 2022). Hydrothermal extraction or autohydrolysis has advantages in the form of short reaction durations, efficient conversion, and mild reaction temperatures. The addition of organic solvents to the process could improve the extraction yield and efficiency of high molecular weight hemicelluloses (Gullón et al. 2012).

Different techniques are described for the purification of hemicelluloses, which include the ethanol precipitation, supercritical CO_2 precipitation, and column chromatography (Sun et al. 2013). The extraction techniques used have a significant impact on its molecular properties, and in this way, affect applicability to different purposes (Peng et al. 2012). Greater polymerization and more complicated branching structures are paramount for their applications as emulsifying agents in food and cosmetic formulations (Rao et al. 2023). Alkali-treated hemicelluloses tend to be modified by carboxymethylation, enhancing water solubility and the ability to interact with other polymers (He et al. 2024).

Films and coatings made from hemicellulose are biodegradable with excellent barriers to oxygen and moisture. For the better performance of the hemicellulose films, researchers prepared the films by adding chitosan, showing improvements in mechanical properties such as tensile strength and elasticity, essential for effective packaging materials (Xu et al. 2019). Researchers also reported novel methods for the application of nanomaterials in the hemicellulose films. Composite films enhance mechanical strength and thermal resistance, provide UV protection, and reduce water vapor and oxygen permeability. They also minimize lipid oxidation and exhibit antibacterial activity against pathogens such as Staphylococcus aureus and Escherichia coli (Liu et al. 2023). In addition, researchers reported adding additives such as citric acid and polyethylene glycol to enhance the mechanical durability and barrier properties (water vapor and oxygen) of hemicellulose films (Zhang et al. 2025).

3.2.5 Soy Protein-Based Materials

Soy protein, a byproduct of the soybean oil industry, is used as a dietary supplement and an ingredient in animal feed. There are three principal categories of commercially processed soy protein products, which include soy flour (SF), soy protein concentrate (SPC), and soy protein isolate (SPI) (Singh et al. 2008). The SF is derived from finely milled soybeans with a protein content of around 50%. The SPC, which comprises around 70% protein, is essentially defatted soy flour from which water-soluble carbohydrates have been removed. The SPI represents the most refined form with 90% protein (dw) (Thrane et al. 2017).

The protein fraction in soybeans consists predominantly of albumins and globulins, which contain 18 amino acids, many of which bear polar functional groups such as carboxyl, amino, and hydroxyl moieties (Qin et al. 2022). Despite these properties, unmodified soy protein is unsuitable as a direct substitute for conventional synthetic polymers (García et al. 1997). Soy protein exhibits a processing temperature close to 200 °C, which approximates its thermal degradation threshold, thereby restricting its thermal processability. Furthermore, soy protein-based materials typically display high stiffness but are also brittle and inflexible. An additional challenge is their pronounced hydrophilicity due to the presence of polar groups and the necessity for plasticizers, which collectively increase their susceptibility to moisture (Singh and Krishnaswamy 2022). To overcome these limitations, strategies such as polymer blending and nanoparticle reinforcement have shown promise, facilitated easier processing and significantly enhanced the functional performance of soy protein-based materials (Kunte et al. 1997). Incorporation of thermally stable nanoscale fillers such as silica, boron nitride, and clays enhances thermal resistance by acting as insulative and barrier layers (Li et al. 2018). Soy protein films exhibit strong gas barrier abilities but poor moisture resistance, which is further compromised by high plasticizer content. The incorporation of nanofillers, particularly montmorillonite, significantly enhances barrier properties (Chen and Zhang 2006). Other nanomaterials such as silica, carbon nanoparticles, and boron nitride also effectively reduce water vapor permeability. Blending soy protein with hydrophobic lipids or resins further improves moisture resistance (Swain et al. 2017).

Soy protein films are primarily fabricated using solution-based and melt-processing techniques. Among these, solution casting is the most prevalent, leveraging the hydrophilic nature of soy protein to disperse it in aqueous media (Brandenburg et al. 1993). This process typically involves blending soy protein suspensions with polymer or nanoparticle dispersions, followed by solvent evaporation (Guerrero et al. 2010).

The incorporation of plasticizers—particularly low molecular weight and highly polar compounds like polyols, amides, and hydroxylamines—improves melt flowability and flexibility by disrupting protein–protein interactions and increasing chain mobility (Brandenburg et al. 1993; Tahir et al. 2024). Plasticizers also influence the morphology of soy protein blends, affecting mechanical performance. Compression-molded films generally outperform solution-cast ones, except in specific blends like soy protein/agar, where casting promotes superior tensile strength due to network formation (Tian et al. 2011).

Mechanical performance is critical for soy protein films, particularly in structural applications. Strength enhancement is effectively achieved through nanofiller incorporation, which depends on particle properties, dispersion quality, and matrix compatibility. While tensile strength and modulus generally increase with filler content, ductility often declines (Li et al. 2018). Polymer blending, especially with biodegradable polymers, further improves the mechanical attributes of soy protein films. Natural polymers (e.g., starch, gelatin, cellulose) offer good compatibility via similar polar groups, facilitating stress transfer (Koshy et al. 2015).

Soy protein/inorganic nanocomposites have garnered attention due to their enhanced mechanical, thermal, and barrier properties. Inorganic fillers, known for superior thermal stability over organic ones, are used to address soy protein's limitations in water resistance and mechanical strength. Acid-treated multiwalled carbon nanotubes also reinforced SPI matrices by forming hydrogen bonds with protein chains (Huang et al. 2020). Silver nanoparticles incorporated SP films displayed uniform dispersion and potent antimicrobial activity against *S. aureus* and *E. coli* (Zhao et al. 2013).

Chemical crosslinking between fillers and SP matrix further amplifies strength and structural integrity. Microfibrillated cellulose (MFC) notably improves toughness and modulus (Ortiz et al. 2018). Nanofibrillar cellulose (NFC) incorporated into soy protein-based aerogels reduces brittleness and boosts compression modulus (Arboleda et al. 2013). Soy protein/natural polymer blends involve biodegradable biopolymers, including proteins, polysaccharides (starch, cellulose, chitosan), and natural rubber. These biopolymers, except natural rubber, are hydrophilic and form strong hydrogen bonds with soy protein, enhancing compatibility. However, their hydrophilicity doesn't significantly resolve soy protein's water sensitivity. Soy protein/synthetic polymer blends require effective interaction and compatibility for optimal performance. The application of the synthetic polymers and their benefit is presented in Table 3.2.

3.2.6 Zein-Based Materials

Zein, the principal storage protein found in the endosperm of *Zea mays*, comprises 80% of the total protein content in maize kernels (Kasaai 2018). As a member of the prolamin family, zein is characterized by its solubility in aqueous alcohols and relative insolubility in water (Shukla and Cheryan 2001). Its amphiphilic nature, arising from a combination of hydrophobic (leucine, proline, alanine) and hydrophilic (glutamine) amino acid residues, imparts unique film-forming and barrier properties (Anderson and Lamsa 2011). Corn gluten (CGM) meal is regarded as a by-product of wet-milling. Zein has garnered renewed interest due to advances in sustainable extraction technologies and the rising demand for biodegradable, bio-functional materials. Currently, zein is utilized across a broad range of sectors, including biodegradable plastics, adhesives, inks, ceramics, cosmetics, edible coatings, and pharmaceutical delivery systems (Kasaai 2018; Lawton 2002).

Table 3.2 Blending of various Synthetic polymers and their benefits

Synthetic polymer	Role/effect in blend with soy protein (SPI)	Key properties/benefits	References
Poly (ethylene oxide)	Forms films with denatured SPI	Ultra-elasticity due to cross-linking interactions	Ji et al. (2012)
Polycaprolactone	Enhances mechanical and water resistance	Improved mechanical strength and water resistance	Zhong and Sun (n.d.)
Polyurethane	Produces semi-IPN blends with diverse properties	Toughened plastics to elastomers, improved flexibility and water resistance	Liu et al. (2008)
Waterborne polyurethane	Enhances mechanical and environmental performance	Better environmental profile, improved flexibility and mechanical strength	Zhang et al. (2012)
Poly (vinyl alcohol)	Offers excellent film properties	High transparency, oxygen barrier, mechanical strength (especially with melt processing)	Su et al. (2010)
Poly (lactic acid)	Improves phase morphology and reduces water absorption	Enhanced mechanical properties and water resistance	González and Alvarez Igarzabal (2013)
Poly(butylene succinate)	Modified with glycerol-plasticized SPI to enhance multiple properties	Improved hydrophobicity, mechanical strength, and biodegradability	Li et al. (2008)

Structurally, zein exists in several isoforms, predominantly α-zein, with lower proportions of β-, γ-, and δ-zeins. The α-Zein (19–25 kDa), the most industrially significant fraction, is highly soluble in 70–95% ethanol and exhibits excellent film-forming capabilities. The β- and γ-zein fractions contain disulfide-linked aggregates that contribute to zein's macromolecular heterogeneity and influence its functional attributes (Momany et al. 2006).

Extraction and purification efficiency depend heavily on feedstock composition and processing conditions. While CGM is the most viable commercial source, other by-products such as distillers' dried grains dry-milled corn (DMC), with soluble (DDGS), and alkali-treated corn vary in protein profile and affect zein yield and purity (Lawton 2002). Conventional extraction typically employs aqueous ethanol, with protein solubilization followed by centrifugation and precipitation. The product is a yellowish powder that often requires multiple purification steps and suffers from modest extraction yields. To enhance sustainability, recent studies have explored greener solvents and solvent-free methods (Shukla and Cheryan 2001).

Zein's α-helical secondary structure is largely conserved across different corn sources, although genotype and environmental conditions (e.g., elevated CO_2) can modulate isoform distribution and functional characteristics (Momany et al. 2006). Zein's ability to form glossy, hydrophobic, and grease-resistant films is attributed to strong intermolecular interactions, including hydrogen bonding, disulfide bridging, and hydrophobic associations. Upon solvent evaporation, zein self-assembles into a fibrillar network with low oxygen permeability and resistance to microbial degradation, making it highly suitable for food packaging (Corradini et al. 2014). Zein films produced via the wet process (solvent casting and evaporation) tend to be brittle, necessitating plasticizers to improve flexibility and mechanical integrity. Alternatively, the dry processing approach exploits zein's thermoplasticity under low-humidity conditions to produce moldable materials (Liu et al. 2019).

Zein's surface activity and interactions have been studied through monolayer and surface plasmon resonance techniques (Wang et al. 2004). Zein also exhibits notable gas barrier and sorption characteristics, comparable to some cellulose-based materials (Ali et al. 2014). The incorporation of phenolic compounds and flavonoids has been shown to enhance mechanical strength and antioxidant activity, broadening zein's application in bioactive and intelligent packaging (Bayer 2021; Kong et al. 2023). Zein's amphiphilic character enables blending with both hydrophilic and hydrophobic polymers to produce biodegradable materials with tunable properties. For example, zein–starch blends exhibit increased tensile strength with higher zein content, though immiscibility may hinder interfacial adhesion (Gao et al. 2017a, b).

Zein–PCL blends tend to exhibit decreased tensile strength but higher stiffness due to limited compatibility (Wu et al. 2012). Conversely, zein–PVP blends display improved mechanical performance, indicating better miscibility (Woods et al. 2009). Zein–nylon films increased strength and water resistance (Selling and Biswas 2012), while incorporation of Pluronic F127 results in flexible films with tunable plasticity and crystallinity depending on concentration (Li et al. 2013).

3.3 Seafood Waste-Based Biodegradable Films

3.3.1 Chitosan-Based Biodegradable Films

The seafood processing sector produces a large amount of residual biomass and by-products, such as heads, tails, skins, scales, viscera, backbones, and shells. All these waste materials are known to be good reservoirs of valuable biopolymers and bioactive components. Chitin and its deacetylated derivative, chitosan, are found to have significant commercial value owing to their multifunctional bioactivities and various agrochemical uses (Aranaz et al. 2021; Kou et al. 2021; Kumar et al. 2004).

Chitin is a linear polysaccharide composed of β-(1→4)-linked 2-acetamido-2-deoxy-D-glucopyranose (N-acetyl-D-glucosamine) units. Typically, one out of every six glucosamine residues remains in a deacetylated form within native chitin (Aranaz et al. 2021). The polymer chain length of chitin varies by biological source-for instance, chitin derived from yeast contains approximately 100 glucosamine units, while chitin from crabs may encompass 5000 to 8000 units (Moussian 2019). Treatment with hot alkali during deproteinization results in a reduction of polymer chain length due to partial depolymerization (Kumar et al. 2004).

Chitin exists in three distinct allomorphic forms: α, β, and γ. In α-chitin, the polymer chains adopt an antiparallel configuration and are predominantly found in arthropods and crustaceans (Kumirska et al. 2010). In contrast, β-chitin features a parallel chain alignment and is typically sourced from marine diatoms. The γ-chitin structure exhibits a more complex and less well-defined organization and remains a topic of ongoing scientific debate (Gardner and Blackwell 1975). Of these, α-chitin is the most thermodynamically stable, and β-chitin can be readily converted to α-chitin through chemical treatments such as lithium thiocyanate exposure or formic acid precipitation (Aranaz et al. 2021; Kou et al. 2021; Kumar et al. 2004).

Chitin is linked to proteins and other biopolymers to constitute a composite matrix. Hence, the purification of pure chitin requires the removal of proteins and mineral content by deproteinization and demineralization, respectively. Both chemical and biological extraction methods depend on the material source, as it greatly influences the purity, structural integrity, and recovery of chitin (No et al. 1989). Chemical extraction is still the most commonly used industrial process for chitin recovery based on the simplicity of operation and quick processing. Shell materials are first thoroughly cleaned, dried, ground into a fine powder, and then pulverized. The process of extraction involves three main steps: demineralization, deproteinization, and decolorization. Demineralization is done through acid treatment (HCl, HNO_3, H_2SO_4, CH_3COOH or HCOOH). To reduce polymer degradation, less severe acid-like acetic acid, EDTA, and sulfurous acid are preferably used. Traditionally, hydrochloric acid is used under controlled levels and agitation for 2–3 h (El Knidri et al. 2018). Then neutralization and drying are necessary preliminary steps.

Deproteinization involves the alkali-mediated hydrolysis of proteins. A wide range of alkaline (NaOH, KOH, Na_2CO_3, $NaHCO_3$, $Ca\ (OH)_2$) is used at concentrations between 0.125 and 0.5 M and a temperature of 160 °C. However, prolonged or intensive alkaline exposure may induce partial deacetylation of chitin and reduce its molecular weight (Abidin et al. 2020; Khanafari et al. 2008).

Enzyme-based methods typically offer improved selectivity and milder reaction conditions, resulting in a purer product (Hoell et al. 2010). Organic acids produced by lactic acid bacteria and other microorganisms facilitate the dissolution of calcium carbonate in the shell matrix, forming soluble calcium salts (Mohan et al. 2022; No et al. 1989). Proteolytic enzymes (alcalase, pepsin, papain, pancreatin, trypsin) derived from microbial, plant, or animal sources are employed to hydrolyze and eliminate protein content from the shell material. Among them, alcalase is favored due to its specificity and effectiveness in simultaneous recovery of chitin, protein hydrolysates, and astaxanthin (Barber et al. 2013).

Studies have demonstrated the efficacy of proteases from bacterial strains such as *Bacillus mojavensis* A21, *B. subtilis* A26, *B. licheniformis* NH1 and MP1, *Vibrio metschnikovii* J1, and *Aspergillus clavatus* ES1 on the chitosan isolation (Khanafari et al. 2008). The protease from *B. mojavensis* A21 achieved a deproteinization rate of 76%, which was further optimized to 88% using response surface methodology by adjusting enzyme-to-substrate ratios, temperature, and incubation time (Wang et al. 2021).

Chitosan is derived from chitin through deacetylation, which involves the removal of acetyl groups, altering its molecular weight via depolymerization (Mohan et al. 2022). Chemical deacetylation can be performed either homogeneously by dispersing chitin in concentrated alkali at 25 °C, followed by cooling, or heterogeneously, using hot concentrated alkali (typically NaOH), achieving deacetylation levels up to 85–99% (El Knidri et al. 2018).

Enzymatic deacetylation presents a greener alternative using chitin deacetylase, which hydrolyzes N-acetamido bonds under mild conditions, preserving polymer integrity (Halder and Mondal 2018). This enzyme, sourced from fungi like *Mucor rouxii*, *Aspergillus nidulans*, and *Colletotrichum lindemuthianum*, is thermally stable up to 50 °C and targets β-(1,4)-linked N-acetyl-D-glucosamine. Though enzymatic methods typically yield lower quantities, they produce higher molecular weight chitosan and facilitate the development of novel biopolymeric products (Liang et al. 2023).

Chitosan is a potential biopolymer for food packaging because it is biodegradable, can form films, and possesses inherent antimicrobial activity (Jiang et al. 2023). Chitosan films have been formulated with additives such as dilute acids, plasticizers (e.g., glycerol), and emulsifiers (e.g., Tween 20, Span 20) to make them more flexible and film-stable, but they are still lacking compared to synthetic films as far as mechanical strength, water resistance, and large-scale use are concerned. For their limitations to be overcome, chitosan is usually combined with other synthetic and natural polymers such as cellulose derivatives, starch, alginate, and pectin for increased mechanical strength, water barrier properties, antioxidant activity, and food preservation. Cellulose and its derivatives (e.g., methylcellulose) enhance durability, elasticity, and transparency (Dutta et al. 2009; Zhao et al. 2023). Chitosan-alginate and chitosan-pectin films support good gas and moisture exchange while maintaining food quality, as demonstrated in fresh produce packaging. Incorporation of other polysaccharides like cyclodextrin, glucomannan, and xylan further strengthens film functionality (Elsabee et al. 2008; Yan et al. 2001).

Animal-derived proteins like casein, gelatin, collagen, and lysozyme enhance the water vapor barrier, mechanical properties, and UV protection of the chitosan films (Ponce and Moreira 2025; Qiao et al. 2017; Sionkowska et al. 2006). Blending with synthetic polymers such as polyvinyl alcohol (PVA) (Kanatt et al. 2012), polylactic acid (PLA) (Fathima et al. 2018), and polyethylene oxide (PEO) (Alhusaiki Alghamdi 2022) improves flexibility, thermal stability, and barrier properties of chitosan films. While this may compromise full biodegradability, it helps reduce persistent plastic use (Fathima et al. 2018).

The addition of nanomaterials significantly improves chitosan film properties. Metallic nanoparticles such as silver, zinc oxide, titanium dioxide, and copper provide strong antimicrobial, thermal, and mechanical reinforcement. Natural biopolymeric nanoparticles, including nano-cellulose, chitin, and lignin, improve structural properties and food safety while maintaining environmental sustainability. Polymer/clay nanocomposites using materials like montmorillonite and Cloisite reduce gas and moisture permeability and enhance tensile strength. Natural polysaccharide-based fillers such as chitin and cellulose whiskers, nano-crystalline cellulose, and lignin offer biodegradable, antimicrobial reinforcement (Dong et al. 2022). These fillers improve strength, barrier properties, and thermal stability, making them viable for food packaging (Hu et al. 2019).

3.4 Dairy-Based Waste Materials

3.4.1 Whey Protein-Based Materials

The dairy industry generates substantial volumes of liquid effluent as a byproduct during the coagulation of casein. Dairy whey (DW) accounts for approximately 85–95% of the original milk volume yet retains nearly 55% of the total nutrient content of whole milk (Dennis 2015).

The protein fraction of DW is a complex assemblage of globular proteins with diverse structural features and biological activities. β-Lactoglobulin (β-LG) (~18.3 kDa) is the predominant whey protein, comprising over 50% of the total protein content in DW (Dennis 2015; Solak and Akin 2012). The α-Lactalbumin (α-LA) 10% of DW proteins (~14 kDa), it displays greater thermal stability compared to β-LG (Jovanović et al. 2005; Smithers 2008). Bovine Serum Albumin (BSA) is a high-molecular-weight protein (~66.3 kDa) known for its remarkable ligand-binding capacity. Immunoglobulins, approximately 0.7 g/L (DW), contribute critical immune-related functionalities (Foegeding et al. 2002). Other Minor proteins are composed of lower quantities of lactoferrin (LF) (Mcintosh et al. 1998; Smithers et al. 1996).

Modern processing technologies have enabled the transformation of DW into commercially viable products by physical and thermal Processing. The DW is often subjected to spray-drying to produce dairy whey powder (DWPD). DW with Reduced Lactose (DWRL)—containing <60% lactose, DW Protein Concentrates (DWPCs), with protein content DWPC34 or DWPC80. DW Protein Isolates (DWPI) are highly purified preparations containing >90% protein and <1% lactose and fat (Madureira et al. 2007). However, advanced membrane filtration techniques, such as ultrafiltration and nanofiltration, are increasingly employed due to their energy efficiency and ability to preserve the native functional structure of proteins (Kilara and Vaghela 2017).

DW proteins are soluble in water and can function well as emulsifiers, allowing for the formation of odorless, colorless, transparent, and flexible edible films (Sabato et al. 2001). DWPI films are normally endowed with a good aroma barrier, low oxygen permeability, but poor moisture barrier functionality (Ustunol and Mert 2004). The most frequent production process is casting and drying of DWPI aqueous solutions (Fang et al. 2002; Huntrakul and Harnkarnsujarit 2020). Industrial film-forming processes like extrusion and compression molding can also be used, allowing solvent-free processing operations (Fang et al. 2002). Mechanical reinforcement and barrier properties of whey protein films can be enhanced appreciably by cross-linking methods (Le Tien et al. 2000; Ustunol and Mert 2004).

Thermal treatment is the most prevalent approach, promoting the formation of disulfide bonds and hydrophobic interactions. Ultraviolet (UV) irradiation induces free radical formation via activation of aromatic rings and unsaturated bonds in amino acids, facilitating covalent cross-link formation (Le Tien et al. 2000; Sabato et al. 2001; Ustunol and Mert 2004). Additionally, γ-irradiation combined with thermal processing has been shown to enhance film functionality through the formation of bityrosine and disulfide cross-links. Chemical agents can reinforce film networks by forming covalent bonds between protein chains, thereby improving mechanical and barrier characteristics. However, concerns about residual toxicity, especially from aldehydes, limit their application in food-contact materials (Azeredo and Waldron 2016). Transglutaminase (TGase) and peroxidase are frequently used to catalyze the formation of isopeptide or dityrosine bonds, respectively, within protein matrices. TGase-mediated cross-linking has been demonstrated to significantly enhance the film's structural integrity, elasticity, and moisture resistance without introducing toxic residues (Faergemand et al. 1998).

To enhance the mechanical and physical properties of hydrocolloidal films, plasticizers are often incorporated. Without plasticizers, DW protein films are brittle and inflexible (Fang et al. 2002). The inclusion of plasticizers imparts flexibility but simultaneously increases water vapor permeability (WVP) (Fang et al. 2002; Galietta et al. 1998). Moreover, increasing plasticizer concentrations typically reduces the tensile strength and Young's modulus while improving elongation capacity. Among the various plasticizers investigated, glycerol has been shown to significantly increase WVP and water solubility, while reducing mechanical resistance, apparent Young's modulus, and glass transition temperature. Plasticizers such as xylitol, sucrose, and propylene glycol have also been examined, and each of them has been found to affect the physical properties of the films produced differently. These results show the relevance of plasticizer concentration and type optimization to match functionality and performance in DW protein-based edible films (Huntrakul and Harnkarnsujarit 2020).

3.5 Animal Food Waste-Based Materials

3.5.1 Collagen-Based Materials

The primary sources of collagen include bones, hides, pig skin, and various other mammalian byproducts. Collagen is the most prevalent protein found in livestock processing byproducts with high protein content. Therefore, extracting collagen from these byproducts would offer the potential to produce valuable protein resources from animal slaughter (Birk and Bruckner 2005; Lee et al. 2001). Type I collagen sourced from bovine is typically extracted from the Achilles tendon, skin, and bones. Type II collagen is predominantly located in cartilage, type III in cardiovascular tissues, and type IV in the basement membrane. Collagen can also be extracted from a range of marine waste like fish, jellyfish, sponges, mollusks such as octopus and cuttlefish, as well as various species of echinoderms (Fratzl 2008; Jafari et al. 2020; Ottani et al. 2001).

Collagen exists in different dimensions and biological functions, depending on the animal tissue of origin. It preferentially takes the right-handed, super twisted triple-helix configuration constructed by α-polypeptide chains. The tropocollagen unit, a monomeric building block, is present in both fibrillar and non-fibrillar forms (Fratzl 2008; Ottani et al. 2001). The adult fibrillar collagen type, represented by type I collagen, has the elongated triple-helical domain as the major component with over 1000 amino acids (Ottani et al. 2001). Non-fibrillar collagen types have interspersed gaps (multiple residues) or small defects (one to three residues) in the helix region. The occurrence of glycine at each third residue position makes it possible for close packing of the helices, positioning it at the center of the collagen triple helix (Birk and Bruckner 2005). The high ratio of Pro and Hyp residues makes the α-chains tend towards a polyproline type II (PPII) helical structure, thus reducing entropic costs during the folding of collagen (Fratzl 2008; Lee et al. 2001; Shoulders and Raines 2009).

Collagen extraction from slaughterhouse-derived by-products typically entails two major phases: pretreatment and extraction (Jafari et al. 2020; Matinong et al. 2022). The pretreatment involves immersion in alkaline solutions to eliminate extraneous substances and non-collagenous proteins, organic solvents for lipid removal, and chelating agents such as EDTA to achieve decalcification (Fratzl 2008). These extracted forms generally preserve the native triple-helical configuration of collagen macromolecules. Proteolytic digestion, particularly enzymatic hydrolysis preferentially, cleaves the non-helical telopeptide regions, enhancing both yield and aqueous solubility (Jafari et al. 2020; Matinong et al. 2022; Van Der Rest and Garrone 1991).

Recently advanced techniques like ultrasound, microwaves, and enzyme-based extractions are also studied (Li et al. 2010; Song et al. 2018). Crude collagen extracts often contain various salts and non-collagenous proteins. To purify collagen, a series of steps, including filtration and centrifugation, are employed. Additionally, purification methods can be customized to isolate collagen within specific molecular weight ranges through techniques such as membrane filtration and chromatographic separation. Partial hydrolysis of collagen results in gelatine, which is a combination of polypeptides of varying molecular weight, usually 15–250 kDa (Matinong et al. 2022).

The strong mechanical properties and barrier properties of collagen films make them attractive for use as edible wrappers (Li et al. 2024). However, some issues with the collagen films need to be addressed. For instance, the collagen film swells rapidly in water and that may cause collagen shell rupturing or detaching from the packed meat during meat loading or cooking (Xu et al. 2020). To address this issue, physical (dehydration, heat, and aging treatment), chemical (glutaraldehyde), or enzymatic (transglutaminase) crosslinking methods are achieved through noncovalent linkages between or within collagen molecules. Covalent bond formation can enhance the physicochemical and mechanical properties of collagen casings and film (Grover et al. 2012).

A novel bioactive film combining collagen, chitosan functionalized with gallic, and ε-polylysine (ε-PL) in varying concentrations has been developed to extend pork shelf life (Zheng et al. 2023). Cross-linking collagen films with quinones enhances their physicochemical characteristics, including reduced water vapor permeability, increased enzymatic degradation resistance, improved mechanical integrity, and superior thermal stability. Cross-linking imparts antioxidant and antimicrobial functionalities to the collagen films, suggesting strong potential for active edible packaging systems (Ran et al. 2025). In further developments, Zhu et al. (2024) has developed a method where CMC was employed as a cross-linking matrix for collagen films. Botanical extract exhibited excellent environmental compatibility, high biodegradability, diminished transparency, and exceptional UV-visible light shielding capabilities (Li et al. 2024).

Collagen-based film reinforced with delphinidin and catalyzed by laccase, offering enhanced mechanical and functional properties. The resultant film improves tensile strength of 41.74 MPa and exhibits potent antioxidant and barrier properties, including efficient free radical scavenging, effective light transmission, and low water vapor permeability. Furthermore, the integration of *Vaccinium oxycoccus* pigment enabled intelligent responsiveness with visible color shifts in response to pH fluctuations (Yin et al. 2023).

3.5.2 Gelatine-Based Materials

Upon hydrolysis, the intra- and intermolecular bonds within collagen are cleaved, yielding a smaller, water-soluble polypeptide known as gelatine. It is characterized by a high molecular weight and comprises various polypeptide chains, namely α, β, and γ-stabilized through hydrogen bonding (Rýglová et al. 2017). The compositional and functional attributes of gelatine are influenced by the source of the raw material, extraction methods, and interspecies variability. Specifically, amino acids such as proline and its hydroxylated derivative, hydroxyproline, significantly enhance the thermal stability of collagen by promoting hydrogen bond formation, thereby increasing Bloom strength and elevating gelatine's melting and gelling temperatures (Liu et al. 2015).

Type A gelatine is conventionally produced via acidic hydrolysis of collagen and is commonly sourced from pigs, poultry, and fish. This process results in gelatine with an isoelectric point between 7 and 9, making it suitable for substrates with lower cross-link densities. Type B gelatine, on the other hand, is obtained through alkaline hydrolysis, typically from bovine origins. Type B gelatine and its corresponding nanoparticles often display increased cross-linking density and exhibit slower degradation kinetics relative to Type A gelatine (Dille et al. 2021; Liu et al. 2015; Rýglová et al. 2017).

In order to improve the mechanical properties and functionality of gelatine films, different chemical and physical treatments have been explored. Enzymes and cross-linking agents are utilized to enhance tensile strength, and the addition of plasticizers, polysaccharides, essential oils, and bioactive compounds can increase elasticity and reduce water solubility and permeability (Said and Sarbon 2022). Natural cross-linkers like genipin, an aglycone derivative isolated from plant origin, have been found to effectively enhance film stiffness, albeit at the cost of flexibility (Zhao and Sun 2018). Addition of polyvinyl alcohol (PVOH) and glycerol as plasticizers has been found to improve the elasticity and durability of gelatine films through the depletion of intermolecular hydrogen bond density. Of these, glycerol has proven the best plasticizer for enhancing film flexibility (Qiao et al. 2024).

Additional functional enhancements have been achieved by integrating chitosan, essential oils, and various plant-based extracts such as anthocyanins and phenolic compounds. These additives confer antimicrobial and antioxidant properties, as well as improved mechanical performance and reduced hydrophilicity (Manzoor et al. 2023). Gelatine-based films intended for packaging applications must withstand mechanical stresses during handling and transport. Therefore, critical parameters such as tensile strength and flexibility are crucial for film integrity (Zhao and Sun 2018). The thickness of these films is influenced by processing conditions and the concentration of solid additives in the filmogenic solution. For example, Alizadeh Sani et al. (2022) observed the incorporation of κ-carrageenan, anthocyanins, and TiO_2 to increase film thickness. Gelatine films with higher protein content and thickness exhibit increased water absorption due to the prevalence of hydrophilic amino acid residues (Lin et al. 2018).

The incorporation of antimicrobial agents into gelatine films is a promising strategy to inhibit microbial proliferation, thereby extending the shelf life of both the packaging material and its contents. For instance, Hanani et al. (2019) demonstrated the antimicrobial efficacy of gelatine films enriched with pomegranate peel powder against *Escherichia coli*, *Staphylococcus aureus*, and *Listeria monocytogenes* using the disc diffusion method. Similarly, embedding antioxidant compounds into the gelatine matrix offers additional benefits, including retarding lipid oxidation and further extending the shelf life of food products.

3.6 Microbial-Based Polymers from Food Waste

3.6.1 Polylactic Acid (PLA)-Based Biodegradable Packaging Materials

Bioplastics primarily consist of polyesters such as polylactic acid (PLA) and polyhydroxyalkanoates (PHAs), which exhibit comparable mechanical, thermal, and physicochemical properties to conventional petrochemical-based polymers (Di Bartolo et al. 2021). Among these, PLA is one of the most extensively studied and commercially viable biopolymers, owing to its biodegradability, biocompatibility, thermoplastic nature, elasticity, rigidity, and excellent moldability (Ibrahim et al. 2021). Structurally, PLA is a biodegradable aliphatic polyester with the formula $(C_3H_4O_2)_n$, and it is synthesized from renewable sources such as maize, starch, sugarcane bagasse, and food waste (Gupta et al. 2007). Being the first commercially successful biopolymer wholly obtained from renewable resources, PLA is widely used in food packaging and is "Generally Recognized as Safe" (GRAS) by the U.S. Food and Drug Administration (FDA) (Shlush and Davidovich-Pinhas 2022).

The fermentative synthesis of lactic acid has garnered considerable attention due to its environmental benefits, cost-efficiency, minimal carbon emissions, and ability to yield optically pure isomers through biocatalysis. Currently, 90% of global LA is produced through microbial fermentation, while only 10% is synthesized chemically (Endres 2019; Lackner 2015; Shruti and Kutralam-Muniasamy 2019). Homofermentative bacteria produce lactic acid predominantly from hexoses, whereas heterofermentative strains generate lactic acid along with byproducts such as acetic acid, ethanol, and carbon dioxide (Di Bartolo et al. 2021; Thakur et al. 2018).

Food waste has proven to be an ideal substrate for LA fermentation, owing to its high carbohydrate and protein content. This waste, rich in fermentable sugars, predominantly originates from fruits, vegetables, bakery items, and animal-derived foods waste (Fig. 3.2). Kitchen waste, including vegetable peels, fruit remnants, and leftover food, supports robust microbial growth for LA synthesis (Kwan et al. 2018). Fermentation at 50 °C using native LAB yielded over 98% optically pure L-LA. Innovations like electro-fermentation and enzymatic pre-treatment have

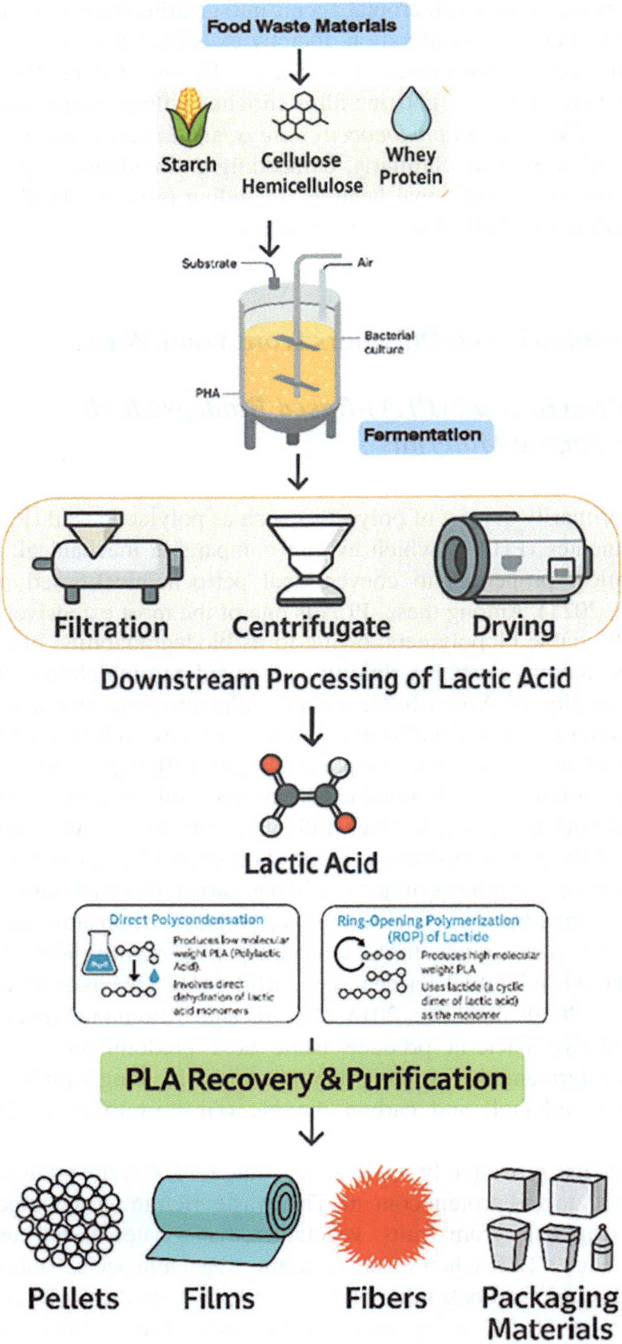

Fig. 3.2 An overview of the Polylactic acid production and film production

further enhanced LA productivity and purity, allowing for a more effective organic waste conversion to high-value lactic acid (Garlotta 2001).

One crucial factor influencing the physical characteristics of PLA is the optical purity of LA, required for synthesizing high-crystalline PLA. Only optically pure D or L isomers produced via microbial fermentation can yield such PLA, unlike racemic lactic acid (Gupta et al. 2007). Optical purity depends on the bacterial strain and fermentation medium used, with Lactobacillus being the primary industrial producer. Over 90% purity is needed for pharmaceutical applications, while 80–90% suffices for food grade. Downstream LA purification is vital and cost-sensitive, involving techniques like adsorption, precipitation, solvent extraction, electrodialysis, molecular distillation, and esterification due to LA's affinity for water (Di Bartolo et al. 2021; Garlotta 2001; Kwan et al. 2018).

PLA offers a viable substitute for conventional petroleum-based plastics such as polystyrene (PS), low- and high-density polyethylene (LDPE, HDPE), and polyethylene terephthalate (PET) (Byun et al. 2010). It has high molecular weight, excellent processability, low water solubility, and biodegradability. Compared to mechanical and barrier properties, PLA closely resembles PET and flexible PVC, offering good tensile strength, odor barrier, and grease resistance. Common PLA applications include films, paperboard coatings, and injection-molded products. It can be chemically recycled back to LA (Byun et al. 2010). PLA packaging is particularly suited for short shelf-life, high-respiration foods like fruits, vegetables, and baked goods (Gao et al. 2017a, b; Turalija et al. 2016).

PLA degradation is influenced by factors such as enzymes, water, ultraviolet (UV) radiation, and microbial activity. The degradation process occurs in two main stages. Initially, water molecules penetrate the PLA matrix and break the ester bonds, reducing molecular weight and forming water-soluble oligomers (Darie-Niţə et al. 2016; Mirkhalaf and Fagerström 2021). In the second stage, oligomers are enzymatically hydrolyzed into lactic acid monomers, which are eventually metabolized into water and carbon dioxide (Darie-Niţə et al. 2016; Jamshidian et al. 2012; Mirkhalaf and Fagerström 2021).

3.6.2 *Polyhydroxyalkanoates*

Polyhydroxyalkanoates (PHAs) are biopolymers synthesized by various microorganisms as intracellular energy and carbon storage compounds. These natural polyesters, composed of 3, 4, 5, and 6 hydroxyalkanoic acids, exhibit thermoplastic properties. Microorganisms can accumulate various intracellular inclusions, both organic and inorganic, typically surrounded by phospholipid or protein layers (Philip et al. 2007). For example, inclusions with iron oxide cores, such as magnetosomes, are classified as inorganic, while those with polyester cores, like PHAs, are considered organic. To date, over 90 genera bacteria have been identified as PHA producers under aerobic and anaerobic conditions. Within bacterial cells,

PHAs are stored in the cytoplasm as granules ranging from 0.2 to 0.5 μm in diameter (Khanna and Srivastava 2005).

PHAs are classified into three groups according to the number of carbon atoms in their monomer side chains: short-chain-length (<5 C), medium-chain-length (5–14 C), and long-chain-length (>14 C) (Tan et al. 2021). Various systems have been developed to convert diverse food waste streams produced during food manufacturing, processing, and consumption into PHAs (Philip et al. 2007). Many food waste components contain complex organic compounds that cannot be directly utilized by PHA-producing microbes (Reddy et al. 2003). Therefore, pre-treatment processes are used to break down these compounds into simpler precursors such as glucose, lactose, acetic acid, or propionic acid. These precursors are then metabolized by selected microbial cultures. Feedstocks like whey, starch, oils, lignocellulose, legumes, and sugar wastes have been explored for PHA production (Raza et al. 2018; Wang et al. 2014).

Cheese industry-derived dairy whey is a viable low-cost carbon feedstock for PHA production. Despite its potential, approximately 50% of whey is discarded or used as animal feed. Studies using organisms like *Haloferax mediterranei*, *Thermus thermophilus* HB8, and genetically modified *Cupriavidus necator* have shown successful PHA synthesis from whey (Amaro et al. 2019).

Starch, a glucose polymer from crops (rice, wheat, potatoes, maize, and cassava), is another viable carbon source for PHA production from food waste (Naitam et al. 2022). The *Cupriavidus* sp. KKU38, isolated from cassava starch wastewater, produced 5.97 g/L biomass with 61.6% PHA content under optimal conditions (Ranganathan et al. 2020). In another study, *Azotobacter croococcum* directly utilized starch without enzymatic hydrolysis, giving 54 g/L dry cell weight with 46% PHA under oxygen-limited conditions (del Mar Sánchez-Peinado et al. 2010). These findings show that starchy food wastes can support high PHA yields, even with limited pre-treatment. Household and industrial waste oils are good carbon substrates for PHA production. *Cupriavidus necator* H16 has shown high PHA yields with palm oil and lard at 6.8 g/L dry cell weight (Kutralam-Muniasamy and Peréz-Guevara 2018).

Lignocellulosic food wastes (bagasse, wheat and rice straw, and bran) contain cellulose, hemicellulose, lignin, and pectin, and require hydrolysis and detoxification before use in PHA production. Sugar industry waste streams, including molasses, sugar beet juice, and bagasse, have been investigated as carbon sources for PHA production. Bagasse, rich in xylose, requires pre-treatment but has shown moderate success with strains like *Burkholderia* and *C. necator* (Naitam et al. 2022). Legume-based food wastes like liquid bean curd and soya waste have been explored as carbon sources for PHA production, with moderate success. *A. latus* cultured on sucrose-supplemented bean curd waste achieved 66.56% PHA content, while *A. eutrophus* showed better results with malt waste than soya, highlighting that the type of food waste significantly influences yield (Sathiyanarayanan et al. 2013).

Solvent extraction is the most common technique for the PHA isolation from the cells. Solvents facilitate cell rupture, solubilization of PHAs in solvents like chloroform or acetone, and precipitation with non-solvents. The floatation method is a variation of solvent extraction, where cells mixed with chloroform self-float after a period, achieving 98% purity and 85% recovery. It also reduces waste and can use eco-friendly solvents (Philip et al. 2007). Digestion methods, both chemical and enzymatic, use agents like sodium hypochlorite or enzymes to break down non-PHA cellular material, allowing for effective recovery. While chemical digestion can lead to polymer degradation, enzymatic methods are more selective and effective (Raza et al. 2018; Sabapathy et al. 2020). Supercritical fluid extraction results in high recovery (up to 89%) and minimal impurities. Aqueous two-phase extraction (ATPE) uses water and non-volatile phases for PHAs isolation, offering an eco-friendly alternative with recovery up to 51% PHA (Raza et al. 2018).

The PHA films exhibit crucial thermal properties, such as melting temperature (177 °C), thermodegradation temperature (227 and 256 °C), and glass transition temperature (−52 to 4 °C), which significantly impact their processing and application (Mirpoor et al. 2023). The PHA crystallinity, which determines brittleness, ranges from highly crystalline to flexible (0–60%). To improve toughness, plasticizers like polyethylene glycol (PEG) and triethyl citrate (TEC) are used, reducing crystallinity and enhancing softness. Mechanical properties of PHAs, including Young's modulus, elongation at break, and tensile strength, vary widely (Pérez-Arauz et al. 2019). For instance, Young's modulus varies between 3.5×10^3 MPa for stiffer PHA and 0.008 MPa for ductile PHA. Elongation at break can range from 2% to 1000%, while tensile strength ranges from 8.8 to 104 MPa (Bhardwaj et al. 2014). In food packaging, PHAs offer good water barrier properties and are hydrophobic, comparable to synthetic thermoplastics like PVC and PET. The PHAs suitable for use in food packaging by UV light barrier properties contribute to product preservation (Sun et al. 2018).

PHA blends and composites for enhanced thermal, mechanical, and barrier properties. PHAs can successfully blend with other polymers like PLA, PE, PP, PBS, and PETG to create hybrid materials with enhanced properties. For example, blending PHAs with PLA can significantly improve tensile properties, impact strength, and flexibility. Plasticizers, such as Lapol 108 and LIM, are also added to improve flexibility, elongation at break, and crystallinity (Bhardwaj et al. 2014; Pérez-Arauz et al. 2019; Sathiyanarayanan et al. 2013; Sun et al. 2018).

3.7 Challenges

Despite rapid advances, converting food-waste biopolymers into market-ready packaging still encounters several intertwined hurdles. First, most promising feedstocks like cellulose nanofibres, starch nanocrystals, chitin/chitosan, or protein isolates demand energy-intensive, high-shear or enzymatic extraction steps and, increasingly, nanotechnology-enabled processing; these raise capital costs and complicate scale-up beyond pilot plants (Chee et al. 2024). Second, films obtained from such hydrophilic matrices are typically inferior to petro-plastics in thermal stability, tensile strength, and, crucially, water and oxygen resistance. In humid supply chains, they can swell, lose integrity, and even foster microbial growth (Thakur et al. 2018). Engineering strategies like surface acetylation, plasma treatment, or the incorporation of hydrophobic fillers can help. However, achieving a defect-free, uniformly dispersed bio-nanocomposite remains technically tricky, with interfacial incompatibility between polar biopolymers and non-polar reinforcements often causing phase separation and opacity (Osterberg et al. 2023). Third, several abundant proteins (e.g., soy, zein) exhibit poor melt-flow behaviour, limiting their compatibility with conventional extrusion or injection-molding lines and demanding bespoke processing windows or reactive compatibilizers (Sabato et al. 2001). Fourth, although the product is biodegradable, the upstream use of organic solvents, synthetic plasticisers, or chemical cross-linkers can introduce toxicity, compromise compostability certifications, and trigger additional regulatory scrutiny (Wu et al. 2021). Finally, active-packaging concepts that embed antimicrobials, antioxidants, or nanometals must be rigorously evaluated for controlled release, migration into food, and long-term efficacy, ensuring consumer safety while preserving shelf life. Addressing these challenges simultaneously through green chemistry routes, multifunctional bio-fillers, and life-cycle-aware design remains the pivotal step toward truly sustainable, industrial-scale packaging sourced from food waste (Westlake et al. 2022).

3.8 Conclusions

Biodegradable packaging derived from food waste offers a transformative solution to two pressing global issues: plastic pollution and food waste. Through advanced extraction, chemical modification, and biopolymer blending, a wide range of biodegradable materials from starch and cellulose to soy protein, chitosan, and microbial polymers can be engineered with desirable properties for food packaging. Despite significant progress, challenges related to performance, cost, processability, and moisture sensitivity remain. Continued research, particularly in nanotechnology, green processing, and scalable fabrication methods, is critical to make these materials competitive with conventional plastics. Integrating waste vaporization with sustainable packaging innovations paves the way for a circular, eco-conscious economy.

References

Abidin NAZ, Kormin F, Abidin NAZ, Anuar NAFM, Bakar MFA (2020) The potential of insects as alternative sources of chitin: an overview on the chemical method of extraction from various sources. Int J Mol Sci 21(14):1–25. MDPI AG. https://doi.org/10.3390/ijms21144978

Adetunji LR, Adekunle A, Orsat V, Raghavan V (2017) Advances in the pectin production process using novel extraction techniques: a review. Food Hydrocoll 62:239–250. Elsevier B.V. https://doi.org/10.1016/j.foodhyd.2016.08.015

Akhter MJ, Sarkar S, Rayhanujjaman M, Kabir MS, Hosain MM (2024) Characterization of mango seed kernel starch: extraction and analysis. Food Chem Adv 5:100806. https://doi.org/10.1016/j.focha.2024.100806

Alhusaiki Alghamdi HM (2022) Structural, morphological, optical, and electrical characteristics of polyethylene oxide/ chitosan-copper oxide nanoparticles for optoelectronic applications. Opt Mater 134. https://doi.org/10.1016/j.optmat.2022.113101

Ali Akbari Ghavimi S, Ebrahimzadeh MH, Solati-Hashjin M, Abu Osman NA (2015) Polycaprolactone/starch composite: fabrication, structure, properties, and applications. J Biomed Mater Res A 103(7):2482–2498. https://doi.org/10.1002/jbm.a.35371

Ali S, Khatri Z, Oh KW, Kim IS, Kim SH (2014) Zein/cellulose acetate hybrid nanofibers: electrospinning and characterization. Macromol Res 22(9):971–977. https://doi.org/10.1007/s13233-014-2136-4

Alizadeh Sani M, Tavassoli M, Salim SA, Azizi-lalabadi M, McClements DJ (2022) Development of green halochromic smart and active packaging materials: TiO2 nanoparticle- and anthocyanin-loaded gelatin/κ-carrageenan films. Food Hydrocoll 124. https://doi.org/10.1016/j.foodhyd.2021.107324

Amaro TMMM, Rosa D, Comi G, Iacumin L (2019) Prospects for the use of whey for polyhydroxyalkanoate (PHA) production. Front Microbiol 10(May). Frontiers Media S.A. https://doi.org/10.3389/fmicb.2019.00992

Anderson TJ, Lamsa BP (2011) Zein extraction from corn, corn products, and coproducts and modifications for various applications: a review. Cereal Chem 88(2):159–173. American Association of Cereal Chemists. https://doi.org/10.1094/CCHEM-06-10-0091

Aranaz I, Alcántara AR, Civera MC, Arias C, Elorza B, Caballero AH, Acosta N (2021) Chitosan: an overview of its properties and applications. Polymers 13(19). MDPI. https://doi.org/10.3390/polym13193256

Arboleda JC, Hughes M, Lucia LA, Laine J, Ekman K, Rojas OJ (2013) Soy protein-nanocellulose composite aerogels. Cellulose 20(5):2417–2426. https://doi.org/10.1007/s10570-013-9993-4

Asfaw WA, Tafa KD, Satheesh N (2023) Optimization of citron peel pectin and glycerol concentration in the production of edible film using response surface methodology. Heliyon 9(3). https://doi.org/10.1016/j.heliyon.2023.e13724

Azeredo HMC, Waldron KW (2016) Crosslinking in polysaccharide and protein films and coatings for food contact - a review. Trends Food Sci Technol 52:109–122. Elsevier Ltd. https://doi.org/10.1016/j.tifs.2016.04.008

Bangar SP, Whiteside WS (2021) Nano-cellulose reinforced starch bio composite films- a review on green composites. Int J Biol Macromol 185:849–860. Elsevier B.V. https://doi.org/10.1016/j.ijbiomac.2021.07.017

Baraheng S, Karrila T (2019) Chemical and functional properties of durian (Durio zibethinus Murr.) seed flour and starch. Food Biosci 30. https://doi.org/10.1016/j.fbio.2019.100412

Barber PS, Shamshina JL, Rogers RD (2013) A "green" industrial revolution: using chitin towards transformative technologies. Pure Appl Chem 85(8):1611–1701. https://doi.org/10.1351/PAC-CON-12-10-14

Bayer IS (2021) Chapter 7: Zein in food packaging. In: Sustainable food packaging technology. https://doi.org/10.1002/9783527820078

Behrooznia Z, Nourmohammadi J (2024) Polysaccharide-based materials as an eco-friendly alternative in biomedical, environmental, and food packaging. Giant 19:100301. Elsevier B.V. https://doi.org/10.1016/j.giant.2024.100301

Bekele M, Satheesh N (2020) Screening of Ethiopian mango cultivars for suitability for preparing jam and determination of pectin, sugar, and acid effects on physico-chemical and sensory properties of mango jam. Sci Afr 7:e00277. https://doi.org/10.1016/j.sciaf.2020.e00277

Bhardwaj U, Dhar P, Kumar A, Katiyar V (2014) Polyhydroxyalkanoates (PHA)-cellulose based nanobiocomposites for food packaging applications. ACS Symp Ser 1162:275–314. https://doi.org/10.1021/bk-2014-1162.ch019

Birk DE, Bruckner P (2005) Collagen suprastructures. Top Curr Chem 247:185–205. https://doi.org/10.1007/b103823

Boonsiriwit A, Lee M, Kim M, Inthamat P, Siripatrawan U, Lee YS (2021) Hydroxypropyl methylcellulose/microcrystalline cellulose biocomposite film incorporated with butterfly pea anthocyanin as a sustainable pH-responsive indicator for intelligent food-packaging applications. Food Biosci 44. https://doi.org/10.1016/j.fbio.2021.101392

Brandenburg AH, Weller CL, Testin RF (1993) Edible films and coatings from soy protein. J Food Sci 58(5):1086–1089

Byun Y, Kim YT, Whiteside S (2010) Characterization of an antioxidant polylactic acid (PLA) film prepared with α-tocopherol, BHT and polyethylene glycol using film cast extruder. J Food Eng 100(2):239–244. https://doi.org/10.1016/j.jfoodeng.2010.04.005

Caffall KH, Mohnen D (2009) The structure, function, and biosynthesis of plant cell wall pectic polysaccharides. Carbohydr Res 344(14):1879–1900. https://doi.org/10.1016/j.carres.2009.05.021

Chandrasekar CM, Krishnamachari H, Farris S, Romano D (2023) Development and characterization of starch-based bioactive thermoplastic packaging films derived from banana peels. Carbohyd Polym Tech App 5. https://doi.org/10.1016/j.carpta.2023.100328

Chee PL, Sathasivam T, Tan YC, Wu W, Leow Y, Lim QRT et al (2024) Nanochitin for sustainable and advanced manufacturing. Nanoscale 16(7):3269–3292. https://doi.org/10.1039/D3NR05533G

Chen P, Zhang L (2006) Interaction and properties of highly exfoliated soy protein/montmorillonite nanocomposites. Biomacromolecules 7(6):1700–1706. https://doi.org/10.1021/bm050924k

Chen Y, Zhao L, He T, Ou Z, Hu Z, Wang K (2019) Effects of mango peel powder on starch digestion and quality characteristics of bread. Int J Biol Macromol 140:647–652. https://doi.org/10.1016/j.ijbiomac.2019.08.188

Cindradewi AW, Bandi R, Park CW, Park JS, Lee EA, Kim JK, Kwon GJ, Han SY, Lee SH (2021) Preparation and characterization of cellulose acetate film reinforced with cellulose nanofibril. Polymers 13(17). https://doi.org/10.3390/polym13172990

Corradini E, Curti PS, Meniqueti AB, Martins AF, Rubira AF, Muniz EC (2014) Recent advances in food-packing, pharmaceutical and biomedical applications of zein and zein-based materials. Int J Mol Sci 15(12):22438–22470. MDPI AG. https://doi.org/10.3390/ijms151222438

Costa JR, Tonon RV, Cabral L, Gottschalk L, Pastrana L, Pintado ME (2020) Valorization of agricultural lignocellulosic plant byproducts through enzymatic and enzyme-assisted extraction of high-value-added compounds: a review. ACS Sustain Chem Eng 8(35):13112–13125. American Chemical Society. https://doi.org/10.1021/acssuschemeng.0c02087

Darie-Niţə RN, Vasile C, Irimia A, Lipşa R, Râpə M (2016) Evaluation of some eco-friendly plasticizers for PLA films processing. J Appl Polym Sci 133(13). https://doi.org/10.1002/app.43223

del Mar Sánchez-Peinado M, Juárez-Jimenez B, González-López J, Pozo C (2010) Society for Microbiology and Biotechnology Biological Treatment of two-phase olive mill wastewater (TPOMW, alpeorujo): polyhydroxyalkanoates (PHAs) production by Azotobacter strains. J Microbiol Biotechnol 20(3):594–601. https://doi.org/10.4014/jmb.0906.06037

Dennis TJ (2015) Whey protein. In: Nutrition: science, issues, and applications, vol 1–2. Bloomsbury Publishing Plc, pp 858–859. https://doi.org/10.4103/2249-5975.99663

Di Bartolo A, Infurna G, Dintcheva NT (2021) A review of bioplastics and their adoption in the circular economy. Polymers 13(8). MDPI AG. https://doi.org/10.3390/polym13081229

Dille MJ, Haug IJ, Draget KI (2021) Gelatin and collagen. In: Handbook of hydrocolloids. Elsevier, pp 1073–1097. https://doi.org/10.1016/b978-0-12-820104-6.00028-0

Diyana ZN, Jumaidin R, Selamat MZ, Ghazali I, Julmohammad N, Huda N, Ilyas RA (2021) Physical properties of thermoplastic starch derived from natural resources and its blends: a review. Polymers 13(9):5–20. MDPI AG. https://doi.org/10.3390/polym13091396

Doelker E (2005) Cellulose derivatives. In: Biopolymers I. Springer Berlin Heidelberg, Berlin, Heidelberg, pp 199–265

Dong W, Su J, Chen Y, Xu D, Cheng L, Mao L, Gao Y, Yuan F (2022) Characterization and antioxidant properties of chitosan film incorporated with modified silica nanoparticles as an active food packaging. Food Chem 373. https://doi.org/10.1016/j.foodchem.2021.131414

Dutta D, Sit N (2024) Comprehensive review on developments in starch-based films along with active ingredients for sustainable food packaging. Sustain Chem Pharm 39. https://doi.org/10.1016/j.scp.2024.101534

Dutta PK, Tripathi S, Mehrotra GK, Dutta J (2009) Perspectives for chitosan based antimicrobial films in food applications. Food Chem 114(4):1173–1182. https://doi.org/10.1016/j.foodchem.2008.11.047

Ebringerová A, Hromádková Z, Heinze T (2005) Hemicellulose. Adv Polym Sci 186:1–67. https://doi.org/10.1007/b136816

El Knidri H, Belaabed R, Addaou A, Laajeb A, Lahsini A (2018) Extraction, chemical modification and characterization of chitin and chitosan. Int J Biol Macromol 120:1181–1189. Elsevier B.V. https://doi.org/10.1016/j.ijbiomac.2018.08.139

Elsabee MZ, Abdou ES, Nagy KSA, Eweis M (2008) Surface modification of polypropylene films by chitosan and chitosan/pectin multilayer. Carbohydr Polym 71(2):187–195. https://doi.org/10.1016/j.carbpol.2007.05.022

Endres HJ (2019) Bioplastics. Adv Biochem Eng Biotechnol 166:427–468. Springer Science and Business Media Deutschland GmbH. https://doi.org/10.1007/10_2016_75

Espitia PJP, Du WX, de Jesús Avena-Bustillos R, Soares ND, McHugh TH (2014) Edible films from pectin: physical-mechanical and antimicrobial properties - a review. Food Hydrocoll 35:287–296. Elsevier B.V. https://doi.org/10.1016/j.foodhyd.2013.06.005

Faergemand M, Otte J, Qvist KB (1998) Cross-linking of whey proteins by enzymatic oxidation. J Agric Food Chem 46(4):1326–1333

Fang Y, Tung MA, Britt IJ, Yada S, Dalgleish DG (2002) Tensile and barrier properties of edible films made from whey proteins. J Food Sci 67:188–193

Fathima PE, Panda SK, Ashraf PM, Varghese TO, Bindu J (2018) Polylactic acid/chitosan films for packaging of Indian white prawn (Fenneropenaeus indicus). Int J Biol Macromol 117:1002–1010. https://doi.org/10.1016/j.ijbiomac.2018.05.214

Foegeding EA, Davis JP, Doucet D, McGuffey MK (2002) Advances in modifying and understanding whey protein functionality. Trends Food Sci Technol 13(5):151–159. https://doi.org/10.1016/S0924-2244(02)00111-5

Fratzl P (2008) Collagen: structure and mechanics, an introduction. In: Collagen: structure and mechanics. Springer US, Boston, pp 1–13

Galietta G, Di Gioia L, Phane Guilbert S, Cuq B (1998) Mechanical and thermomechanical properties of films based on whey proteins as affected by plasticizer and crosslinking agents. J Dairy Sci 81:3123–3130

Gao H, Fang X, Chen H, Qin Y, Xu F, Jin TZ (2017a) Physiochemical properties and food application of antimicrobial PLA film. Food Control 73:1522–1531. https://doi.org/10.1016/j.foodcont.2016.11.017

Gao P, Wang F, Gu F, Liang J, Li N, Ludescher RD (2017b) Preparation and characterization of zein thermo-modified starch films. Carbohydr Polym 157:1254–1260. https://doi.org/10.1016/j.carbpol.2016.11.004

García MC, Torre M, Marina ML, Laborda F (1997) Composition and characterization of Soyabean and related products. Crit Rev Food Sci Nutr 37(4):361–391. https://doi.org/10.1080/10408399709527779

Gardner KH, Blackwell J (1975) Refinement of the structure of @-chitin. Biopolymers 14:1581–1595

Garlotta D (2001) A literature review of poly (lactic acid). J Polym Environ 9(2):63–84

González A, Alvarez Igarzabal CI (2013) Soy protein - poly (lactic acid) bilayer films as biodegradable material for active food packaging. Food Hydrocoll 33(2):289–296. https://doi.org/10.1016/j.foodhyd.2013.03.010

Grover CN, Gwynne JH, Pugh N, Hamaia S, Farndale RW, Best SM, Cameron RE (2012) Crosslinking and composition influence the surface properties, mechanical stiffness and cell reactivity of collagen-based films. Acta Biomater 8(8):3080–3090. https://doi.org/10.1016/j.actbio.2012.05.006

Grzybek P, Dudek G, van der Bruggen B (2024) Cellulose-based films and membranes: a comprehensive review on preparation and applications. Chem Eng J 495. Elsevier B.V. https://doi.org/10.1016/j.cej.2024.153500

Guerrero P, Retegi A, Gabilondo N, De La Caba K (2010) Mechanical and thermal properties of soy protein films processed by casting and compression. J Food Eng 100(1):145–151. https://doi.org/10.1016/j.jfoodeng.2010.03.039

Gullón P, Romaní A, Vila C, Garrote G, Parajó JC (2012) Potential of hydrothermal treatments in lignocellulose biorefineries. Biofuel Bioprod Bior 6(2):219–232. https://doi.org/10.1002/bbb.339

Gupta B, Revagade N, Hilborn J (2007) Poly(lactic acid) fiber: an overview. Prog Polym Sci 32(4):455–482. https://doi.org/10.1016/j.progpolymsci.2007.01.005

Halder SK, Mondal KC (2018) Microbial valorization of chitinous bioresources for chitin extraction and production of chito-oligomers and N-acetylglucosamine: trends, perspectives and prospects. Microbial Biotechnol 2:69–107. Springer Singapore. https://doi.org/10.1007/978-981-10-7140-9_4

Hanani ZAN, Yee FC, Nor-Khaizura MAR (2019) Effect of pomegranate (Punica granatum L.) peel powder on the antioxidant and antimicrobial properties of fish gelatin films as active packaging. Food Hydrocoll 89:253–259. https://doi.org/10.1016/j.foodhyd.2018.10.007

He Y, Liu Y, Zhang M (2024) Hemicellulose and unlocking potential for sustainable applications in biomedical, packaging, and material sciences: a narrative review. Int J Biol Macromol 280. Elsevier B.V. https://doi.org/10.1016/j.ijbiomac.2024.135657

Heinze T (2015) Cellulose: structure and properties. Adv Polym Sci 271:1–52. https://doi.org/10.1007/12_2015_319

Hoell IA, Vaaje-Kolstad G, Eijsink VGH (2010) Structure and function of enzymes acting on chitin and chitosan. Biotechnol Genet Eng Rev 27(1):331–366. https://doi.org/10.1080/02648725.2010.10648156

Hu X, Jia X, Zhi C, Jin Z, Miao M (2019) Improving the properties of starch-based antimicrobial composite films using ZnO-chitosan nanoparticles. Carbohydr Polym 210:204–209. https://doi.org/10.1016/j.carbpol.2019.01.043

Huang X, Li R, Zeng L, Li X, Xi Z, Wang K, Li Y (2020) A multifunctional carbon nanotube reinforced nanocomposite modified via soy protein isolate: a study on dispersion, electrical and mechanical properties. Carbon 161:350–358. https://doi.org/10.1016/j.carbon.2020.01.069

Huntrakul K, Harnkarnsujarit N (2020) Effects of plasticizers on water sorption and aging stability of whey protein/carboxy methyl cellulose films. J Food Eng 272. https://doi.org/10.1016/j.jfoodeng.2019.109809

Ibrahim NI, Shahar FS, Hameed Sultan MT, Md Shah AU, Azrie Safri SN, Mat Yazik MH (2021) Overview of bioplastic introduction and its applications in product packaging. Coatings 11(11). https://doi.org/10.3390/COATINGS11111423

Jafari H, Lista A, Siekapen MM, Ghaffari-Bohlouli P, Nie L, Alimoradi H, Shavandi A (2020) Fish collagen: extraction, characterization, and applications for biomaterials engineering. Polymers 12(10):1–37. MDPI AG. https://doi.org/10.3390/polym12102230

Jahurul MHA, Azzatul FS, Sharifudin MS, Norliza MJ, Hasmadi M, Lee JS, Patricia M, Jinap S, Ramlah George MR, Firoz Khan M, Zaidul ISM (2020) Functional and nutritional properties of rambutan (Nephelium lappaceum L.) seed and its industrial application: a review. Trends Food Sci Technol 99:367–374. Elsevier Ltd. https://doi.org/10.1016/j.tifs.2020.03.016

Jamshidian M, Tehrany EA, Imran M, Akhtar MJ, Cleymand F, Desobry S (2012) Structural, mechanical and barrier properties of active PLA-antioxidant films. J Food Eng 110(3):380–389. https://doi.org/10.1016/j.jfoodeng.2011.12.034

Janjarasskul T, Krochta JM (2010) Edible packaging materials. Annu Rev Food Sci Technol 1(1):415–448. https://doi.org/10.1146/annurev.food.080708.100836

Ji J, Li B, Zhong WH (2012) An ultraelastic poly(ethylene oxide)/soy protein film with fully amorphous structure. Macromolecules 45(1):602–606. https://doi.org/10.1021/ma202347v

Jiang A, Patel R, Padhan B, Palimkar S, Galgali P, Adhikari A, Varga I, Patel M (2023) Chitosan based biodegradable composite for antibacterial food packaging application. Polymers 15(10). MDPI. https://doi.org/10.3390/polym15102235

Jovanović S, Barać M, Maćej O (2005) Whey proteins-properties and possibility of application. Mljekarstvo 55(3):215–233

Kanatt SR, Rao MS, Chawla SP, Sharma A (2012) Active chitosan-polyvinyl alcohol films with natural extracts. Food Hydrocoll 29(2):290–297. https://doi.org/10.1016/j.foodhyd.2012.03.005

Kasaai MR (2018) Zein and zein -based nano-materials for food and nutrition applications: a review. Trends Food Sci Technol 79:184–197. Elsevier Ltd. https://doi.org/10.1016/j.tifs.2018.07.015

Khamsucharit P, Laohaphatanalert K, Gavinlertvatana P, Sriroth K, Sangseethong K (2018) Characterization of pectin extracted from banana peels of different varieties. Food Sci Biotechnol 27(3):623–629. https://doi.org/10.1007/s10068-017-0302-0

Khanafari A, Marandi R, Sanatei S (2008) Recovery of chitin and chitosan from shrimp waste by chemical and microbial methods. J Environ Health Sci Eng 5(1). www.SID.ir

Khanna S, Srivastava AK (2005) Recent advances in microbial polyhydroxyalkanoates. Process Biochem 40(2):607–619. https://doi.org/10.1016/j.procbio.2004.01.053

Kilara A, Vaghela MN (2017) Whey proteins. In: Proteins in food processing, 2nd edn. Elsevier, pp 93–126. https://doi.org/10.1016/B978-0-08-100722-8.00005-X

Klangmuang P, Sothornvit R (2016) Barrier properties, mechanical properties and antimicrobial activity of hydroxypropyl methylcellulose-based nanocomposite films incorporated with Thai essential oils. Food Hydrocoll 61:609–616. https://doi.org/10.1016/j.foodhyd.2016.06.018

Kong J, Ge X, Sun Y, Mao M, Yu H, Chu R, Wang Y (2023) Multi-functional pH-sensitive active and intelligent packaging based on highly cross-linked zein for the monitoring of pork freshness. Food Chem 404. https://doi.org/10.1016/j.foodchem.2022.134754

Koshy RR, Mary SK, Thomas S, Pothan LA (2015) Environment friendly green composites based on soy protein isolate - a review. Food Hydrocoll 50:174–192. Elsevier B.V. https://doi.org/10.1016/j.foodhyd.2015.04.023

Kou SG, Peters LM, Mucalo MR (2021) Chitosan: a review of sources and preparation methods. Int J Biol Macromol 169:85–94. Elsevier BV. https://doi.org/10.1016/j.ijbiomac.2020.12.005

Kumar MNVR, Muzzarelli RAA, Muzzarelli C, Sashiwa H, Domb AJ (2004) Chitosan chemistry and pharmaceutical perspectives. Chem Rev 104(12):6017–6084. https://doi.org/10.1021/cr030441b

Kumar P, Tanwar R, Gupta V, Upadhyay A, Kumar A, Gaikwad KK (2021) Pineapple peel extract incorporated poly(vinyl alcohol)-corn starch film for active food packaging: preparation, characterization and antioxidant activity. Int J Biol Macromol 187:223–231. https://doi.org/10.1016/j.ijbiomac.2021.07.136

Kumar S, Konwar J, Purkayastha MD, Kalita S, Mukherjee A, Dutta J (2023) Current progress in valorization of food processing waste and by-products for pectin extraction. Int J Biol Macromol 239. https://doi.org/10.1016/j.ijbiomac.2023.124332

Kumirska J, Czerwicka M, Kaczyński Z, Bychowska A, Brzozowski K, Thöming J, Stepnowski P (2010) Application of spectroscopic methods for structural analysis of chitin and chitosan. Mare Drugs 8(5):1567–1636. MDPI AG. https://doi.org/10.3390/md8051567

Kunte LA, Gennadios A, Cuppett SL, Hanna MA, Weller CL (1997) Cast films from soy protein isolates and fractions. Cereal Chem 74(2):115–118

Kutralam-Muniasamy G, Peréz-Guevara F (2018) Genome characteristics dictate poly-R-(3)-hydroxyalkanoate production in Cupriavidus necator H16. World J Microbiol Biotechnol 34(6). Springer Netherlands. https://doi.org/10.1007/s11274-018-2460-5

Kwan TH, Hu Y, Lin CSK (2018) Techno-economic analysis of a food waste valorisation process for lactic acid, lactide and poly(lactic acid) production. J Clean Prod 181:72–87. https://doi.org/10.1016/j.jclepro.2018.01.179

Lackner M (2015) Bioplastics. In: Kirk-Othmer encyclopedia of chemical technology. Wiley, pp 1–41. https://doi.org/10.1002/0471238961.koe00006
Lawton JW (2002) Zein: a history of processing and use. Cereal Chem 79(1):1–18
Le Tien C, Letendre M, Ispas-Szabo P, Mateescu MA, Delmas-Patterson G, Yu HL, Lacroix M (2000) Development of biodegradable films from whey proteins by cross-linking and entrapment in cellulose. J Agric Food Chem 48(11):5566–5575. https://doi.org/10.1021/jf0002241
Lee CH, Singla A, Lee Y (2001) Biomedical applications of collagen. Int J Pharm 221. www.elsevier.com/locate/ijpharm
Lerdlattaporn R, Phalakornkule C, Trakulvichean S, Songkasiri W (2021) Implementing circular economy concept by converting cassava pulp and wastewater to biogas for sustainable production in starch industry. Sustain Environ Res 31(1). https://doi.org/10.1186/s42834-021-00093-9
Li YD, Zheng JB, Wang XL, Yang KK, Wang YZ (2008) Structure and properties of soy protein/poly(butylene succinate) blends with improved compatibility. Biomacromolecules 9(11):3157–3164. https://doi.org/10.1021/bm800745p
Li DF, Mu CD, Zhang Q, Zhou YL, Lin W (2010) Effects of microwave irradiation on collagen denaturation. J Food Biochem 34(6):1319–1331. https://doi.org/10.1111/j.1745-4514.2010.00334.x
Li J, Li Y, Lee TC, Huang Q (2013) Structure and physical properties of zein/pluronic F127 composite films. J Agric Food Chem 61(6):1309–1318. https://doi.org/10.1021/jf3043055
Li Y, Ji J, Wang Y, Li R, Zhong WH (2018) Soy protein-treated nanofillers creating adaptive interfaces in nanocomposites with effectively improved conductivity. J Mater Sci 53(11):8653–8665. https://doi.org/10.1007/s10853-018-2121-y
Li L, Luo X, Liu Y, Teng M, Liu X, Zhang X, Liu X (2024) Edible packaging revolution: enhanced functionality with natural collagen aggregates. Food Hydrocoll 156. https://doi.org/10.1016/j.foodhyd.2024.110331
Liang B, Song W, Xing R, Liu S, Yu H, Li P (2023) The source, activity influencing factors and biological activities for future development of chitin deacetylase. Carbohydr Polym 321. Elsevier Ltd. https://doi.org/10.1016/j.carbpol.2023.121335
Lin J, Wang Y, Pan D, Sun Y, Ou C, Cao J (2018) Physico-mechanical properties of gelatin films modified with lysine, arginine and histidine. Int J Biol Macromol 108:947–952. https://doi.org/10.1016/j.ijbiomac.2017.11.015
Lindström T (2017) Aspects on nanofibrillated cellulose (NFC) processing, rheology and NFC-film properties. Curr Opin Colloid Interface Sci 29:68–75. Elsevier Ltd. https://doi.org/10.1016/j.cocis.2017.02.005
Liu D, Tian H, Zhang L, Chang PR (2008) Structure and properties of blend films prepared from castor oil-based polyurethane/soy protein derivative. Ind Eng Chem Res 47(23):9330–9336. https://doi.org/10.1021/ie8009632
Liu D, Nikoo M, Boran G, Zhou P, Regenstein JM (2015) Collagen and gelatin. Annu Rev Food Sci Technol 6:527–557. Annual Reviews Inc. https://doi.org/10.1146/annurev-food-031414-111800
Liu Z, Cao X, Ren S, Wang J, Zhang H (2019) Physicochemical characterization of a zein prepared using a novel aqueous extraction technology and tensile properties of the zein film. Ind Crop Prod 130:57–62. https://doi.org/10.1016/j.indcrop.2018.12.071
Liu B, Li J, Liu L, Yao M, Liang J, Qin C, Liang C, Huang C, Zhang Z, Yao S (2022) Efficient separation of eucalyptus hemicellulose and improvement of the stability of the remaining components by 1-amino-2-naphthol-4-sulfonic acid pretreatment. Ind Crop Prod 187. https://doi.org/10.1016/j.indcrop.2022.115406
Liu G, Shi K, Sun H (2023) Research progress in hemicellulose-based nanocomposite film as food packaging. Polymers 15(4). MDPI. https://doi.org/10.3390/polym15040979
Lu Y, He Q, Fan G, Cheng Q, Song G (2021) Extraction and modification of hemicellulose from lignocellulosic biomass: a review. Green Process Synth 10(1):779–804. De Gruyter Open Ltd. https://doi.org/10.1515/gps-2021-0065
Lupidi G, Pastore G, Marcantoni E, Gabrielli S (2023) Recent developments in chemical derivatization of microcrystalline cellulose (MCC): pre-treatments, functionalization, and applications. Molecules 28(5). MDPI. https://doi.org/10.3390/molecules28052009

Madureira AR, Pereira CI, Gomes AMP, Pintado ME, Xavier Malcata F (2007) Bovine whey proteins - overview on their main biological properties. Food Res Int 40(10):1197–1211. https://doi.org/10.1016/j.foodres.2007.07.005

Manzoor A, Yousuf B, Pandith JA, Ahmad S (2023) Plant-derived active substances incorporated as antioxidant, antibacterial or antifungal components in coatings/films for food packaging applications. Food Biosci 53. Elsevier Ltd. https://doi.org/10.1016/j.fbio.2023.102717

Matinong AME, Chisti Y, Pickering KL, Haverkamp RG (2022) Collagen extraction from animal skin. Biology 11(6). MDPI. https://doi.org/10.3390/biology11060905

Mcintosh GH, Royle PJ, Le Leu RK, Regester GO, Johnson MA, Grinsted RL, Kenward RS, Smithers GW (1998) Whey proteins as functional food ingredients? Int Dairy J 8:425–434

Mirkhalaf SM, Fagerström M (2021) The mechanical behavior of polylactic acid (PLA) films: fabrication, experiments and modelling. Mech Time-Depend Mater 25(2):119–131. https://doi.org/10.1007/s11043-019-09429-w

Mirpoor SF, Patanè GT, Corrado I, Giosafatto CVL, Ginestra G, Nostro A, Foti A, Gucciardi PG, Mandalari G, Barreca D, Gervasi T, Pezzella C (2023) Functionalization of Polyhydroxyalkanoates (PHA)-based bioplastic with Phloretin for active food packaging: characterization of its mechanical, antioxidant, and antimicrobial activities. Int J Mol Sci 24(14). https://doi.org/10.3390/ijms241411628

Mohan K, Ganesan AR, Ezhilarasi PN, Kondamareddy KK, Rajan DK, Sathishkumar P, Rajarajeswaran J, Conterno L (2022) Green and eco-friendly approaches for the extraction of chitin and chitosan: a review. Carbohydr Polym 287. Elsevier Ltd. https://doi.org/10.1016/j.carbpol.2022.119349

Mohnen D (2008) Pectin structure and biosynthesis. Curr Opin Plant Biol 11(3):266–277. https://doi.org/10.1016/j.pbi.2008.03.006

Momany FA, Sessa DJ, Lawton JW, Selling GW, Hamaker SAH, Willett JL (2006) Structural characterization of α-zein. J Agric Food Chem 54(2):543–547. https://doi.org/10.1021/jf058135h

Moro TMA, Ascheri JLR, Ortiz JAR, Carvalho CWP, Meléndez-Arévalo A (2017) Bioplastics of native starches reinforced with passion fruit peel. Food Bioprocess Technol 10(10):1798–1808. https://doi.org/10.1007/s11947-017-1944-x

Moussian B (2019) Chitin: structure, chemistry and biology. Adv Exp Med Biol 1142:5–18. Springer New York LLC. https://doi.org/10.1007/978-981-13-7318-3_2

Mukprasirt A, Sajjaanantakul K (2004) Physico-chemical properties of flour and starch from jackfruit seeds (Artocarpus heterophyllus Lam.) compared with modified starches. Int J Food Sci Technol 39(3):271–276. https://doi.org/10.1111/j.1365-2621.2004.00781.x

Naitam MG, Kaushik R, Tomar GS, Singh S, Nain L (2022) Agro-industrial waste as potential renewable feedstock for biopolymer poly-Hydroxyalkanoates (PHA) production. Enz Eng 11:190. https://doi.org/10.35841/2329-6674.22.11.190

No HK, Samuel J, Meyers P, Lee KS (1989) Isolation and characterization of chitin from crawfish shell waste. J Agric Food Chem 37(3):575–579

Ortiz CM, Salgado PR, Dufresne A, Mauri AN (2018) Microfibrillated cellulose addition improved the physicochemical and bioactive properties of biodegradable films based on soy protein and clove essential oil. Food Hydrocoll 79:416–427. https://doi.org/10.1016/j.foodhyd.2018.01.011

Osterberg M, Henn KA, Farooq M, Valle-Delgado JJ (2023) Biobased nanomaterials— the role of interfacial interactions for advanced materials. Chem Rev 123(5):2200–2241. https://doi.org/10.1021/acs.chemrev.2c00492

Ottani V, Raspanti M, Ruggeri A (2001) Collagen structure and functional implications. www.elsevier.com/locate/micron

Pareek S (2015) Nutritional and biochemical composition of lychee (Litchi chinensis Sonn.) cultivars. In: Nutritional composition of fruit cultivars. Elsevier, pp 395–418. https://doi.org/10.1016/B978-0-12-408117-8.00017-9

Pasini Cabello SD, Takara EA, Marchese J, Ochoa NA (2015) Influence of plasticizers in pectin films: microstructural changes. Mater Chem Phys 162:491–497. https://doi.org/10.1016/j.matchemphys.2015.06.019

Peng F, Peng P, Xu F, Sun RC (2012) Fractional purification and bioconversion of hemicelluloses. Biotechnol Adv 30(4):879–903. https://doi.org/10.1016/j.biotechadv.2012.01.018

Pérez-Arauz AO, Aguilar-Rabiela AE, Vargas-Torres A, Rodríguez-Hernández AI, Chavarría-Hernández N, Vergara-Porras B, López-Cuellar MR (2019) Production and characterization of biodegradable films of a novel polyhydroxyalkanoate (PHA) synthesized from peanut oil. Food Packag Shelf Life 20. https://doi.org/10.1016/j.fpsl.2019.01.001

Philip S, Keshavarz T, Roy I (2007) Polyhydroxyalkanoates: biodegradable polymers with a range of applications. J Chem Technol Biotechnol 82(3):233–247. https://doi.org/10.1002/jctb.1667

Ponce AG, Moreira MR (2025) Casein and chitosan polymers: use in antimicrobial packaging. In: Antimicrobial food packaging. Elsevier, pp 635–652. https://doi.org/10.1016/B978-0-323-90747-7.00041-7

Punia Bangar S, Sharma N, Kaur H, Kaur M, Sandhu KS, Maqsood S, Ozogul F (2022) A review of sapodilla (Manilkara zapota) in human nutrition, health, and industrial applications. Trends Food Sci Technol 127:319–334. Elsevier Ltd. https://doi.org/10.1016/j.tifs.2022.05.016

Qiao C, Ma X, Zhang J, Yao J (2017) Molecular interactions in gelatin/chitosan composite films. Food Chem 235:45–50. https://doi.org/10.1016/j.foodchem.2017.05.045

Qiao D, Huang Y, Hou X, Ye F, Wu K, Jiang F, Zhao G, Zhang B, Xie F (2024) Enhancing thermal stability and mechanical resilience in gelatin/starch composites through polyvinyl alcohol integration. Carbohydr Polym 344. https://doi.org/10.1016/j.carbpol.2024.122528

Qin P, Wang T, Luo Y (2022) A review on plant-based proteins from soybean: health benefits and soy product development. J Agric Food Res 7. https://doi.org/10.1016/j.jafr.2021.100265

Rajeshkumar L (2021) Biodegradable polymer blends and composites from renewable resources. In: Biodegradable polymers, blends and composites. Elsevier, pp 527–549. https://doi.org/10.1016/B978-0-12-823791-5.00015-6

Raji Z, Khodaiyan F, Rezaei K, Kiani H, Hosseini SS (2017) Extraction optimization and physicochemical properties of pectin from melon peel. Int J Biol Macromol 98:709–716. https://doi.org/10.1016/j.ijbiomac.2017.01.146

Ran R, He YQ, Zhang X, Zeng X, Guo X, Luo X, Li G (2025) Development of multi-functional collagen films for maintaining and visually monitoring shrimp freshness. Food Hydrocoll 160. https://doi.org/10.1016/j.foodhyd.2024.110734

Ranganathan S, Dutta S, Moses JA, Anandharamakrishnan C (2020) Utilization of food waste streams for the production of biopolymers. Heliyon 6(9) Elsevier Ltd. https://doi.org/10.1016/j.heliyon.2020.e04891

Rao J, Lv Z, Chen G, Peng F (2023) Hemicellulose: structure, chemical modification, and application. Prog Polym Sci 140. Elsevier Ltd. https://doi.org/10.1016/j.progpolymsci.2023.101675

Raza ZA, Abid S, Banat IM (2018) Polyhydroxyalkanoates: characteristics, production, recent developments and applications. Int Biodeterior Biodegradation 126:45–56. Elsevier Ltd. https://doi.org/10.1016/j.ibiod.2017.10.001

Reddy CSK, Ghai R, Kalia V (2003) Polyhydroxyalkanoates: an overview. Bioresour Technol 87(2):137–146. https://doi.org/10.1016/S0960-8524(02)00212-2

Rivera–González G, Amaya–Guerra CA, de la Rosa–Millán J (2019) Physicochemical characterisation and in vitro starch digestion of avocado seed flour (Persea americana V. Hass) and its starch and fibrous fractions. Int J Food Sci Technol 54(7):2447–2457. https://doi.org/10.1111/ijfs.14160

Rýglová Š, Braun M, Suchý T (2017) Collagen and its modifications—crucial aspects with concern to its processing and analysis. Macromol Mater Eng 302(6). Wiley-VCH Verlag. https://doi.org/10.1002/mame.201600460

Sabapathy PC, Devaraj S, Meixner K, Anburajan P, Kathirvel P, Ravikumar Y, Zabed HM, Qi X (2020) Recent developments in Polyhydroxyalkanoates (PHAs) production – a review. Bioresour Technol 306. Elsevier Ltd. https://doi.org/10.1016/j.biortech.2020.123132

Sabato SF, Ouattara B, Yu H, D'Aprano G, Le Tien G, Mateescu MA, Lacroix M (2001) Mechanical and barrier-properties of cross-linked soy and whey protein based films. J Agric Food Chem 49(3):1397–1403. https://doi.org/10.1021/jf0005925

Said NS, Sarbon NM (2022) Physical and mechanical characteristics of gelatin-based films as a potential food packaging material: a review. Membranes 12(5). MDPI. https://doi.org/10.3390/membranes12050442

Sathiyanarayanan G, Saibaba G, Seghal Kiran G, Selvin J (2013) A statistical approach for optimization of polyhydroxybutyrate production by marine Bacillus subtilis MSBN17. Int J Biol Macromol 59:170–177. https://doi.org/10.1016/j.ijbiomac.2013.04.040

Selling GW, Biswas A (2012) Blends of Zein and Nylon-6. J Polym Environ 20(3):631–637. https://doi.org/10.1007/s10924-012-0426-5

Sethulakshmi AG, Saravanakumar MP (2024) Sustainable papaya plant waste and green tea residue composite films integrated with starch and gelatin for active food packaging applications. Int J Biol Macromol 260. https://doi.org/10.1016/j.ijbiomac.2023.129153

Shah U, Naqash F, Gani A, Masoodi FA (2016) Art and science behind modified starch edible films and coatings: a review. Compr Rev Food Sci Food Saf 15(3):568–580. https://doi.org/10.1111/1541-4337.12197

Shlush E, Davidovich-Pinhas M (2022) Bioplastics for food packaging. Trends Food Sci Technol 125:66–80. Elsevier Ltd. https://doi.org/10.1016/j.tifs.2022.04.026

Shoulders MD, Raines RT (2009) Collagen structure and stability. Annu Rev Biochem 78:929–958. https://doi.org/10.1146/annurev.biochem.77.032207.120833

Shruti VC, Kutralam-Muniasamy G (2019) Bioplastics: missing link in the era of microplastics. Sci Total Environ 697. Elsevier B.V. https://doi.org/10.1016/j.scitotenv.2019.134139

Shukla R, Cheryan M (2001) Zein: the industrial protein from corn. Ind Crop Prod 13. www.elsevier.com/locate/indcrop

Singh P, Krishnaswamy K (2022) Sustainable zero-waste processing system for soybeans and soy by-product valorization. Trends Food Sci Technol 128:331–344. Elsevier Ltd. https://doi.org/10.1016/j.tifs.2022.08.015

Singh P, Kumar R, Sabapathy SN, Bawa AS (2008) Functional and edible uses of soy protein products. Compr Rev Food Sci Food Saf 7(1):14–28

Sionkowska A, Wisniewski M, Skopinska J, Poggi GF, Marsano E, Maxwell CA, Wess TJ (2006) Thermal and mechanical properties of UV irradiated collagen/chitosan thin films. Polym Degrad Stab 91(12):3026–3032. https://doi.org/10.1016/j.polymdegradstab.2006.08.009

Smithers GW (2008) Whey and whey proteins-From “gutter-to-gold”. Int Dairy J 18(7):695–704. https://doi.org/10.1016/j.idairyj.2008.03.008

Smithers GW, John Ballard F, Copeland AD, De Silva KJ, Dionysius DA, Francis GL, Goddard C, Grieve PA, Mcintosh GH, Mitchell IR, Pearce RJ, Regester GO (1996) New opportunities from the isolation and utilization of whey proteins. J Dairy Sci 79(8):1454–1459. https://doi.org/10.3168/jds.S0022-0302(96)76504-9

Soares Rodrigues CI, Jackson JJ, Montross MD (2016) A molar basis comparison of calcium hydroxide, sodium hydroxide, and potassium hydroxide on the pretreatment of switchgrass and miscanthus under high solids conditions. Ind Crop Prod 92:165–173. https://doi.org/10.1016/j.indcrop.2016.08.010

Solak BB, Akin N (2012) Health benefits of whey protein: a review. J Food Sci Eng 2:129

Song KM, Jung SK, Kim YH, Kim YE, Lee NH (2018) Development of industrial ultrasound system for mass production of collagen and biochemical characteristics of extracted collagen. Food Bioprod Process 110:96–103. https://doi.org/10.1016/j.fbp.2018.04.001

Sothornvit R, Krochta JM (2000) Plasticizer effect on oxygen permeability of β-lactoglobulin films. J Agric Food Chem 48(12):6298–6302. https://doi.org/10.1021/jf000836l

Spiridon I, Popa VI (2008) Hemicelluloses: major sources, properties and applications. In: Monomers, polymers and composites from renewable resources, pp 289–304. Elsevier

Streimikyte P, Viskelis P, Viskelis J (2022) Enzymes-assisted extraction of plants for sustainable and functional applications. Int J Mol Sci 23(4). MDPI. https://doi.org/10.3390/ijms23042359

Su JF, Yuan XY, Huang Z, Xia WL (2010) Properties stability and biodegradation behaviors of soy protein isolate/poly (vinyl alcohol) blend films. Polym Degrad Stab 95(7):1226–1237. https://doi.org/10.1016/j.polymdegradstab.2010.03.035

Sun XF, Wang HH, Jing ZX, Mohanathas R (2013) Hemicellulose-based pH-sensitive and biodegradable hydrogel for controlled drug delivery. Carbohydr Polym 92(2):1357–1366. https://doi.org/10.1016/j.carbpol.2012.10.032

Sun S, Liu P, Ji N, Hou H, Dong H (2018) Effects of various cross-linking agents on the physicochemical properties of starch/PHA composite films produced by extrusion blowing. Food Hydrocoll 77:964–975. https://doi.org/10.1016/j.foodhyd.2017.11.046

Swain SK, Sethy PK, Pattanayak AJ (2017) Synthesis of soy protein based biocomposites for packaging applications. In: Green biocomposites: design and applications., 9783319493817. Springer International Publishing, pp 143–166. https://doi.org/10.1007/978-3-319-49382-4_7

Tafa KD, Satheesh N, Abera W (2023) Mechanical properties of tef starch based edible films: development and process optimization. Heliyon 9(2). https://doi.org/10.1016/j.heliyon.2023.e13160

Tahir M, Li A, Moore M, Ford E, Theyson T, Seyam AFM (2024) Development of eco-friendly soy protein fiber: a comprehensive critical review and prospects. Fibers 12(4). Multidisciplinary Digital Publishing Institute (MDPI). https://doi.org/10.3390/fib12040031

Tan D, Wang Y, Tong Y, Chen GQ (2021) Grand challenges for industrializing Polyhydroxyalkanoates (PHAs). Trends Biotechnol 39(9):953–963. Elsevier Ltd. https://doi.org/10.1016/j.tibtech.2020.11.010

Thakur S, Chaudhary J, Sharma B, Verma A, Tamulevicius S, Thakur VK (2018) Sustainability of bioplastics: opportunities and challenges. Curr Opin Green Sustain Chem 13:68–75. Elsevier B.V. https://doi.org/10.1016/j.cogsc.2018.04.013

Thrane M, Paulsen PV, Orcutt MW, Krieger TM (2017) Soy protein: impacts, production, and applications. In: Sustainable protein sources. Elsevier Inc, pp 23–45. https://doi.org/10.1016/B978-0-12-802778-3.00002-0

Tian H, Xu G, Yang B, Guo G (2011) Microstructure and mechanical properties of soy protein/agar blend films: effect of composition and processing methods. J Food Eng 107(1):21–26. https://doi.org/10.1016/j.jfoodeng.2011.06.008

Trache D, Hussin MH, Haafiz MKM, Thakur VK (2017) Recent progress in cellulose nanocrystals: sources and production. Nanoscale 9(5):1763–1786. Royal Society of Chemistry. https://doi.org/10.1039/c6nr09494e

Turalija M, Bischof S, Budimir A, Gaan S (2016) Antimicrobial PLA films from environment friendly additives. Compos Part B 102:94–99. https://doi.org/10.1016/j.compositesb.2016.07.017

Ustunol Z, Mert B (2004) Water solubility, mechanical, barrier, and thermal properties of cross-linked whey protein isolate-based films. J Food Sci 69. www.ift.org

Van Der Rest M, Garrone R (1991) Collagen family of proteins. www.fasebj.org

Wadhwa M, Bakshi MPS, Makkar HPS (2016) Wastes to worth: value added products from fruit and vegetable wastes. CABI Rev 2015:1–25

Walters CM, Boott CE, Nguyen TD, Hamad WY, Maclachlan MJ (2020) Iridescent cellulose nanocrystal films modified with hydroxypropyl cellulose. Biomacromolecules 21(3):1295–1302. https://doi.org/10.1021/acs.biomac.0c00056

Wang Q, Wang JF, Geil PH, Padua GW (2004) Zein adsorption to hydrophilic and hydrophopic surfaces investigates by surface plasmon resonance. Biomacromolecules 5(4):1356–1361. https://doi.org/10.1021/bm049965r

Wang Y, Yin J, Chen GQ (2014) Polyhydroxyalkanoates, challenges and opportunities. Curr Opin Biotechnol 30:59–65. Elsevier Ltd. https://doi.org/10.1016/j.copbio.2014.06.001

Wang J, Li X, Chen H, Lin B, Zhao L (2021) Heterologous expression and characterization of a high-efficiency Chitosanase from Bacillus mojavensis SY1 suitable for production of chitosan oligosaccharides. Front Microbiol 12. https://doi.org/10.3389/fmicb.2021.781138

Westlake JR, Tran MW, Jiang Y, Zhang X, Burrows AD, Xie M (2022) Biodegradable active packaging with controlled release: principles, progress, and prospects. ACS Food Sci Technol 2(8):1166–1183. https://doi.org/10.1021/acsfoodscitech.2c00070

Woods KK, Selling GW, Cooke PH (2009) Compatible blends of zein and polyvinylpyrrolidone. J Polym Environ 17(2):115–122. https://doi.org/10.1007/s10924-009-0128-9

Wu F, Wei J, Liu C, O'Neill B, Ngothai Y (2012) Fabrication and properties of porous scaffold of zein/PCL biocomposite for bone tissue engineering. Compos Part B 43(5):2192–2197. https://doi.org/10.1016/j.compositesb.2012.02.040

Wu F, Misra M, Mohanty AK (2021) Challenges and new opportunities on barrier performance of biodegradable polymers for sustainable packaging. Prog Polym Sci 117:101395. https://doi.org/10.1016/j.progpolymsci.2021.101395

Xiao Y, Chen N, He Q, Zeng WC (2025) Longan seeds used as novel resources to prepare resistant starch: structure, properties and digestion. Int J Biol Macromol 310. https://doi.org/10.1016/j.ijbiomac.2025.143451

Xie F, Zhang H, Xia Y, Ai L (2020) Effects of tamarind seed polysaccharide on gelatinization, rheological, and structural properties of corn starch with different amylose/amylopectin ratios. Food Hydrocoll 105. https://doi.org/10.1016/j.foodhyd.2020.105854

Xu J, Xia R, Zheng L, Yuan T, Sun R (2019) Plasticized hemicelluloses/chitosan-based edible films reinforced by cellulose nanofiber with enhanced mechanical properties. Carbohydr Polym 224. https://doi.org/10.1016/j.carbpol.2019.115164

Xu J, Liu F, Wang T, Goff HD, Zhong F (2020) Fabrication of films with tailored properties by regulating the swelling of collagen fiber through pH adjustment. Food Hydrocoll 108. https://doi.org/10.1016/j.foodhyd.2020.106016

Yan X-L, Khor E, Lim L-Y (2001) Chitosan-alginate films prepared with chitosans of different molecular weights. J Biomed Mater Res 58(4):358–365

Yang X, Nisar T, Hou Y, Gou X, Sun L, Guo Y (2018) Pomegranate peel pectin can be used as an effective emulsifier. Food Hydrocoll 85:30–38. https://doi.org/10.1016/j.foodhyd.2018.06.042

Yildirim-Yalcin M, Tornuk F, Toker OS (2022) Recent advances in the improvement of carboxymethyl cellulose-based edible films. Trends Food Sci Technol 129:179–193. Elsevier Ltd. https://doi.org/10.1016/j.tifs.2022.09.022

Yin S, Zhang Y, Zhang X, Tao K, Li G (2023) High-strength collagen/delphinidin film incorporated with Vaccinium oxycoccus pigment for active and intelligent food packaging. Collagen Leather 5(1). https://doi.org/10.1186/s42825-023-00118-6

Zhang M, Song F, Wang XL, Wang YZ (2012) Development of soy protein isolate/waterborne polyurethane blend films with improved properties. Colloids Surf B: Biointerfaces 100:16–21. https://doi.org/10.1016/j.colsurfb.2012.05.031

Zhang Y, Rempel C, Liu Q (2014) Thermoplastic starch processing and characteristics-a review. Crit Rev Food Sci Nutr 54(10):1353–1370. https://doi.org/10.1080/10408398.2011.636156

Zhang R, Chai J, Guo W, Lu P, Wu R (2025) Improved paper barrier properties based on coating by citric acid crosslinking of hemicellulose. Int J Biol Macromol 304. https://doi.org/10.1016/j.ijbiomac.2025.140892

Zhao Y, Sun Z (2018) Effects of gelatin-polyphenol and gelatin–genipin cross-linking on the structure of gelatin hydrogels. Int J Food Prop 20:S2822–S2832. https://doi.org/10.1080/10942912.2017.1381111

Zhao S, Yao J, Fei X, Shao Z, Chen X (2013) An antimicrobial film by embedding in situ synthesized silver nanoparticles in soy protein isolate. Mater Lett 95:142–144. https://doi.org/10.1016/j.matlet.2012.12.106

Zhao Y, Zhang Y, Dong H, Wu W, Yang X, He Q (2023) Functional biopolymers for food packaging: formation mechanism and performance improvement of chitosan-based composites. Food Biosci 54. Elsevier Ltd. https://doi.org/10.1016/j.fbio.2023.102927

Zheng T, Tang P, Yang C, Ran R, Li G (2023) Development of active packaging films based on collagen/gallic acid-grafted chitosan incorporating with ε-polylysine for pork preservation. Food Hydrocoll 140. https://doi.org/10.1016/j.foodhyd.2023.108590

Zhong Z, Sun XS (n.d.) Properties of soy protein isolate/polycaprolactone blends compatibilized by methylene diphenyl diisocyanate q. www.elsevier.nl/locate/polymer

Zhou Y, Hu Y, Tan Z, Zhou T (2024) Cellulose extraction from rice straw waste for biodegradable ethyl cellulose films preparation using green chemical technology. J Clean Prod 439. https://doi.org/10.1016/j.jclepro.2024.140839

Zhu K, Yu Z, Li J, Chiou BS, Chen M, Zhong F, Liu F (2024) Sodium carboxymethyl cellulose as a stabilizer for fabricating mineralized collagen films with improved wet mechanical properties. Food Hydrocoll 150. https://doi.org/10.1016/j.foodhyd.2023.109676

Part II
Environmental Remediation and Cultural Restoration

Chapter 4
Microbes and Bioremediation: An Emerging Approach for Environmental Cleanness

Simpal Kumari and Atul Kumar Srivastava

Abstract The persistent accumulation of recalcitrant pollutants in the environment presents serious ecological and public health concerns due to their long-term stability, toxicity, and bioaccumulation potential. Among the various remediation strategies, microbial bioremediation stands out as a sustainable and eco-friendly approach that utilizes the enzymatic capabilities of microorganisms to convert toxic compounds into less harmful or inert byproducts. This chapter provides a comprehensive overview of key aspects related to microbial bioremediation. First, it discusses the major classes of persistent environmental contaminants, including heavy metals, pesticides, hydrocarbons, and synthetic dyes. Next, it explores the underlying biodegradation mechanisms employed by bacteria, fungi, and algae, with schematic illustrations to elucidate their metabolic pathways. A comparative analysis of degradation efficiencies across different microbial species is presented in tabular format, offering insights into their effectiveness under varied environmental conditions. Additionally, special emphasis is placed on emerging microbial strains with enhanced degradation potential, such as genetically engineered bacteria and extremophiles, which show promise in tackling complex and resistant pollutants. These novel strains contribute to the development of efficient bioremediation strategies tailored for diverse ecosystems. The chapter concludes by emphasizing the need for continued research and technological integration to optimize microbial bioremediation approaches. Overall, the insights presented aim to support the advancement of practical and scalable bioremediation technologies, contributing to global efforts toward environmental sustainability and pollution mitigation.

S. Kumari
Department of Microbiology, Faculty of Science & Technology, Dr. Shakuntala Misra National Rehabilitation University, Lucknow, Uttar Pradesh, India

A. K. Srivastava (✉)
Key Laboratory of Microbiology, College of Agriculture, Guizhou University, Guiyang, People's Republic of China

N. K. Arora, J. Mishra (eds.), *Biotechnological Solutions for a Sustainable Future*, Microorganisms for Sustainability 55,
https://doi.org/10.1007/978-981-95-3989-5_4

Keywords Recalcitrant pollutants · Bioremediation · Microorganism · Environmental sustainability · Toxic pollutants

4.1 Introduction

Physical and biological processes within the environment are adversely affected by pollution. One major environmental concern is the deterioration of water quality due to anthropogenic activities, such as mining and the discharge of hazardous metal effluents from steel mills, battery manufacturers, and power generation facilities (Gaur et al. 2022). Pollutants like petroleum, plastics, and trace metals contribute significantly to environmental degradation. Heavy metals, naturally occurring in the earth's crust, are particularly persistent and difficult to break down. Found as ores in rocks and extracted as minerals, these metals can enter the environment in large quantities through human activities. Once released, they remain hazardous for extended periods (Masindi et al. 2021). Among these pollutants, persistent organic pollutants (POPs) have drawn global concern due to their high toxicity, environmental persistence, and ability to bioaccumulate. These compounds can travel long distances and transfer through food chains, posing serious health risks even at very low concentrations (Gaur et al. 2022). Initially, 12 POPs were identified, including certain pesticides, industrial chemicals, and byproducts. By 2017, this number had increased to 16, as adopted by 181 parties (Madaj et al. 2018). POPs are known to cause immunotoxicity by altering immune responses (Suzuki et al. 2020).

According to the World Health Organization (WHO 2022), approximately 12.6 million deaths annually are attributed to environmental exposures, including contaminated drinking water, poor air quality, and toxic chemical exposure. In India, the Central Pollution Control Board (CPCB 2021) reported over 351 polluted river stretches, primarily due to untreated industrial effluents and municipal sewage. These polluted waters contribute to significant public health issues such as gastrointestinal infections, skin diseases, and cancer. Even in industrialized nations with advanced waste management systems, persistent pollutants like per- and polyfluoroalkyl substances (PFAS), chlorinated compounds, and pharmaceutical residues continue to contaminate natural resources. In the United States, the Environmental Protection Agency (EPA 2023) estimates that over 40 million people are exposed to drinking water containing PFAS above safe levels. Additionally, global urbanization and intensified agriculture have led to increased discharges of nitrogen- and phosphorus-rich waste into aquatic ecosystems, accelerating eutrophication and biodiversity loss. These widespread environmental issues have significant health and ecological implications, underscoring the urgent need for innovative and sustainable remediation strategies. Conventional cleanup methods such as landfilling, recycling, pyrolysis, and incineration are often inefficient, costly, and can generate harmful byproducts (Paul et al. 2005). Despite numerous efforts, pollution remains a persistent global challenge. Therefore, alternative strategies are essential for safeguarding human and environmental health. Biological decontamination methods, particularly those utilizing microorganisms, are gaining preference because microbes can degrade a wide range of pollutants without producing toxic intermediates (Furukawa 2003). As the demand for sustainable ecosystem restoration grows, microbial-based bioremediation has emerged as a promising approach for eliminating POPs from the environment (Kumar et al. 2016). Bioremediation involves the use of microorganisms to break down environmental pollutants, as illustrated in detail in a further section. Some microbes can thrive in extreme conditions from as low as −196 °F to as high as 1200 °F, making them well-suited for treating hazardous waste. Carbon is a critical nutrient for these microorganisms. Diverse microbial species such as *Alcaligenes*, *Xanthobacter*, *Achromobacter*, *Pseudomonas*, *Bacillus*, and *Corynebacterium*, isolated from various environments, have been utilized in bioremediation efforts (Sharma et al. 2021).

This approach typically employs indigenous microbial communities, and its success depends on several factors, including microbial diversity and environmental conditions (Malla et al. 2018; Sharma et al. 2022). Thus, understanding and predicting the behavior of toxins in the environment requires considering a wide array of variables when designing bioremediation strategies. To enhance our understanding of bioremediation mechanisms and to develop new solutions, researchers are increasingly employing tools from metabolomics, transcriptomics, genomics, proteomics, and fluxomics. Advances in next-generation sequencing (NGS) and in silico modelling have empowered environmental microbiologists to identify and study active microbial populations in contaminated areas previously a "black box" (Maphosa et al. 2010). Recent progress in molecular biology, genomics, and systems microbiology has further enabled the development of targeted and efficient bioremediation strategies (Zhou et al. 2023b). Innovative techniques such as bioaugmentation, bio stimulation, and microbial consortia have shown success in remediating contaminated environments with minimal ecological disruption (Gupta et al. 2022). Moreover, integrating microbes with emerging technologies like nanomaterials, biosensors, and bioreactors offers substantial potential for real-time monitoring and large-scale applications (Singh et al. 2024). As environmental challenges intensify with climate change and population growth, microbial bioremediation represents a dynamic and sustainable solution for environmental cleanup and the restoration of ecological balance.

4.2 Types of Persistent Organic Pollutants

Various types of organic pollutants present in the environment are difficult to remove. Among them, three major categories: pesticides (e.g., Aldrin, Dieldrin, Eldrin, Mirex, Chlordane, Heptachlor, Toxaphene, Hexachlorobenzene, and Dichlorodiphenyltrichloroethane [DDT]), hydrocarbons (e.g., Benzene, Toluene, Ethylbenzene, Xylene [BTEX], and polycyclic aromatic hydrocarbons [PAHs]), and industrial chemicals (e.g., polychlorinated biphenyls [PCBs], polychlorinated dibenzo-p-dioxins [PCDDs], and polychlorinated dibenzofurans [PCDFs]) are particularly hazardous to human health (Fig. 4.1). These pollutants can enter the food web and lead to serious, life-threatening diseases.

4.2.1 Pesticide

A pesticide is a chemical agent, or a mixture of chemical agents, used to destroy or repel pests and weeds. In Egypt, the first recorded use of pesticides dates back to around 1550 BCE (Aktar et al. 2009). By 1900, more than 3000 pesticide formulations were in use, including 165 fungicides, 300 insecticides, 290 herbicides, and

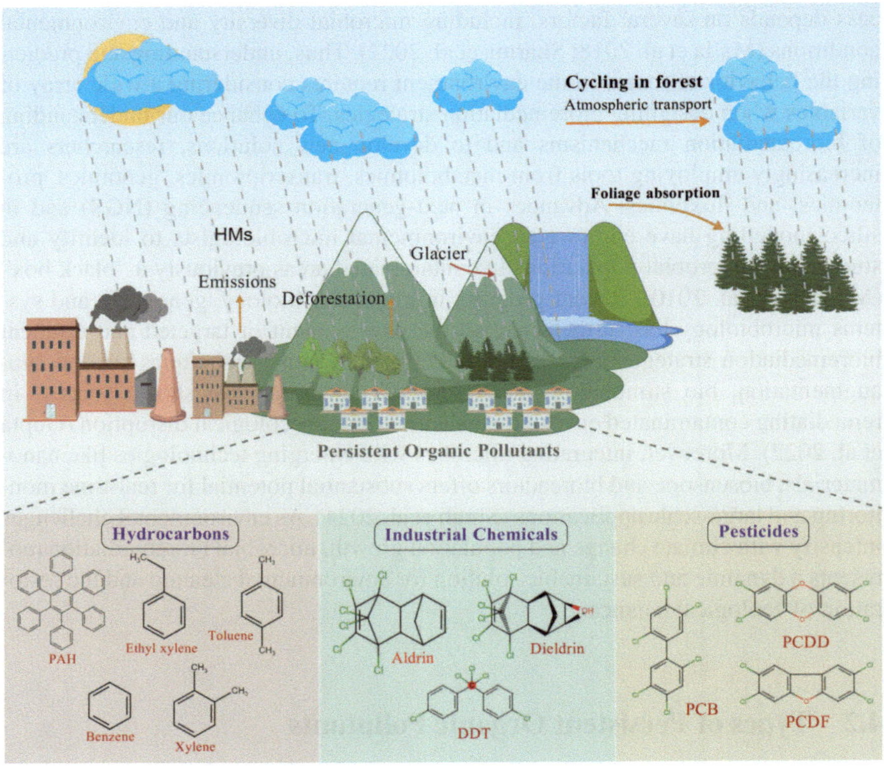

Fig. 4.1 Illustration of the environmental pathways and sources of persistent organic pollutants (POPs) entering forest ecosystems. Industrial activities release emissions containing heavy metals (HMs) and POPs, which are transported through atmospheric deposition and rainfall. Deforestation and glacier melting further contribute to pollutant mobilization. POPs undergo long-range transport and cycling in forest ecosystems via foliage absorption and atmospheric processes. The bottom panel categorizes key POPs into three groups: Hydrocarbons (e.g., PAH, benzene, toluene, xylene), Industrial Chemicals (e.g., aldrin, dieldrin, DDT), and Pesticides (e.g., PCB, PCDD, PCDF), highlighting their structural diversity and environmental persistence

various other chemicals (Carvalho 2017). The global annual consumption of pesticides is approximately 2 billion kilograms (Popp et al. 2013). Despite their revolutionary role in boosting agricultural productivity, the indiscriminate use of synthetic organic chemicals (SOCs) has led to the widespread contamination of surface and groundwater resources (Zhou et al. 2022; Sharma and Singh 2024). Runoff from agricultural fields is the primary source of water body pollution, while domestic and industrial chemical compounds contribute to a lesser extent.

Many pesticides have half-lives ranging from days to years, which leads to their persistence in the environment, resulting in bioaccumulation and biomagnification through food chains. Once these compounds enter the food chain, they tend to accumulate in the fatty tissues of humans and animals, causing various adverse health effects (Venkatesan and Halden 2014). It has been demonstrated that exposure to dichlorodiphenyltrichloroethane (DDT) in colon cancer cells inhibits apoptosis, a process regulated by the Wnt/β-catenin and PI3K/AKT signaling pathways (Santiago et al. 2021).

4.2.2 Hydrocarbons

These are organic compounds composed entirely of carbon and hydrogen. They are primarily derived from fossil fuels such as petroleum, natural gas, and coal. Hydrocarbons are widely used in industrial processes and are found in fuels, lubricants, plastics, pesticides, and solvents. However, their extensive use and improper disposal have led to significant environmental contamination, posing serious risks to human health. Hydrocarbons can be classified into two types: aliphatic and aromatic hydrocarbons. Aliphatic hydrocarbons include straight or branched-chain compounds such as methane, ethane, and octane while aromatic hydrocarbons contain one or more benzene rings, such as benzene, toluene, and polycyclic aromatic hydrocarbons (PAHs). Hydrocarbons can enter the environment through various anthropogenic activities and subsequently infiltrate the food chain via the following mechanisms:

- *Oil spills and industrial waste*: Major oil spills and industrial discharges release hydrocarbons into water bodies and soil. Aquatic organisms absorb these contaminants directly, which then bioaccumulate through the food chain (Kingston 2002).
- *Atmospheric deposition*: Hydrocarbons emitted from vehicle exhaust and industrial combustion can settle on soil and vegetation, contaminating agricultural produce via atmospheric deposition (Menzie et al. 1992).
- *Soil and groundwater contamination*: Leaking underground storage tanks and improper waste disposal introduce hydrocarbons into soil and groundwater. Plants absorb these pollutants through their roots, and animals feeding on these plants may accumulate them (Cerniglia 1992).

- *Bioaccumulation in aquatic organisms*: Fish and shellfish living in contaminated waters can accumulate hydrocarbons in their tissues. These are later consumed by humans, thereby entering the human food chain (Baumard et al. 1998).
- *Food packaging and processing*: Hydrocarbons used in packaging materials and food processing oils may migrate into food, particularly under high temperatures or prolonged storage (Franz and Welle 2002).

Hydrocarbons, especially aromatic types such as PAHs, are associated with several adverse health effects:

- Certain PAHs, such as benzopyrene, are classified as carcinogenic to humans. Prolonged exposure can lead to cancers of the lung, liver, and skin (IARC 1998).
- Exposure to volatile hydrocarbons (e.g., toluene, xylene) may result in dizziness, memory loss, and long-term cognitive impairment (Dorsey 2000).
- Hydrocarbons can disrupt endocrine function, potentially leading to reproductive issues, developmental delays, and hormonal imbalances (Diamanti-Kandarakis et al. 2009).
- Chronic exposure can result in hepatic and renal dysfunction due to the bioactivation of these compounds in the liver (Srogi 2007).
- Some studies suggest that hydrocarbons may weaken the immune system, reducing resistance to infections (Knafla et al. 2006).

4.2.3 Industrial Chemicals

Industrial chemicals are substances used or produced during manufacturing processes across various sectors, including agriculture, textiles, mining, pharmaceuticals, petroleum, and plastics. While many of these chemicals are essential to modern life, their release into the environment can lead to serious consequences—especially when they enter the food chain and pose health risks to humans. Common industrial chemicals of concern include:

- Heavy metals: lead, mercury, cadmium, arsenic
- Persistent organic pollutants (POPs): polychlorinated biphenyls (PCBs), dioxins, and organochlorine pesticides (e.g., DDT)
- Plasticizers: bisphenol A (BPA), phthalates
- Per and polyfluoroalkyl substances (PFAS): found in non-stick coatings, fire retardants, etc.
- Volatile organic compounds (VOCs): benzene, toluene, xylene

4.3 Pathways into the Food Chain

Pathways into the food chain describe the critical routes through which various substances—including essential nutrients, harmful contaminants, or persistent pollutants—enter ecosystems and subsequently transfer between organisms at different trophic levels. This movement is fundamental to ecosystem dynamics but poses significant risks when toxic substances are involved. Key pathways include direct absorption from the environment by plants and primary producers, consumption by herbivores, and transfer to higher-level predators, including humans. Aquatic contamination serves as a major entry point, where industrial effluents are discharged into rivers, lakes, or oceans, introducing chemicals like heavy metals and organic pollutants directly into the water. Aquatic organisms, such as fish and shellfish, absorb these toxins through their gills and digestive systems, leading to bioaccumulation within their tissues. These contaminants are then passed on and often amplified as they move up the food chain via predation (WHO 2010). Similarly, soil contamination arises from agricultural practices such as the application of sewage sludge, chemical fertilizers, or the use of contaminated irrigation water. These practices lead to the accumulation of heavy metals (e.g., cadmium, lead) and synthetic chemicals (e.g., pesticides) in agricultural soils. Crops grown in these soils readily absorb these contaminants through their root systems, introducing them directly into the human and animal diet when consumed (Nagajyoti et al. 2010). Airborne deposition provides another significant pathway; pollutants emitted from industrial sources like factories, power plants, and incinerators—including dioxins, furans, and heavy metals—can travel long distances before settling onto farmland, pastures, and water surfaces through atmospheric deposition. These contaminants deposit on plant leaves and fruits or seep into soils, where they are taken up by plants or ingested by grazing animals, thereby entering terrestrial and aquatic food chains (Zhang et al. 2015). Furthermore, human activities during food processing and packaging introduce contaminants. Chemicals used in processing, such as cleaning agents, lubricants, or preservatives, can inadvertently remain on or migrate into food. Additionally, chemicals from packaging materials, notably endocrine disruptors like Bisphenol A (BPA) from plastics or PFAS from coatings, can leach into food contents, especially under conditions of high temperature, acidity, or prolonged storage (Geueke et al. 2018). Once ingested by animals, many persistent organic pollutants (POPs), such as PCBs and dioxins, exhibit a strong tendency to accumulate in fatty tissues due to their lipophilic nature and resistance to metabolic breakdown. This bioaccumulation results in increasingly higher concentrations of these toxins in organisms at the top of the food chain. Consequently, humans consuming contaminated animal products—including meat, dairy, eggs, and particularly fatty fish—face the risk of ingesting these chemicals at levels that can pose significant toxicological threats over time (Van den Berg et al. 2006). Understanding these diverse pathways is crucial for assessing exposure risks and developing strategies to protect food safety and ecosystem health.

4.3.1 Health Risks Associated with Industrial Chemicals

Industrial chemicals present significant health hazards through multiple exposure pathways—inhalation, dermal absorption, and ingestion. While acute exposure can cause immediate effects like poisoning, chemical burns, or respiratory distress, the greater concern lies in *chronic exposure*, which induces complex molecular and cellular damage leading to severe, often irreversible health outcomes. Workers in manufacturing and residents near industrial zones face elevated risks due to sustained contact. The diverse toxicological mechanisms include:

- *Neurotoxicity*: Heavy metals such as lead (Pb) and mercury (Hg) readily cross the blood-brain barrier, inducing neuronal damage and oxidative stress. This is particularly detrimental during development, impairing cognitive function, memory, and learning in children (Grandjean and Landrigan 2006).
- *Endocrine disruption*: Chemicals like bisphenol A (BPA) and phthalates structurally mimic endogenous hormones (e.g., estrogens) or interfere with hormone synthesis/signaling. This disruption is mechanistically linked to reproductive dysfunction (infertility), developmental disorders, metabolic dysregulation (obesity), and altered pubertal timing (Diamanti-Kandarakis et al. 2009).
- *Carcinogenicity*: Persistent bioaccumulative toxins, including dioxins, polychlorinated biphenyls (PCBs), and specific volatile organic compounds (VOCs), can initiate or promote carcinogenesis via DNA adduct formation, genotoxicity, or chronic inflammation. Epidemiological studies associate these agents with increased incidence of liver, breast, prostate, and hematopoietic cancers (IARC 1998).
- *Organ-specific toxicity*: Many chemicals exhibit tropism for vital organs. Cadmium (Cd), for instance, accumulates in renal cortex tubules via metallothionein binding, inducing oxidative stress and apoptosis, leading to progressive nephrotoxicity and kidney failure (Jarup 2003). Similar mechanisms affect the liver (e.g., solvent-induced hepatotoxicity) and lungs (e.g., fibrosis from particulate inhalation).
- *Immunotoxicity*: Certain persistent organic pollutants (POPs) suppress immune function by altering lymphocyte proliferation, cytokine production, or thymic development. This compromises host defense, increasing susceptibility to infections and potentially reducing vaccine efficacy (Dietert et al. 2010).
- *Developmental toxicity*: Prenatal and early-life exposure is critically vulnerable. Phthalates and mercury are established *teratogens* and developmental toxicants, causing structural birth defects, neurodevelopmental delays, epigenetic changes, and measurable reductions in childhood IQ, even at low exposure levels (Bellinger 2013).

Mitigating these multifaceted risks requires stringent exposure controls, evidence-based regulatory frameworks, and substitution with safer alternatives where feasible.

4.4 Principle of Bioremediation

Bioremediation is the process by which living organisms degrade or transform hazardous organic contaminants into less toxic or non-toxic substances. Microorganisms and their enzymes break down hazardous compounds to obtain nutrients and other chemicals necessary for their optimal functioning (Fig. 4.2). Microbial enzymes play a crucial role at every stage of the metabolic process (Chen et al. 2016a, b). Enzymes such as oxidoreductases, lyases, transferases, and hydrolases possess both specific and non-specific substrate affinities, enabling them to degrade a wide range of pollutants. Microorganisms capable of degrading various pollutants such as chloroaromatics, polycyclic aromatics, biphenyls, polychlorinated biphenyls (PCBs), and spilled oils—have been isolated, with the goal of harnessing their metabolic potential for the bioremediation of contaminated sites (Tripathi and Verma 2004; Paul et al. 2005). The degradation process can occur under aerobic conditions, which require molecular oxygen and enzymes like monooxygenases and dioxygenases that act as both terminal electron acceptors and co-substrates. Alternatively, it

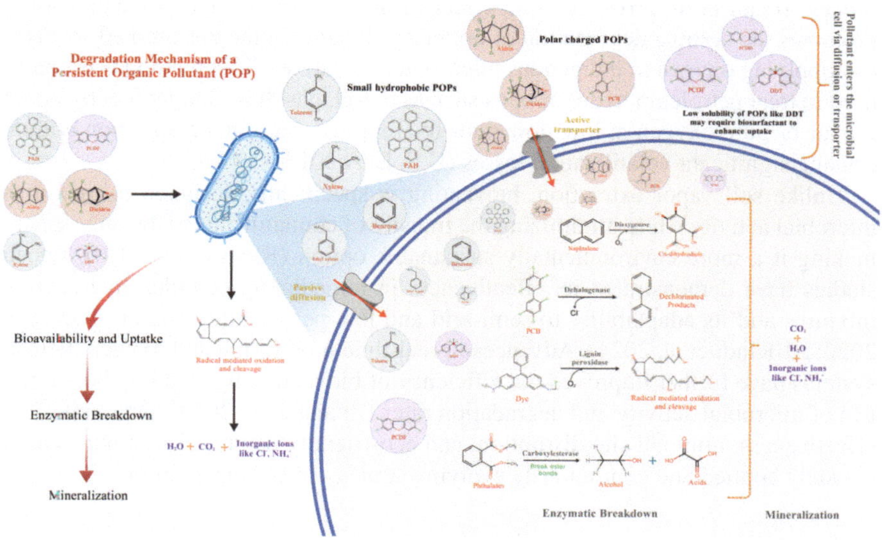

Fig. 4.2 Schematic representation of the degradation mechanism of persistent organic pollutants (POPs) by microorganisms. The process involves three key phases: (1) Bioavailability and Uptake of POPs such as DDT, PCBs, PAHs, and phthalates enter microbial cells via passive diffusion or active transporters, with uptake efficiency influenced by hydrophobicity and charge; biosurfactants may enhance the solubility of less bioavailable compounds. (2) Enzymatic Breakdown by intracellular enzymes like dioxygenases, dehalogenases, peroxidases, and carboxylesterases catalyze the transformation of POPs into intermediate metabolites or dechlorinated products. (3) Mineralization includes the final degradation results in non-toxic end products such as CO_2, H_2O, and inorganic ions (e.g., Cl^-, NH_4^+), completing the detoxification process

can occur under anaerobic conditions, which rely on the presence of electron-withdrawing groups to initiate reductive attack (Gaur et al. 2018).

These microbial processes result in the production of carbon dioxide, water, and inorganic salts by either modifying the compounds' functional groups or breaking down their fundamental structures. The rate of microbial growth and the efficiency of the biodegradation process can be enhanced by manipulating environmental parameters. Biodegradation is a key component of bioremediation, converting harmful organic pollutants into non-toxic or naturally occurring inorganic compounds such as carbon dioxide and water that are safe for humans, animals, aquatic life, and plants (Priyadarshanee and Das 2021). The following methods are considered when employing bioremediation technology for the elimination of persistent organic pollutants (POPs):

4.4.1 Bioventing

Bioventing is a process that involves aerating soil and water to stimulate the in situ biodegradation of organic contaminants and enhance the overall bioremediation process (Gaur et al. 2018). It is an effective in situ bioremediation technique that promotes the aerobic degradation of organic pollutants in the unsaturated soil zone by supplying oxygen to indigenous microbial populations. This method is primarily used to treat petroleum hydrocarbons such as diesel, gasoline, and jet fuel by injecting air or pure oxygen at low flow rates to stimulate microbial respiration without causing significant volatilization losses (U.S. EPA 2013).

Unlike soil vapor extraction, bioventing is specifically designed to maximize microbial activity while minimizing the transfer of contaminants to the atmosphere, making it a more environmentally sustainable option (Khan et al. 2020). Recent studies have demonstrated its effectiveness in remediating complex hydrocarbon mixtures and its adaptability to semi-arid and low-permeability soils (Singh et al. 2022; Al-Kindi et al. 2024). Advances in real-time monitoring and oxygen delivery systems have further improved the efficiency of bioventing by enabling better control of microbial activity and degradation rates (Zhang et al. 2023). Due to its cost-effectiveness, minimal site disruption, and eco-friendly nature, bioventing remains a widely applied and continuously evolving strategy in bioremediation.

4.4.2 Biostimulation

Biostimulation involves modifying the contaminated environment to support the growth of native soil bacteria, often by adjusting pH levels and supplementing limiting nutrients to optimize the carbon: nitrogen:phosphorous (C: N: P) ratio (Bala et al. 2022). It is a widely adopted bioremediation strategy that enhances indigenous microbial activity by optimizing environmental conditions to accelerate the degradation of organic pollutants. This technique typically includes the addition of nutrients (e.g., nitrogen and phosphorus), electron acceptors, or amendments such as biosurfactants and oxygen-releasing compounds to stimulate native microbial populations capable of degrading contaminants like petroleum hydrocarbons, chlorinated solvents, and pesticides (Gomez and Sartaj 2020). Biostimulation has shown particular promise in treating contaminated soils and aquifers, especially where microbial activity is limited by nutrient deficiencies or suboptimal redox conditions (Kumar et al. 2021b).

Recent advancements focus on the use of slow-release nutrient formulations and nanomaterials to sustain microbial stimulation and enhance contaminant bioavailability, thereby increasing treatment efficacy (Abbas et al. 2023). Moreover, field-scale applications have demonstrated their cost-effectiveness and environmental compatibility compared to more invasive remediation methods (Alonso-Gutiérrez et al. 2024). As interest in sustainable remediation grows, biostimulation remains a key component of integrated bioremediation frameworks due to its adaptability, low environmental impact, and compatibility with in situ treatment strategies.

4.4.3 Bioaugmentation

Bioaugmentation involves the deliberate introduction of specific microbial strains or biocatalysts under controlled conditions to degrade persistent organic pollutants (POPs). It is an advanced bioremediation technique particularly effective at sites where native microbial populations lack the metabolic capabilities or diversity required to degrade complex or recalcitrant compounds, such as chlorinated solvents, polycyclic aromatic hydrocarbons (PAHs), and heavy oil fractions (Tyagi et al. 2020). This method can utilize either autochthonous (site-derived) or allochthonous (externally sourced) microorganisms, including genetically modified strains engineered for enhanced biodegradation (Santos et al. 2022). Recent studies emphasize the importance of microbial consortia and their synergistic interactions in improving degradation rates and resilience under site-specific stress conditions (Gupta et al. 2023). Advances in microbial encapsulation, delivery systems, and genomic tools have further increased the efficiency and adaptability of bioaugmentation strategies (Khan et al. 2024). As regulatory frameworks evolve and sustainable remediation practices gain traction, bioaugmentation is becoming increasingly recognized as a targeted and environmentally friendly solution for the remediation of complex contaminated sites.

4.5 Microbes Involved in Biodegradation

Microorganisms like bacteria, fungi, and algae possess enzymatic systems capable of breaking down complex organic molecules into simpler, less harmful substances. This biological process not only detoxifies contaminated environments but also helps restore ecological balance by recycling organic matter back into the ecosystem.

4.5.1 Bacteria

The bacterial population plays a crucial role in the degradation of organic pollutants at contaminated sites and significantly contributes to environmental sustainability and the geochemical cycle. Bacteria are central to both natural and engineered remediation processes, making them a cornerstone of bioremediation strategies. Their remarkable metabolic versatility enables them to enzymatically transform or mineralize a wide range of contaminants including hydrocarbons, pesticides, chlorinated solvents, and dyes under both aerobic and anaerobic conditions (Megharaj et al. 2017). Species such as *Pseudomonas*, *Bacillus*, *Rhodococcus*, and *Alcaligenes* have been extensively studied for their ability to degrade petroleum hydrocarbons and other persistent organic pollutants (Kumar et al. 2020) (Table 4.1). These microorganisms utilize pollutants as sources of carbon and energy, often breaking them down into harmless byproducts such as carbon dioxide and water. Recent advancements in genomics, metagenomics, and systems biology have facilitated the discovery of novel degradative pathways and regulatory mechanisms, broadening the scope of bacterial biodegradation to include emerging contaminants such as pharmaceuticals and microplastics (Zhou et al. 2023a; Li et al. 2025). Furthermore, bioengineering approaches are being developed to enhance bacterial degradation efficiency under extreme environmental conditions, supporting their use in diverse ecosystems. As our understanding of bacterial ecology and functionality deepens, bacterial biodegradation continues to emerge as a sustainable and effective tool for mitigating environmental pollution.

4.5.2 Fungi

Fungi are common organisms that play an essential role in every ecosystem. These microscopic eukaryotes can grow on a wide variety of substrates and often function over extended periods. Fungi encompass a broad group including filamentous fungi, molds, and yeasts that are capable of treating various types of industrial wastewater. They represent a powerful and increasingly studied avenue in bioremediation, owing to their unique enzymatic capabilities and tolerance to harsh environmental conditions. Filamentous fungi, particularly white-rot species such as *Phanerochaete*

Table 4.1 Bioremediation of hazardous organic pollutants by bacteria

Name of bacteria	Pollutant type	Pollutants name	Bioremediation mechanism	References
Sphingomonas spp.	Chlorinated compounds	Pentachlorophenol, PCB	Reductive dechlorination	White et al. (1996)
Mycobacterium spp.	Polycyclic aromatic hydrocarbons (PAHs)	Phenanthrene, Fluoranthene	Intracellular degradation	Kazunga and Aitken (2000)
Arthrobacter spp.	Pesticides, Herbicides	Lindane, Atrazine	Enzymatic degradation, cometabolism	Strong et al. (2002)
Burkholderia cepacia	Herbicides and pesticides	2,4-D, Atrazine	Enzymatic degradation	Coenye and Vandamme (2003)
Alcanivorax borkumensis	Alkanes (Marine oil spills)	n-alkanes, branched alkanes	Alkane monooxygenase activity	Yakimov et al. (2007)
Bacillus subtilis	Petroleum hydrocarbons	Diesel, Kerosene	Surfactant production, biodegradation	Das and Mukherjee (2007)
Rhodococcus erythropolis	Alkanes, Aromatic compounds	Crude oil, BTEX compounds	Cometabolism	Martínková et al. (2009)
Acinetobacter calcoaceticus	Phenols, Hydrocarbons	Phenol, Diesel oil	Oxygenase-mediated degradation	Wang et al. (2010)
Dehalococcoides spp.	Chlorinated solvents	Trichloroethylene (TCE), PCE	Reductive dehalogenation	Löffler et al. (2013)
P. putida	Hydrocarbons	Toluene, Xylene, Benzene	Oxidation via dioxygenases	Singh et al. (2014)
B. megaterium	Phenols	Phenol, Cresol	Oxidative degradation	Patel et al. (2020)
P. aeruginosa	Hydrocarbons	Crude oil, Toluene	Biosurfactant production, oxidation	Khan et al. (2021)
Dehalococcoides mccartyi	Chlorinated solvents	TCE, PCE	Reductive dehalogenation	Lee et al. (2022)
Rhodococcus qingshengii	PAHs	Naphthalene, Fluorene	Cometabolic degradation	Zhao et al. (2023)
Sphingobium yanoikuyae	Pesticides	Atrazine, Lindane	Enzymatic degradation	Nguyen et al. (2024)

chrysosporium, *Trametes versicolor*, and *Pleurotus ostreatus*, produce extracellular oxidative enzymes like lignin peroxidase, manganese peroxidase, and laccase. These enzymes enable the degradation of structurally complex organic pollutants, including polycyclic aromatic hydrocarbons (PAHs), synthetic dyes, pesticides, and endocrine-disrupting chemicals (Singh and Chen 2019; Harms et al. 2021). Unlike many bacteria, fungi can degrade both aliphatic and aromatic compounds through metabolic pathways, often without utilizing the pollutant as a primary carbon source.

Recent studies have demonstrated fungi's ability to survive in low-nutrient and extreme environments, making them highly suitable for remediating contaminated soils, sediments, and industrial effluents (Al-Hawash et al. 2022). The integration of molecular tools and omics technologies has enhanced our understanding of fungal biodegradation mechanisms and accelerated the discovery of novel strains with improved capabilities (Kumar et al. 2024). With their broad substrate range, high enzyme secretion potential, and remarkable adaptability, fungi continue to play an increasingly vital role in sustainable bioremediation practices (Table 4.2).

4.5.3 Algae

Algae have emerged as a promising, eco-friendly strategy for bioremediation due to their ability to metabolize a wide range of organic pollutants while simultaneously contributing to oxygen production and nutrient recovery. Microalgae and cyanobacteria such as *Chlorella vulgaris*, *Scenedesmus obliquus*, and *Spirulina platensis* have demonstrated the capacity to degrade hydrocarbons, dyes, pharmaceuticals, pesticides, and phenolic compounds through enzymatic pathways and biosorption mechanisms (Elumalai et al. 2020; Manzoor et al. 2021). Algae-based systems are particularly advantageous in aquatic environments, where they can operate in both suspended and immobilized forms, often in combination with bacterial consortia in phycoremediation approaches. Recent research has highlighted the use of genetically engineered microalgae to enhance the degradation of persistent organic pollutants and heavy metals, thereby expanding their application in wastewater treatment and polluted ecosystems (Mohanraj et al. 2023; Wang et al. 2025). Furthermore, algae can sequester CO_2 and generate valuable biomass, offering co-benefits for bioenergy production and integration into circular economy models. Owing to their autotrophic nature, metabolic plasticity, and low operational costs, algae represent a sustainable and multifunctional platform for biodegradation in modern bioremediation frameworks (Table 4.3).

4.6 Advanced Tools for Bioremediation

4.6.1 Synthetic Biology for Precision Bioremediation

The inherent limitations of natural microbiomes in degrading persistent synthetic pollutants—such as PFAS, chlorinated solvents, and plastics—have driven the development of engineered microbial solutions. Leveraging synthetic biology, researchers now design microbes with de novo or enhanced catabolic pathways,

Table 4.2 Bioremediation of hazardous organic pollutants by fungi

Fungal species	Pollutant type	Pollutants name	Bioremediation mechanism	References
Phanerochaete chrysosporium	PAHs	Benzo[a]pyrene, Phenanthrene	Ligninolytic enzyme degradation	Singh et al. (2015)
Trametes versicolor	Dyes	Reactive Black 5, Congo Red	Laccase-mediated degradation	Chen et al. (2016b)
Aspergillus niger	Hydrocarbons	Diesel, Toluene	Biosurfactant production, oxidation	Yadav et al. (2017)
Pleurotus ostreatus	Pesticides	Malathion, DDT	Enzymatic degradation	Martínez et al. (2018)
F. solani	PCBs	Aroclor 1242	Oxidative degradation	Ghosh et al. (2019)
Mucor racemosus	Phenols	Phenol, Cresol	Peroxidase-mediated degradation	Zhang et al. (2020a)
Penicillium chrysogenum	Pharmaceuticals	Ibuprofen, Diclofenac	Mycoremediation	Kumar et al. (2021a)
Cladosporium cladosporioides	Hydrocarbons	Crude oil, Kerosene	Oxidation via extracellular enzymes	Ali et al. (2022)
Rhizopus oryzae	Dyes	Methylene Blue, Crystal Violet	Adsorption and enzymatic degradation	Tariq et al. (2023)
Lentinula edodes	Herbicides	Atrazine	Ligninolytic enzyme activity	Wang et al. (2015)
Aspergillus flavus	PAHs	Fluoranthene, Pyrene	Enzymatic oxidation	Lee et al. (2016)
Trichoderma harzianum	Pesticides	Chlorpyrifos	Biodegradation through hydrolysis	Das et al. (2017)
Myrothecium verrucaria	Dyes	Direct Red 28	Peroxidase enzyme system	Nawaz et al. (2018)
Paecilomyces variotii	Phenols	Phenol	Biodegradation by oxidative enzymes	Patel et al. (2019)
Ganoderma lucidum	Pharmaceuticals	Carbamazepine	Mycoremediation via laccases	Sun et al. (2020)

real-time biosensing capabilities, and improved environmental fitness for targeted bioremediation (Cao et al. 2022; French et al. 2021; Wang et al. 2020). Key strategies include: (1) Heterologous Expression: Introducing foreign degradative enzymes (e.g., PETase/MHETase from *Ideonella sakaiensis* into *E. coli* or *Pseudomonas*) to enable plastic hydrolysis (Yoshida et al. 2016; Tanasupawat et al. 2020); (2) De Novo Pathway Construction: Assembling novel metabolic routes in chassis strains (e.g., engineering *Rhodococcus* with P450 monooxygenases for PFAS defluorination) (Huang and Jaffé 2019); (3) Enzyme Optimization: Using directed evolution (e.g., laccase mutagenesis for enhanced dye decolorization) (Mate and Alcalde 2017); and (4) Strain Integration: Combining multiple pathways in robust hosts, exemplified by the *P. putida* "Superbug" engineered with hydroxylase and dehalogenase genes for pentachlorophenol mineralization (van Dillewijn et al. 2008).

Table 4.3 Bioremediation of hazardous organic pollutants by algae

Algal species	Pollutant type	Pollutants name	Bioremediation mechanism	References
Chlorella vulgaris	Dyes	Methylene Blue, Congo Red	Biosorption and biodegradation	Zhou et al. (2015)
Scenedesmus obliquus	Pharmaceuticals	Ibuprofen, Paracetamol	Enzymatic degradation	Kumar et al. (2016)
Spirulina platensis	Phenols	Phenol, Bisphenol A	Adsorption and enzymatic oxidation	Sharma et al. (2017)
Nannochloropsis oculata	Hydrocarbons	Diesel, Crude oil	Biosorption and metabolic transformation	Yadav et al. (2018)
Chlamydomonas reinhardtii	Pesticides	Atrazine	Uptake and biodegradation	Liu et al. (2019)
Anabaena variabilis	Herbicides	2,4-D	Enzymatic degradation	Singh et al. (2020)
Oscillatoria limnetica	PAHs	Naphthalene	Metabolic degradation	Chen et al. (2021)
Spirulina maxima	Textile dyes	Crystal Violet	Biosorption	Zhang et al. (2022)
Botryococcus braunii	Petroleum hydrocarbons	Kerosene, Toluene	Adsorption and metabolic breakdown	Ali et al. (2023)
Dunaliella salina	Plasticizers	Phthalates	Biosorption and enzymatic transformation	Patel et al. (2024)
Chlorella pyrenoidosa	Heavy oils	Used motor oil	Cell wall absorption	Fernandez et al. (2025)
Nostoc muscorum	PAHs	Anthracene	Bioaccumulation and enzymatic degradation	Lee et al. (2015)
Tetraselmis suecica	Phenolic compounds	Cresol	Bioadsorption and transformation	Wang et al. (2016)
Haematococcus pluvialis	Pesticides	Chlorpyrifos	Metabolic conversion	Nguyen et al. (2017)
Microcystis aeruginosa	Dyes	Direct Red 28	Enzymatic reduction	Iqbal et al. (2018)

4.6.2 *Biosensor Integration: Real-Time Pollution Monitoring*

Engineered microbes can act as living biosensors by combining their ability to degrade pollutants with the detection of contaminants. Fluorescent and bioluminescent reporters, such as the *lux* gene for bioluminescence or *gfp* for green fluorescent protein, are commonly fused to pollutant-responsive promoters. For example, *E. coli* engineered with toluene-sensitive promoters emit a visible glow in the presence of benzene derivatives (van der Meer and Belkin 2010). Colorimetric outputs also play a role; enzymes like β-galactosidase generate distinct color changes when specific pollutants, such as arsenic, are present (Stocker et al. 2003). These biosensing systems have practical applications, including portable devices that use freeze-dried engineered yeast to detect heavy metals in water.

4.6.3 Tolerance Enhancement

Tolerance enhancement is crucial for microbial survival in harsh conditions commonly found at contaminated sites, which may involve extreme pH, high salinity, or accumulation of toxic metabolites. To cope with such stressors, microbes employ various strategies. One approach is the overexpression of stress-resistance genes, such as heat-shock proteins like Hsp70 or osmolytes like betaine, which help stabilize cellular functions under stress (Yancey 2005; Lindquist and Craig 1988). Another method involves modifications to the cell membrane, where changes in the lipid bilayer composition can enhance resistance to toxic solvents, as seen in *Pseudomonas* species during toluene degradation (Sikkema et al. 1995). Additionally, CRISPR-Cas9 technology enables precise gene editing, such as knocking out genes that heighten stress sensitivity—for example, removing prophage elements in *Deinococcus radiodurans* to improve radiation resistance (Jeong et al. 2017).

4.6.4 Engineered Consortia

Complex pollutant mixtures, such as oil spills, often require the cooperative action of multiple microbial species to achieve effective degradation. This microbial teamwork involves a division of labor, where one microbial strain may specialize in degrading aromatic hydrocarbons while another targets alkane. Coordination among these microbes is regulated through quorum sensing, a communication system that uses signaling molecules like acyl-homoserine lactones to synchronize gene expression across different species. In synthetic ecology approaches, combinations like *Bacillus subtilis*, known for forming robust biofilms, and *Sphingomonas*, an efficient degrader of polycyclic aromatic hydrocarbons (PAHs), are used together to enhance soil bioremediation outcomes (Chen and Huang 2020).

4.6.5 Electro-Bioremediation & Microbial Electrochemical Technologies (METs)

Electro-bioremediation combines microbial metabolism with electrochemical processes to enhance the breakdown of pollutants in soil, sediment, and wastewater. Microbial electrochemical technologies (METs) leverage the ability of electroactive bacteria to interact with electrodes, enabling simultaneous contaminant degradation and energy recovery (Logan et al. 2019a). These systems are particularly effective for recalcitrant compounds, including chlorinated solvents, petroleum hydrocarbons, and heavy metals, where traditional bioremediation falls short.

4.6.6 Microbial Fuel Cells (MFCs): Bioelectricity-Driven Degradation

Microbial fuel cells (MFCs) utilize electrochemically active microbes, known as exoelectrogens, to oxidize organic or inorganic contaminants at the anode, while simultaneously generating usable electricity. In the anode chamber, bacteria such as *Geobacter sulfurreducens* and *Shewanella oneidensis* degrade pollutants like phenols or benzene and transfer electrons to the anode (Logan et al. 2006; Rabaey and Verstraete 2005). For instance, *Geobacter* metabolizes acetate according to the reaction:

$$CH_3COO^- + 2H_2O \rightarrow 2CO_2 + 7H^+ + 8e^- -$$

These electrons travel through an external circuit to the cathode, where they reduce terminal electron acceptors. Oxygen may be reduced to water, or chlorinated pollutants like trichloroethylene (TCE) to ethene. An example of such a cathodic reaction is the reduction of perchlorate:

$$ClO_4^- + 8H^+ + 8e^- \rightarrow Cl^- + 4H_2O$$

MFCs have promising applications, including wastewater treatment, where they degrade textile effluent contaminants while producing electricity (typically 0.1–1 W/m^2), and sediment remediation through benthic MFCs that remove sulfides and hydrocarbons in marine environments.

4.6.7 Electrokinetic Bioremediation

Electrokinetics involves the application of low-voltage electric fields (1–5 V/cm) to soil and water systems to enhance the mobilization and degradation of contaminants. This technique mobilizes pollutants through electrophoresis, where charged species like Cr^{6+} and Pb^{2+} migrate toward oppositely charged electrodes, and electroosmosis, where water flow transports dissolved contaminants such as BTEX compounds to designated treatment zones (Virkutyte et al. 2002). Additionally, electrokinetics stimulates microbial activity by enabling electrodes to serve as electron donors or acceptors, thereby accelerating the metabolic rates of microorganisms like *Dehalococcoides* for chlorinated solvent degradation and *Desulfovibrio* for sulfate reduction. pH changes near the electrodes can further enhance enzyme activities such as those of dehalogenases. In a notable case study, electrokinetic treatment in clay soil led to a 300% increase in *Dehalococcoides* activity, resulting in complete dechlorination of trichloroethylene (TCE) to ethene (Yang et al. 2025).

Recent innovations have integrated microbial fuel cells (MFCs) with electrokinetic techniques to enhance remediation efficiency through synergistic mechanisms. In bioelectrokinetic remediation (BEKR), anodes facilitate the degradation of organic pollutants, while cathodes support the reduction of heavy metals such as hexavalent chromium (Cr^{6+}) to trivalent chromium (Cr^{3+}) (Huang et al. 2019). Simultaneously, electric fields generated in the system promote the movement of essential nutrients, such as lactate, toward anaerobic zones to stimulate microbial degradation (Zhang et al. 2020b). In microbial electrolysis cells (MECs), the application of a small external voltage (~0.3 V) enables hydrogen production at the cathode, which aids in reductive dechlorination processes (Logan et al. 2019b). These advanced methods offer several advantages over conventional techniques, including energy-neutral or even energy-positive operations due to electricity generation by MFCs, precise targeting of pollutants through electrode placement, and the versatility to treat complex contaminant mixtures, such as polycyclic aromatic hydrocarbons (PAHs) combined with heavy metals (Rabaey and Verstraete 2005).

4.6.8 Future Prospects

The field of microbial bioremediation is poised for groundbreaking advances, propelled by interdisciplinary innovation and mounting environmental urgency. Emerging strategies span five key frontiers. First, de novo pathway design will tackle stubborn pollutants like PFAS and plastics using synthetic biology, aided by AI-guided protein folding to create novel enzymes such as advanced PETases (Jumper et al. 2021). Space bioremediation is also on the horizon, with NASA exploring engineered microbes for waste degradation during space missions (Zea et al. 2020). Second, smart bioremediation systems will integrate CRISPR-based biosensors with IoT networks, enabling real-time monitoring and deployment of autonomous remediation tools (Wen et al. 2020; Badr and Almehdar 2021). Machine learning will optimize microbial consortia tailored to specific contamination scenarios. Third, electro-biohybrid platforms will enhance microbial activity via nanobioelectrodes and closed-loop electrokinetic systems powered by solar energy, enabling metal recovery through microbial nanoparticle synthesis (Sharma et al. 2020; Li et al. 2022). Fourth, climate-proofing bioremediation involves engineering extremophiles capable of thriving in harsh conditions like high heat or salinity (Canganella and Wiegel 2011). Carbon-negative remediation strategies will pair pollutant degradation with methane-to-bioplastic conversion. Finally, policy and socioeconomic integration will support field deployment of engineered microbes with biocontainment features, while circular economy collaborations will scale up waste-to-resource innovations (Saidi et al. 2021). Addressing challenges such as GMO containment, contaminant complexity, and high costs will require solutions like synthetic auxotrophy, microbial swarms, and energy-recovering bioelectrochemical systems.

4.7 Conclusion

Bioremediation represents a forefront strategy in environmental biotechnology, leveraging the inherent metabolic versatility of microorganisms to transform and detoxify hazardous contaminants in ecosystems. This technology harnesses complex microbial enzymatic pathways to mineralize or transform a wide spectrum of organic pollutants into less toxic or benign end-products, offering an eco-friendly and sustainable alternative to conventional physico-chemical remediation methods. Despite significant advances in understanding microbial metabolism and genetic engineering approaches to enhance biodegradative capacities, the effective remediation of recalcitrant and highly toxic pollutants remains a formidable challenge. Current limitations arise largely because many hazardous contaminants exhibit chemical structures that are resistant to microbial enzymatic attack, and the majority of indigenous microbial consortia capable of degrading these compounds are either unculturable under laboratory conditions or lack the necessary catabolic pathways. This microbial unculturability restricts access to the full metabolic potential existing in natural environments. Consequently, environmental biotechnologists are now pivoting towards integrative and innovative strategies such as metagenomics, synthetic biology, and microbial consortia engineering to overcome these bottlenecks. By employing high-throughput sequencing and omics-based approaches, researchers can decode the vast genetic repertoire of environmental microbiomes, identifying novel catabolic genes and metabolic networks involved in pollutant degradation. Synthetic biology enables the design and construction of tailor-made microbial strains or microbial communities with enhanced degradative functions and resilience under extreme environmental conditions. Additionally, the application of bioaugmentation, coupled with biostimulation techniques, optimizes in situ conditions, thereby promoting the proliferation and activity of pollutant-degrading microorganisms. In summary, the future of bioremediation hinges on multidisciplinary innovation, integrating molecular biology, environmental engineering, and systems ecology to unlock the full potential of microbial metabolism. Such advanced biotechnological interventions promise to revolutionize the remediation of persistent environmental pollutants, paving the way towards safer, more effective, and sustainable ecosystem restoration.

References

Abbas SZ, Rafatullah M, Ismail N (2023) Nano-enhanced biostimulation for the bioremediation of contaminated environments: current trends and future prospects. Sci Total Environ 856:159102

Aktar W, Sengupta D, Chowdhury A (2009) Impact of pesticides use in agriculture: their benefits and hazards. Interdiscip Toxicol 2(1):1–12

Al-Hawash AB, Zhang Y, Ma F (2022) Fungal biodegradation of persistent organic pollutants: mechanisms and recent advances. Environ Pollut 300:118956

Ali MI, Khan S, Ahmed A (2022) *Cladosporium cladosporioides* as a crude oil-degrading fungus: role of extracellular enzymes in bioremediation. J Environ Manag 310:114763

Ali MI, Ahmad M, Khan S (2023) Biodegradation of petroleum hydrocarbons using Botryococcus braunii and its potential in bioremediation. Bioresour Technol Rep 21:101150

Al-Kindi S, Al-Wahaibi Y, Al-Bahry S (2024) Optimization of bioventing for hydrocarbon degradation in arid environments. J Environ Manag 341:118222

Alonso-Gutiérrez J, Vilchez-Vargas R, Leiva C (2024) Field-scale biostimulation of hydrocarbon degradation in a marine sediment chronically impacted by oil pollution. Mar Pollut Bull 200:115806

International Agency for Research on Cancer (1998) Monographs on the evaluation of carcinogenic risks tohumans. Some antineoplastic and immunosuppressive agents, 26.https://cir.nii.ac.jp/crid/1370001114204490117

Dorsey A (2000) Toxicological profile fortoluene. The Agency

Badr A, Almehdar H (2021) CRISPR-Cas-based tools for environmental biosensing. Biosensors 11(2):45

Bala S, Garg D, Thirumalesh BV, Sharma M, Sridhar K, Inbaraj BS, Tripathi M (2022) Recent strategies for bioremediation of emerging pollutants: a review for a green and sustainable environment. Toxics 10(8):484. https://doi.org/10.3390/toxics10080484

Baumard P, Budzinski H, Garrigues P, Narbonne JF (1998) Polycyclic aromatic hydrocarbons in sediments and mussels of the western Mediterranean Sea. Environ Toxicol Chem 17(5):765–776

Bellinger DC (2013) Prenatal exposures to environmental chemicals and children's neurodevelopment: an update. Safe Health Work 4(1):1–11

Canganella F, Wiegel J (2011) Extremophiles: from abyssal to terrestrial ecosystems and possibly beyond. Naturwissenschaften 98(4):253–279. https://doi.org/10.1007/s00114-010-0762-8

Cao M, Zhang Y, Zhao Y, Wang C, Liu W (2022) Synthetic biology approaches for biodegradation of recalcitrant environmental pollutants. Biotechnol Adv 59:107973. https://doi.org/10.1016/j.biotechadv.2022.107973

Carvalho FP (2017) Pesticides, environment, and food safety. Food Energy Secur 6(2):48–60

Cerniglia CE (1992) Biodegradation of polycyclic aromatic hydrocarbons. Biodegradation 3(2–3):351–368

Chen A, Huang Y (2020) Acyl-homoserine lactone-based quorum sensing affects phenanthrene removal by Novosphingobium pentaromativorans through altering cell surface properties. Int Biodeterior Biodegrad 147:104841. https://doi.org/10.1016/j.ibiod.2020.104841

Chen BY, Ma CM, Han K, Yueh PL, Qin LJ, Hsueh CC (2016a) Influence of textile dye and decolorized metabolites on microbial fuel cell-assisted bioremediation. Bioresour Technol 200:1033–1038. https://doi.org/10.1016/j.biortech.2015.10.079

Chen C, Zhang S, Liu X (2016b) Laccase-mediated decolorization of synthetic dyes by Trametes versicolor. J Environ Sci 44:260–268

Chen M, Ma Y, Chen H (2021) Biodegradation of polycyclic aromatic hydrocarbons by cyanobacteria and their removal mechanisms. Ecotoxicol Environ Saf 208:111697

Coenye T, Vandamme P (2003) Diversity and significance of Burkholderia species occupying diverse ecological niches. Environ Microbiol 5(9):719–729

CPCB (2021) River stretches for restoration of water quality. Central Pollution Control Board, Ministry of Environment, Forest and Climate Change, Government of India

Das K, Mukherjee AK (2007) Crude petroleum-oil biodegradation efficiency of *Bacillus subtilis and Pseudomonas aeruginosa* strains isolated from a petroleum-oil contaminated soil. Bioresour Technol 98(7):1339–1345

Das N, Sengupta D, Mukherjee S (2017) Biodegradation of chlorpyrifos by *Trichoderma harzianum*: a promising approach for mycoremediation. Environ Monit Assess 189:387

Diamanti-Kandarakis E et al (2009) Endocrine-disrupting chemicals: an Endocrine Society scientific statement. Endocr Rev 30(4):293–342

Dietert RR et al (2010) Immunological endpoints for environmental health risk assessment and research in the 21st century. Toxicology 271(2–3):131–140

Elumalai S, Suresh A, Ananthi T (2020) Bioremediation potential of microalgae: a review on mechanisms and applications. Environ Sci Pollut Res 27(35):43594–43613

EPA (2023) PFAS strategic roadmap: EPA's commitments to action 2021–2024. Environmental Protection Agency

Fernandez R, Lopez A, Martinez R (2025) Role of Chlorella pyrenoidosa in the removal of used motor oil from wastewater: a case study. Sci Total Environ 858:160972

Franz R, Welle F (2002) Migration of hydrocarbons into foodstuffs from packaging materials. Food Addit Contam 19(S1):44–56

French KE, Terry LA, Ralph PJ (2021) Engineered microbial consortia and bioremediation: a synthetic biology approach to pollution mitigation. Environ Pollut 269:116066. https://doi.org/10.1016/j.envpol.2020.116066

Furukawa K (2003) 'Super bugs' for bioremediation. Trends Biotechnol 21(5):187–190. https://doi.org/10.1016/S0167-7799(03)00071-6

Gaur N, Narasimhulu K, Y P (2018) Recent advances in the bioremediation of persistent organic pollutants and its effect on environment. J Clean Prod 198:1602–1631. https://doi.org/10.1016/j.jclepro.2018.07.026

Gaur N, Dutta D, Singh A, Dubey R, Kamboj DV (2022) Recent advances in the elimination of persistentorganic pollutants by photocatalysis. Front Environ Sci, 10, p.872514

Geueke B et al (2018) Food packaging in the circular economy: overview of chemical safety aspects. Food Addit Contam Part A Chem Anal Control Expo Risk Assess 35(8):1503–1518

Ghosh P, Rathinasabapathi B, Ma LQ (2019) Degradation of PCBs by *Fusarium solani* through oxidative mechanisms. Chemosphere 221:698–705

Gomez DE, Sartaj M (2020) Biostimulation for in situ bioremediation of petroleum hydrocarbon-contaminated soils: a review. J Environ Manag 255:109896

Grandjean P, Landrigan PJ (2006) Developmental neurotoxicity of industrial chemicals. Lancet 368(9553):2167–2178

Gupta S, Yadav P, Choudhary S (2022) Advances in microbial consortia for bioremediation: a holistic approach to environmental cleanup. Ecotoxicol Environ Saf 236:113448

Gupta S, Yadav P, Choudhary S (2023) Microbial consortia in bioaugmentation: a comprehensive review on synergistic interactions for enhanced biodegradation. Ecotoxicol Environ Saf 248:114402

Harms H, Schlosser D, Wick LY (2021) Untapped potential: exploiting fungi in bioremediation of hazardous chemicals. Nat Rev Microbiol 19(3):177–192

Huang S, Jaffé PR (2019) Defluorination of perfluorooctanoic acid (PFOA) by acidimicrobium sp. strain A6. Environ Sci Technol 53(19):11410–11419. https://doi.org/10.1021/acs.est.9b03175

Huang X, Liu Y, Guo J (2019) Bioelectrokinetic remediation of hexavalent chromium-contaminated soil using dual electrodes. Environ Sci Pollut Res 26(21):21724–21734. https://doi.org/10.1007/s11356-019-05639-0

IARC (International Agency for Research on Cancer) (2012) Monographs on the evaluation of carcinogenic risks to humans, vol 100F

Iqbal HMN, Kyazze G, Keshavarz T (2018) Biodegradation of azo dye direct red 28 using *Microcystis aeruginosa* and its possible enzymatic pathway. Biocatal Agric Biotechnol 13:129–134

Jarup L (2003) Hazards of heavy metal contamination. Br Med Bull 68(1):167–182

Jeong SW, Kim JH, Park SJ, Lim HM (2017) Genome editing of *Deinococcus radiodurans* using CRISPR-Cas9. Microb Cell Factories 16(1):1–9. https://doi.org/10.1186/s12934-017-0765-6

Jumper J, Evans R, Pritzel A et al (2021) Highly accurate protein structure prediction with AlphaFold. Nature 596(7873):583–589. https://doi.org/10.1038/s41586-021-03819-2

Kazunga C, Aitken MD (2000) Products from the incomplete metabolism of pyrene by *Mycobacterium* spp. Appl Environ Microbiol 66(5):1917–1922

Khan FI et al (2020) Bioremediation of petroleum contaminated sites: a case study of bioventing technology. Environ Technol Innov 17:100557

Khan Z, Rehman A, Shakoori AR (2021) Crude oil degradation by biosurfactant-producing *Pseudomonas aeruginosa* strain isolated from oil-contaminated soil. Biodegradation 32(2):143–154

Khan A, Li J, Wang L (2024) Novel delivery systems and genome editing tools in bioaugmentation: emerging trends and future directions. Curr Opin Environ Sci Health 37:100520

Kingston PF (2002) Long-term environmental impact of oil spills. Spill Sci Technol Bull 7(1–2):53–61

Knafla A, Phillipps K, Brecher RW, Petrovic S (2006) Human health risk assessment for petroleum hydrocarbon fraction 3 (C16–C34) in soil. Hum Ecol Risk Assess 12(5):1074–1090

Kumar KS, Dahms HU, Won EJ, Lee JS, Shin KH (2016) Microalgae–A promising tool for heavy metal remediation. Ecotoxicol Environ Saf 113:329–352

Kumar M, Singh L, Kumar A (2020) Bacterial degradation of hydrocarbons: strategies and prospects for bioremediation. Environ Technol Innov 20:101074

Kumar R, Meena M, Sharma V (2021a) Mycoremediation of pharmaceutical pollutants using *Penicillium chrysogenum*: mechanistic insights and efficiency. Environ Technol Innov 22:101458

Kumar R, Singh P, Yadav AN (2021b) Biostimulation: an eco-friendly strategy for enhancing microbial degradation of pollutants. Environ Sustain 4(3):497–506

Kumar R, Sharma M, Yadav AN (2024) Fungal biotechnology for environmental cleanup: genomic insights and future applications. Crit Rev Biotechnol 44(1):22–38

Lee CH, Lee JH, Shin KS (2015) Biodegradation of anthracene by Nostoc muscorum: kinetics and metabolic pathways. J Hazard Mater 296:135–143

Lee H, Kim YM, Cho MH (2016) Biodegradation of PAHs by *Aspergillus flavus* and characterization of its enzymatic system. Environ Technol 37(14):1806–1814

Lee SH, Kim JY, Choi YJ (2022) Genomic insights into organohalide respiration by Dehalococcoides mccartyi in trichloroethylene-contaminated environments. Environ Res 210:112912

Li Y, Wang L, Sun D (2022) Electro-biohybrid systems for sustainable bioremediation and resource recovery. Bioresour Technol 346:126595. https://doi.org/10.1016/j.biortech.2021.126595

Li X, Wang Y, Huang B (2025) Microbial degradation of microplastics: recent progress and future challenges. Sci Total Environ 903:166789

Lindquist S, Craig EA (1988) The heat-shock proteins. Annu Rev Genet 22:631–677. https://doi.org/10.1146/annurev.ge.22.120188.003215

Liu Y, Xu H, Liu L (2019) Biodegradation of atrazine using green microalgae *Chlamydomonas reinhardtii*. Environ Sci Pollut Res 26(14):14281–14289

Löffler FE, Yan J, Ritalahti KM, Adrian L, Edwards EA, Konstantinidis KT et al (2013) *Dehalococcoides mccartyi* gen. nov., sp. nov., obligately organohalide-respiring anaerobic bacteria relevant to halogen cycling and bioremediation, belong to a novel bacterial class, Dehalococcoidia classis nov., within the phylum Chloroflexi. Syst Appl Microbiol 36(7):404–415

Logan BE, Hamelers B, Rozendal R, Schröder U, Keller J, Freguia S et al (2006) Microbial fuel cells: methodology and technology. Environ Sci Technol 40(17):5181–5192. https://doi.org/10.1021/es0605016

Logan BE, Rabaey K, Keller J (2019a) Microbial electrochemical technologies: a new paradigm for biological wastewater treatment. Nat Rev Microbiol 17(5):307–319. https://doi.org/10.1038/s41579-019-0173-3

Logan BE, Rossi R, Ragab A, Saikaly PE (2019b) Electroactive microorganisms in bioelectrochemical systems. Nat Rev Microbiol 17(5):307–319. https://doi.org/10.1038/s41579-019-0153-7

Madaj R, Sobiecka E, Kalinowska H (2018) Lindane, kepone and pentachlorobenzene: chloropesticidesbanned by Stockholm convention. Int J Environ Sci Technol, 15(2):471–480

Malla MA, Dubey A, Yadav S, Kumar A, Hashem A, Abd_Allah EF (2018) Understanding and designing the strategies for the microbe-mediated remediation of environmental contaminants using omics approaches. Front Microbiol 9:1132. https://doi.org/10.3389/fmicb.2018.01132

Manzoor M, Ahmad M, Shahid M (2021) Microalgae-based bioremediation of pollutants: advances and challenges. Bioresour Technol 337:125489

Maphosa F, de Vos WM, Smidt H (2010) Exploiting the ecogenomics toolbox for environmental diagnosticsof organohalide-respiring bacteria. Trends Biotechnol, 28(6):308–316

Martínez ÁT, Ruiz-Dueñas FJ, Camarero S (2018) Biodegradation of pesticides by *Pleurotus ostreatus*: enzymatic pathways and potential applications. Bioresour Technol 265:50–59

Martínková L, Uhnáková B, Pátek M, Nešvera J, Pátek M (2009) Biodegradation potential of the genus *Rhodococcus*. Environ Int 35(1):162–177

Masindi V, Mkhonza P, Tekere M (2021) Sources of heavy metals pollution. In Remediation of heavy metals: 419–454. Cham: Springer International Publishing

Mate DM, Alcalde M (2017) Laccase: a multi-purpose biocatalyst at the forefront of biotechnology. Microb Biotechnol 10(6):1457–1467. https://doi.org/10.1111/1751-7915.12422

Megharaj M, Ramakrishnan B, Venkateswarlu K, Sethunathan N, Naidu R (2017) Bioremediation approaches for organic pollutants: a critical perspective. Environ Int 105:27–45

Menzie CA, Potocki BB, Santodonato J (1992) Exposure to carcinogenic PAHs in the environment. Environ Sci Technol 26(7):1278–1284

Mohanraj J, Subramanian B, Sridhar S (2023) Engineered microalgae for enhanced removal of emerging contaminants in water. J Clean Prod 409:137180

Nagajyoti PC, Lee KD, Sreekanth TVM (2010) Heavy metals, occurrence and toxicity for plants: a review. Environ Chem Lett 8(3):199–216

Nawaz M, Zahir ZA, Khan MY (2018) Decolorization of textile dyes by *Myrothecium verrucaria* via peroxidase production. Int Biodeterior Biodegradation 132:12–18

Nguyen TT, Lee DJ, Chang JS (2017) Removal of chlorpyrifos by green microalga Haematococcus pluvialis: role of metabolic conversion. Algal Res 26:220–227

Nguyen TT, Tran HT, Le CD (2024) Biodegradation of atrazine and lindane by *Sphingobium yanoikuyae*: enzymatic and metabolic analysis. Environ Technol Innov 25:103621

Patel H, Shah V, Kunjadia P (2019) Biodegradation of phenol using *Paecilomyces variotii*: a study on oxidative enzymes. Biodegradation 30(3–4):279–290

Patel V, Sharma A, Pandey D (2020) Biodegradation of phenolic compounds using *Bacillus megaterium*: enzymatic insights and detoxification pathway. Environ Sci Pollut Res 27(16):19810–19820

Patel S, Singh M, Mishra A (2024) Biodegradation of phthalates by microalgae: a sustainable approach for plasticizer pollution. Environ Pollut 330:121694

Paul D, Pandey G, Pandey J, Jain RK (2005) Accessing microbial diversity for bioremediation andenvironmental restoration. TRENDS Biotechnol, 23(3), pp.135–142

Popp J, Pető K, Nagy J (2013) Pesticide productivity and food security. A review. Agron Sustain Dev 33(1):243–255

Priyadarshanee M, Das S (2021) Biosorption and removal of toxic heavy metals by metal tolerating bacteria for bioremediation of metal contamination: a comprehensive review. J Environ Chem Eng 9(4):104686. https://doi.org/10.1016/j.jece.2021.104686

Rabaey K, Verstraete W (2005) Microbial fuel cells: novel biotechnology for energy generation. Trends Biotechnol 23(6):291–298. https://doi.org/10.1016/j.tibtech.2005.04.008

Saidi M, Rahimi M, Shirzad R (2021) Bioplastic production from methane by methanotrophs: a promising sustainable strategy. Environ Sci Pollut Res 28:28946–28960. https://doi.org/10.1007/s11356-021-12852-0

Santiago YR, Nava-Castro KE, Montor JM (2021) Environmental pollution as a risk factor to develop colorectal cancer: the role of endocrine-disrupting chemicals in the inflammatory process. In: Immunotherapy in resistant cancer: from the lab bench work to its clinical perspectives, vol 2. Academic Press, pp 131–148. https://doi.org/10.1016/B978-0-12-821725-0.00007-4

Santos VL, Lopes JMM, Freire DMG (2022) Advances in the use of microbial bioaugmentation for the remediation of polluted environments. Environ Technol Innov 26:102293

Sharma M, Singh R (2024) Environmental contamination by synthetic organic chemicals: challenges and microbial remediation strategies. J Environ Manag 345:118765

Sharma B, Dangi AK, Shukla P (2017) Contemporary enzyme-based technologies for bioremediation: a review. J Environ Manag 210:10–22

Sharma Y, Ghosh PC, Kaushik A (2020) Electroactive microbial remediation technologies: progress and prospects. Sci Total Environ 705:135830. https://doi.org/10.1016/j.scitotenv.2019.135830

Sharma P, Pandey AK, Kim SH, Singh SP, Chaturvedi P, Varjani S (2021) Critical review on microbial community during in-situ bioremediation of heavy metals from industrial wastewater. Environ Technol Innov 24:101826. https://doi.org/10.1016/j.eti.2021.101826

Sharma P, Singh SP, Iqbal HMN, Tong YW (2022) Omics approaches in bioremediation of environmental contaminants: an integrated approach for environmental safety and sustainability. Environ Res 211:113102. https://doi.org/10.1016/j.envres.2022.113102

Sikkema J, de Bont JA, Poolman B (1995) Mechanisms of membrane toxicity of hydrocarbons. Microbiol Rev 59(2):201–222. https://doi.org/10.1128/mr.59.2.201-222.1995

Singh H, Chen Y (2019) White-rot fungi in the biodegradation of environmental pollutants. Int Biodeterior Biodegradation 134:1–21

Singh A, Kuhad RC, Ward OP (2014) Biological remediation of petroleum pollutants. In: Advances in applied bioremediation. Springer, pp 1–20

Singh H, Sharma R, Rana V (2015) Biodegradation of polycyclic aromatic hydrocarbons by white-rot fungus *Phanerochaete chrysosporium*. Environ Sci Pollut Res 22(9):7128–7135

Singh R, Kumar M, Mittal A, Mehta PK (2020) Microbial degradation of herbicides. In: Microbial biodegradation and bioremediation. Elsevier, pp 213–237

Singh H, Kumar P, Yadav AN (2022) Microbial bioventing: a sustainable solution for remediation of hydrocarbon-contaminated soils. Environ Sustain 5(1):89–97

Singh R, Kumar A, Thakur R (2024) Emerging technologies in microbial bioremediation: integration of nanotechnology and biotechnology. Environ Technol Innov 29:102991

Srogi K (2007) Monitoring of environmental exposure to polycyclic aromatic hydrocarbons: a review. Environ Chem Lett 5(4):169–195

Stocker J, Balluch D, Gsell M, Harms H, Feliciano J, Daunert S et al (2003) Development of a set of simple bacterial biosensors for quantitative and rapid field measurements of arsenite and arsenate in potable water. Environ Sci Technol 37(20):4743–4750. https://doi.org/10.1021/es030363p

Strong LC, Rosendahl C, Johnson G, Sadowsky MJ, Wackett LP (2002) Arthrobacter aurescens TC1 metabolizes diverse s-triazine ring compounds. Appl Environ Microbiol 68(12):5973–5980

Sun S, Hu M, Li Y (2020) Removal of carbamazepine by *Ganoderma lucidum*: role of laccase in pharmaceutical degradation. J Clean Prod 260:121031

Suzuki T, Hidaka T, Kumagai Y, Yamamoto M (2020) Environmental pollutants and the immune response.Nat Immunol, 21(12), pp.1486–1495

Tanasupawat S, Moonmangmee S, Tanasupawat D, Yukphan P (2020) Advances in microbial PET degradation by synthetic biology. Curr Opin Biotechnol 66:242–250. https://doi.org/10.1016/j.copbio.2020.09.005

Tariq M, Khan F, Abbas M (2023) Biodegradation and decolorization of synthetic dyes by *Rhizopus oryzae*: a dual action via adsorption and enzymatic degradation. Biotechnol Rep 38:e00791

Tripathi G, Verma P (2004) Endosulfan-mediated biochemical changes in the freshwater fish *Clarias batrachus*. Biomed Environ Sci 17:47–56

Tyagi M, da Fonseca MMR, de Carvalho CCCR (2020) Bioaugmentation and biostimulation strategies to improve the effectiveness of bioremediation processes. Biodegradation 31(4):377–393

U.S. EPA (2013) Engineering issue: in situ treatment technologies for contaminated soil. Office of Solid Waste and Emergency Response

Van den Berg M et al (2006) The 2005 World Health Organization revaluation of human and mammalian toxic equivalency factors for dioxins and dioxin-like compounds. Toxicol Sci 93(2):223–241

van der Meer JR, Belkin S (2010) Where microbiology meets microengineering: design and applications of reporter bacteria. Nat Rev Microbiol 8(7):511–522. https://doi.org/10.1038/nrmicro2392

van Dillewijn P, Wittich RM, Caballero A, Ramos JL (2008) Bioremediation of 2,4,6-trichlorophenol and pentachlorophenol by a recombinant *Pseudomonas putida* strain. Appl Environ Microbiol 74(20):6631–6637. https://doi.org/10.1128/AEM.00871-08

Venkatesan AK, Halden RU (2014) Brominated flame retardants in US biosolids from the EPA nationalsewage sludge survey and chemical persistence in outdoor soil mesocosms. Water res, 55, pp.133–142

Virkutyte J, Sillanpää M, Latostenmaa P (2002) Electrokinetic soil remediation—critical overview. Sci Total Environ 289(1–3):97–121. https://doi.org/10.1016/S0048-9697(01)01051-3

Wang S, Hu Y, Guo G (2010) Biodegradation of phenol and diesel oil by *Acinetobacter calcoaceticus*. J Hazard Mater 179(1–3):1067–1070

Wang X, Guo W, Zhang Y (2015) Atrazine degradation by *Lentinula edodes*: role of ligninolytic enzymes. Environ Sci Pollut Res 22(17):12878–12886

Wang X, Gao Y, Li C (2016) Removal of cresol by marine microalga *Tetraselmis suecica* through bioadsorption and transformation. Mar Pollut Bull 111(1–2):411–417

Wang B, Yu T, Liu Y, Zhang Z (2020) Advances in synthetic biology for enhanced bioremediation. Crit Rev Biotechnol 40(3):300–318. https://doi.org/10.1080/07388551.2020.1730862

Wang L, Zhang Y, Li H (2025) Algal bioremediation: advances in applications and technologies for sustainable environmental cleanup. Sci Total Environ 902:166750

Wen KY, Cameron DE, Collins JJ (2020) A synthetic biology framework for programming eukaryotic transcription functions. Cell Syst 10(6):484–499. https://doi.org/10.1016/j.cels.2020.04.009

White DC, Ringelberg DB, Sutton S (1996) Reductive dechlorination of pentachlorophenol by *Sphingomonas* spp. Appl Environ Microbiol 62(11):4089–4093

WHO (2022) Preventing disease through healthy environments: a global assessment of the burden of disease from environmental risks. World Health Organization

WHO (World Health Organization) (2010) Preventing disease through healthy environments: exposure to highly hazardous pesticides: a major public health concern. World Health Organization

Yadav AN, Verma P, Kumar S (2017) Hydrocarbon-degrading potential of *Aspergillus niger* isolated from petroleum-contaminated soil. 3 Biotech 7(4):246

Yadav KK, Gupta N, Kumar V, Khan SA, Kumar A (2018) A review of emerging trends in bioremediation of petroleum hydrocarbon contaminants. J Environ Manag 225:351–370

Yakimov MM, Timmis KN, Golyshin PN (2007) Obligate oil-degrading marine bacteria. Curr Opin Biotechnol 18(3):257–266

Yancey PH (2005) Organic osmolytes as compatible, metabolic and counteracting cytoprotectants in high osmolarity and other stresses. J Exp Biol 208(15):2819–2830

Yang S-Y, Lai C-Y, Zhao H-P (2025) Trichloroethylene detoxification in low-permeability soil via electrokinetic-enhanced bioremediation technology: long-term feasibility and spatial-temporal patterns. J Hazard Mater 484:136743

Yoshida S, Hiraga K, Takehana T, Taniguchi I, Yamaji H, Maeda Y et al (2016) A bacterium that degrades and assimilates poly(ethylene terephthalate). Science 351(6278):1196–1199. https://doi.org/10.1126/science.aad6359

Zea L, Larsen M, Estante F et al (2020) Potential of space biotechnology for the development of bioremediation processes. Front Microbiol 11:488

Zhang Y et al (2015) Atmospheric deposition of polycyclic aromatic hydrocarbons (PAHs) in agricultural soils around Beijing, China. Environ Pollut 206:23–30

Zhang Y, Li J, Wang X (2020a) Phenol and cresol removal using Mucor *racemosus* through peroxidase-mediated degradation. J Hazard Mater 399:123097
Zhang Y, Angelidaki I, Lu Y (2020b) Enhanced lactate transport and microbial degradation in bio-electrokinetic remediation systems. Bioresour Technol 295:122259. https://doi.org/10.1016/j.biortech.2019.122259
Zhang Y, Wang J, Wang J (2022) Biosorption of crystal violet dye from aqueous solution by *Spirulina maxima*: optimization and kinetics. Environ Res 204:111984
Zhang J, Li Z, Chen Y (2023) Advances in oxygen delivery systems for enhanced bioventing performance. Biotechnol Adv 61:108105
Zhao Y, Liu Y, Xu X (2023) Cometabolic degradation of polycyclic aromatic hydrocarbons by *Rhodococcus qingshengii* and its potential in bioremediation. J Environ Chem Eng 11(1):109815
Zhou Q, Zhang J, Fu J, Shi J, Jiang G (2015) Biomonitoring: an appealing tool for assessment of metal pollution in the aquatic ecosystem. Anal Lett 48(9):1410–1432
Zhou Y, Li X, Xu P (2022) Global pesticide pollution and human health: current scenario and future strategies. Environ Pollut 306:119421
Zhou Y, Li C, Xu P (2023a) Advances in bacterial bioremediation of pharmaceutical pollutants: genomic insights and metabolic pathways. J Hazard Mater 450:131120
Zhou Y, Li C, Xu P (2023b) Genomic and systems-level insights into microbial biodegradation of emerging contaminants. J Hazard Mater 450:131120

Chapter 5
Beyond Conventional Cleanup: The Transformative Potential of Advanced Bioremediation

V. I. Gwa, E. J. Ekefan, and F. A. Lum

Abstract The escalating environmental contamination from industrial activities, oil exploration, and plastic waste necessitates innovative and sustainable remediation approaches. Advanced bioremediation techniques offer eco-friendly solutions by harnessing biological systems to detoxify pollutants. This chapter explores three cutting-edge strategies: phytoremediation of heavy metals, mycoremediation for oil spills, and Clustered Regularly Interspaced Short Palindromic Repeats (CRISPR) based genetically engineered microbes for plastic degradation. Phytoremediation employs hyperaccumulator plants to extract, stabilize, or volatilize toxic heavy metals such as lead, cadmium, and arsenic from contaminated soils and water bodies. Mycoremediation utilizes the enzymatic and absorptive capabilities of fungi, particularly white-rot species, to degrade complex hydrocarbons and restore oil-contaminated environments. Furthermore, the integration of genetic engineering, particularly CRISPR-Cas systems, has enabled the development of tailor-made microbial strains capable of degrading recalcitrant plastics like polyethylene terephthalate (PET) and polyethylene (PE), and polystyrene (PS), thereby accelerating decomposition processes that would otherwise take a prolonged time. These engineered microbes offer enhanced specificity, metabolic efficiency, and adaptability, presenting a promising solution to the global plastic waste crisis. These techniques collectively represent a paradigm shift in environmental biotechnology, offering scalable, low-cost, and sustainable alternatives to conventional physicochemical remediation methods. However, challenges such as ecological risks, biosafety concerns, and field scalability remain to be addressed. Future research should focus on

V. I. Gwa (✉)
Department of Agronomy, Faculty of Agriculture, Federal University of Lafia, Lafia, Nigeria
e-mail: iorungwa@agric.fulafia.edu.ng

E. J. Ekefan
Department of Crop and Environmental Protection, Joseph Sarwuan Tarka University, Makurdi, Benue, Nigeria

F. A. Lum
Kings University, Odeomu, Osun, Nigeria

N. K. Arora, J. Mishra (eds.), *Biotechnological Solutions for a Sustainable Future*, Microorganisms for Sustainability 55,
https://doi.org/10.1007/978-981-95-3989-5_5

optimizing these approaches through interdisciplinary integration, risk assessment frameworks, and field validation to ensure environmental and public safety.

Keywords Bioremediation · CRISPR · Heavy metals · Mycoremediation · Oil spills · Phytoremediation · Plastic degradation

5.1 Introduction

The rapid pace of industrialization and urbanization, coupled with unsustainable human activities, has led to the widespread release of hazardous pollutants—including pesticides, heavy metals, hydrocarbons, and synthetic organic compounds—posing severe threats to ecosystems, biodiversity, and human health. Conventional remediation techniques like incineration, chemical treatment, and landfilling frequently face significant limitations, such as high costs, substantial energy demands, risks of secondary contamination, and limited effectiveness in complex environmental settings (Sharma et al. 2018). In response to these challenges, advanced bioremediation techniques have emerged as environmentally benign, economical, and sustainable alternatives for cleaning up contaminated sites.

Bioremediation harnesses the natural metabolic capabilities of microorganisms, plants, fungi, and algae to degrade, detoxify, or immobilize environmental contaminants. Strategies can be broadly categorized as in situ (e.g., bioventing, biosparging, phytoremediation) or ex situ (e.g., composting, land farming, bioreactors), selected based on site-specific conditions and pollutant characteristics (Azubuike et al. 2016; Jacob et al. 2018). However, traditional bioremediation often encounters constraints related to microbial efficiency, pollutant bioavailability, and adverse environmental conditions (Sharma 2020).

Recent breakthroughs in biotechnology, synthetic biology, and systems ecology are driving the development of enhanced bioremediation approaches designed to overcome these limitations. This expanding toolkit includes phytoremediation for heavy metals, mycoremediation for oil spills, and CRISPR-based genetic engineering of microbes for plastic degradation—all offering environmental safety and adaptability across diverse contaminants and ecosystems (Dangi et al. 2019). Complementary strategies further enriching this field include genetically engineered microorganisms (GEMs), nano-bioremediation, bioaugmentation, biosurfactants, and omics-based approaches (metagenomics, transcriptomics, proteomics) for comprehensive process understanding and optimization (Kumar et al. 2023).

Phytoremediation specifically utilizes plants, particularly hyperaccumulators, to extract, stabilize, or detoxify pollutants like cadmium (Cd), lead (Pb), mercury (Hg), and arsenic (As) from soil and water (Verbruggen et al. 2009). Its efficacy is enhanced through rhizosphere engineering, symbiotic microbial inoculation, and its application in soils co-contaminated with heavy metals and persistent organic pollutants. Synergistic effects improving metal immobilization and soil restoration are observed when phytoremediation is combined with bioaugmentation using rhizospheric microbes (Anza et al. 2019). Genetic modifications, such as the overexpression of metal transporter genes and antioxidant enzymes in model plants like *Arabidopsis thaliana* and *Brassica juncea*, are significantly advancing metal uptake and tolerance capabilities (Yadav et al. 2022).

Mycoremediation employs fungi, especially white-rot species like *Pleurotus ostreatus* and *Phanerochaete chrysosporium*, to degrade and detoxify pollutants, particularly petroleum hydrocarbons and polycyclic aromatic hydrocarbons (PAHs) (Akpasi et al. 2023). Fungi produce potent extracellular enzymes (e.g., laccases, peroxidases, lipases) capable of breaking down complex hydrocarbons (Shanmugapriya et al. 2019). Genomic tools have enabled the identification and manipulation of key microbial genes for xenobiotic degradation, leading to the development of fungal-bacterial consortia and immobilized fungal biocatalysts with enhanced catabolic pathways (Rezaei and Moghimi 2024). Furthermore, nanobioremediation—the integration of nanotechnology (e.g., metal-oxide nanoparticles, carbon-based nanostructures) with bioremediation—shows promise in increasing pollutant bioavailability and boosting microbial degradation rates (Kumar et al. 2021).

Addressing the persistent challenge of plastic pollution like Polyethylene (PE), Polyethylene Terephthalate Polyethylene (PET) and Polystyrene (PS) biodegradation offers a viable solution where natural microbial degradation is often slow. CRISPR-Cas9 genome editing has enabled the development of microbial strains with significantly enhanced enzymatic capacities for plastic degradation (Palit et al. 2025). For instance, CRISPR-mediated overexpression of PETase and MHETase enzymes from *Ideonella sakaiensis* has dramatically accelerated PET breakdown (Yoshida et al. 2021). Synthetic biology approaches further facilitate the design of optimized microbial consortia and metabolic pathways to convert plastic monomers into valuable bioproducts, offering scalable, circular solutions for plastic waste management (Wu et al. 2025).

The convergence of plant science, fungal biotechnology, and genetic engineering represents a transformative shift in environmental remediation. This chapter focuses specifically on three pivotal advanced bioremediation strategies: phytoremediation for heavy metals, mycoremediation for oil spills, and CRISPR-based genetic engineering of microbes for plastic degradation. These approaches exemplify environmentally benign, adaptable, and sustainable solutions to diverse pollution challenges, offering comprehensive insights into the future of effective pollution management.

5.1.1 Heavy Metal Pollution, Human Health and Hyperaccumulator Plant

Heavy metal contamination represents one of the most significant environmental and public health challenges of the modern era, with profound implications for human health through multiple exposure pathways, while simultaneously offering opportunities for environmental restoration through the strategic use of hyperaccumulator plants in phytoremediation technologies. Heavy metals pose severe health risks to humans through their ability to cause oxidative stress, cellular damage, and systemic toxicity, even at relatively low concentrations (Chollom et al. 2023; Jomova et al. 2025). These toxic elements, including arsenic (As), mercury (Hg), cadmium (Cd), chromium (Cr), copper (Cu), lead (Pb), tin (Sn) nickel (Ni), and zinc (Zn), can disrupt essential biological processes by displacing vital metals in enzyme systems and pigment synthesis (Balali-Mood et al. 2021). The mechanisms of heavy metal toxicity are multifaceted, involving the generation of reactive oxygen species (ROS) that cause oxidative stress, leading to damage of cellular components, including DNA, proteins, and lipids (Table 5.1). Heavy metals interfere with antioxidant defence mechanisms, enzyme inactivation, and cellular processes such as growth, proliferation, differentiation, and apoptosis (Henkler et al. 2010). These metals can cause both carcinogenic and non-carcinogenic health effects, with As, Cd, Cr, and Ni classified as Group 1 carcinogens by the International Agency for Research on Cancer (Kim et al. 2015; Manić et al. 2022). Bioaccumulation through the food chain presents a critical pathway for human exposure, as heavy metals accumulate

Table 5.1 Major toxic metals, affected organ systems, molecular targets, and associated health risks

Metal	Primary organ systems	Key molecular targets	Consensus risk level	Reference
Arsenic	Skin, liver, lung, bladder	Inhibits PDH complex; induces ROS via arsenite-thiol binding	Group 1 carcinogen	Platanias (2009) and Jomova et al. (2025)
Cadmium	Kidney, bone, prostate	Competes with Ca^{2+}; up-regulates metallothionein; induces Nrf2 signalling dysregulation	Probable carcinogen	Yang and Shu (2015) and Yan and Allen (2021)
Lead	CNS, cardiovascular	Mimics Ca^{2+} in voltage-gated channels; disrupts dopamine release	Developmental neurotoxin	Engwa et al. (2019)
Chromium (VI)	Respiratory, renal	DNA cross-linking after Cr-DNA adducts	Group 1 carcinogen	Macfie et al. (2010)
Copper	Liver, brain	Exceeds ceruloplasmin capacity → Fenton chemistry	Risk at >10 mg day^{-1}	Spasojević et al. (2010)
Zinc	GI tract, immune	Competes with Cu/Zn-SOD, causing immune suppression at >200 mg day^{-1}	Moderate	Osredkar and Sustar (2011)
Mercury	CNS, renal, foetal brain	Binds seleno-enzymes; disrupts microtubule assembly	High	Bizerea et al. (2018)
Nickel	Respiratory, skin	Histone modification → epigenetic silencing	Group 1 carcinogen	Sun et al. (2013)

in plant tissues and subsequently enter the human food supply. Studies have demonstrated that heavy metals can concentrate in edible parts of crops, posing significant health risks to consumers through dietary intake. The bioaccumulation process is particularly concerning because these metals are persistent in the environment and can biomagnify as they move up the food chain.

Phytoremediation emerges as a sustainable solution to heavy metal contamination, utilizing specialized plants known as hyperaccumulators that can absorb, concentrate, and tolerate high levels of toxic metals (Zhakypbek et al. 2024). These plants, numbering approximately 450 taxa globally, possess unique physiological and biochemical mechanisms that enable them to accumulate abnormally high concentrations of potentially toxic trace elements (Skuza et al. 2022). The selection of non-edible hyperaccumulators is crucial for safe and effective phytoremediation, as using edible crops could introduce heavy metals into the food chain and pose health risks to humans and animals. This strategic approach ensures that the remediation process does not compromise food safety while effectively removing contaminants from the environment. There are various types of phytoremediation techniques that serve specific remediation purposes:

5.1.2 Phytoextraction (Phytoaccumulation)

Phytoextraction, also known as phytoaccumulation, photoabsorption, or phytosequestration, is the process by where plants uptake metals via their roots and transport them to aboveground shoots (Lasat 2002). Harvesting these shoots (and sometimes roots) removes the contaminants from the soil or water for disposal or recycling (Pivetz 2001). This approach is often considered more environmentally friendly than conventional remediation methods.

There have been numerous plant species identified capable of phytoextraction. For instance, studies on sludge-grown plants identified several native species accumulating iron, with *Amaranthus spinosus* L. specifically accumulating Zn and Mn in its shoots (Chinmayee et al. 2012). Most tested plants accumulated Fe, Mn, and Zn primarily in leaves, except *A. spinosus* L. and *Ricinus communis* (Lajayer et al. 2019). Ornamental plants also show potential for HM phytoextraction in contaminated habitats. High-biomass crops like *Brassica juncea* (Rani et al. 2023), *Helianthus annuus*, *Cannabis sativa*, *Nicotiana tabacum*, and *Zea mays* (Meers et al. 2005; Herzig et al. 2014) are effective, especially when bioavailability is enhanced using chelating agents like EDTA. Grasses are also suitable due to their short life cycle, rapid growth, high biomass, and tolerance to abiotic stress (Sladkovska et al. 2022). For example, *Trifolium alexandrinum* is effective for Cd, Pb, Cu, and Zn phytoextraction because of its fast growth, pollution tolerance, high biomass, and capacity for multiple harvests (Abuzaid et al. 2021). Aquatic plants also demonstrate efficacy. A meta-analysis of Cu, Zn, and Cd absorption capacities has been reported in the Tamaricaceae, Frankeniaceae, Poaceae, and Chenopodiaceae families of aquatic plants (Li et al. 2015). Phytoextraction of metals (Fe, Zn, Cr, Ni,

Cd) from municipal wastewater using various plants has also been documented (Irshad et al. 2015). Makarova et al. (2021) demonstrated that iron chelate, plant growth regulators, and potassium salt of hydroxyethylidene diphosphonic acid (K2HEDP) significantly improved Cd phytoextraction by *Trifolium repens*. Furthermore, recovering HMs from harvested biomass offers a sustainable decontamination and recycling strategy. However, phytoextraction has limitations. It is often site-specific and time-consuming, requiring repeated harvests over years or decades. A significant concern is the potential for contaminants to enter the food chain if accumulator plants are consumed, as highlighted by Ugya et al. (2019), who found significant Cd and Hg accumulation in vegetables grown on soils irrigated with refinery effluent. Certain plant species, such as the fern *Pteris vittata* (Chinese brake fern), that are mainly As accumulators can also metabolize other toxic metals. For instance, in a study, *Pteris vittata* L., known arsenic hyperaccumulators, are discovered for the first time to also hyperaccumulate toxic thallium (Tl) in Tl-As co-contaminated soils from China. The ferns accumulated extremely high levels of both As (up to 11,110 mg/kg) and Tl (up to 111 mg/kg) from the soil, showing high bioaccumulation factors (BCF Tl = 28.4 > As = 7.8), offering significant potential for phytoremediation of Tl-As co-contaminated soils globally (Wei et al. 2020).

5.1.3 Phytostabilization

Phytostabilization entails the development of vegetative cover over contaminated areas to limit contaminant mobility in the vadose zone by facilitating root uptake or promoting immobilization processes within the rhizosphere, thereby minimizing the risk of contaminant dispersion beyond the site (Radziemska et al. 2017). Metal-tolerant plants are mainly used in phytostabilization to immobilize heavy metals in contaminated soils, reducing their environmental bioavailability (Sharma et al. 2023). In a study Luo et al. (2019) showed the viability of phytostabilizing zinc smelting slag by combining four woody plants with organic supplements. According to the findings, growing plants directly in the slag from zinc smelting enhanced nutrient accumulation and reduced the bioavailability of heavy metals (Cu, Zn, and Cd). Another study by Saran et al. (2020) reported the potential of Helianthus petiolaris for phytostabilization by thriving in soils with 1000 mg/kg Pb^{2+} and 50 mg/kg Cd^{2+}. It accumulated over three times the soil Cd concentration in its aerial parts and translocated notable amounts of Pb to the shoots when exposed to soil contaminated with up to 500 mg/kg Pb. In a different analysis, Galal et al. (2017) revealed the potential of emerging macrophytes like *Vossia cuspidata* as a phytoremediator to absorb heavy metals from contaminated water sources. The concentrations of the examined HMs in the different plant tissues of *V. cuspidate* showed significant seasonal change, with HMs building up in larger quantities in below-ground sections compared to aboveground shoots.

Vegetal grass sown on mine tailings successfully stabilized Pb and Zn pollutants, lowering leaching and erosion while increasing vegetative cover Wang et al. (2020).

These regions can thereby develop green spaces and provide ecosystem services. Soil amendments like organic matter (e.g., composted manure), treated biosolids, microbial inoculants, lime, and biochar can significantly improve the efficiency of phytostabilization by modifying soil properties and promoting plant-metal interactions (Sigua et al. 2019; Ke et al. 2021; Kacprzak et al. 2014). Commonly utilized plants for this technique include *Festuca arundinacea* (tall fescue), *Vetiveria zizanioides* (vetiver grass), and *Populus* spp. (poplar) (Vangronsveld et al. 2009). Its key advantages are preventing erosion and metal leaching, making it ideal where plant uptake is undesired (e.g., near food crops or water sources). However, its primary limitations are the inability to remove metals permanently and the requirement for ongoing monitoring.

5.1.4 Phytovolatilization

Phytovolatilization is a phytoremediation process wherein plants absorb contaminants (primarily certain heavy metals/metalloids or organic compounds) from the soil, transform them into volatile species within their tissues, and subsequently release these gaseous forms into the atmosphere through transpiration via the foliar system (Awa and Hadibarata 2020). The process involves root uptake, enzymatic conversion to volatile compounds (often less toxic or more mobile), and foliar emission. It is particularly applicable to elements like Se, Hg, and As. Plants, notably species within the Brassicaceae family like *Brassica juncea* (Indian mustard) and *Brassica napus* (rapeseed), metabolize inorganic selenium (e.g., selenate, selenite) into organic selenoamino acids (selenocysteine, selenomethionine), which are further converted to dimethylselenide (DMSe), a volatile and less toxic gas (Awa and Hadibarata 2020; Bañuelos and Dhillon 2011). Few plants can absorb ionic mercury (Hg^{2+}) via roots or foliage. Within plant cells, Hg^{2+} is enzymatically reduced (e.g., by mercuric reductase) to elemental mercury (Hg^0), which is highly volatile and diffuses out through the leaves (Rajendran et al. 2024).

The primary advantage of phytovolatilization is its ability to remove contaminants from the soil without the need for physical harvesting and disposal of contaminated plant biomass. However, a significant disadvantage is that contaminants are not destroyed but transferred to the atmosphere, posing risks of secondary contamination via deposition elsewhere (Awa and Hadibarata 2020). This technique is only applicable to contaminants readily converted into volatile forms, such as Se and mercury Hg. While limited As methylation can occur, phytovolatilization is generally inefficient and not a primary removal pathway for As compared to phytoextraction. It is ineffective for non-volatile heavy metals like Pb or Cd. Careful assessment is crucial to ensure atmospheric concentrations of volatilized contaminants remain within safe regulatory limits and do not pose unacceptable risks to human health or ecosystems downwind (Egendorf et al. 2020).

5.1.5 Phytofiltration

In phytofiltration, pollutants, particularly heavy metals and radionuclides, are absorbed, adsorbed, or precipitated from contaminated water using plant roots (rhizofiltration), shoots (caulofiltration), or seedlings (blastofiltration) (Hanus-Fajerska and Kozminska 2016). Phytofiltration has more utility in treating groundwater and wastewater (Hanus-Fajerska and Kozminska 2016). Heavy metals are either absorbed by the roots or adsorbed onto the root surface during rhizofiltration. To further reduce the transport of heavy metals to subterranean water, root exudates can alter the pH of the rhizosphere, which causes heavy metals to precipitate on plant roots (Verma et al. 2006). It requires plants with deep, fibrous roots and can be done in hydroponic or floating systems. *Helianthus annuus* has the potential to remove Pb and As from contaminated water, as shown by Alam et al. (2023), with removal efficiencies of up to 65%. African Medicago sativa (alfalfa) shoots demonstrated strong potential for phytofiltration of heavy metals from aqueous solutions. Batch and column experiments revealed rapid and efficient binding of metals like Cd(II), Cr(III), Pb(II), and Zn(II), with optimal uptake at pH ~5. Silica-immobilized alfalfa shoots effectively removed metals under flow conditions, with over 90% recovery for most ions using mild acid treatment (Gardea-Torresdey et al. 1998). Rhizofiltration not only removes heavy metals but also significantly reduces pathogens. A pilot system using Phragmites australis and Kyllinga nemoralis in South Africa outperformed an unplanted control, lowering BOD, COD, turbidity, and suspended solids (Odinga et al. 2019). The planted section showed higher metal uptake and removed 45–98% of pathogens, including 77% of *Ascaris lumbricoides* eggs (Odinga et al. 2019). This highlights the dual effectiveness of macrophyte-based rhizofiltration for treating co-contaminated municipal wastewater. Limitations of phytofiltration include the need for regular water flow and upkeep, as well as the requirement that plants be harvested and disposed of appropriately.

5.1.6 Phytodesalination

Soil salinization poses an escalating threat to agricultural productivity, particularly in arid and semi-arid regions. Phytodesalination, utilizing salt-tolerant halophyte plants, offers a promising biological strategy to mitigate this problem (Ahmadi et al. 2022). Halophytes possess unique physiological adaptations enabling them to remediate saline soils and water bodies. They absorb excess salts, accumulating them within specialized tissues or vacuoles, or excreting them via salt glands (Manousaki and Kalogerakis 2011). This process effectively reduces soil electrical conductivity (EC), improves soil structure, and aids in the recovery of degraded land (Rahman et al. 2021). The success of phytodesalination hinges critically on selecting appropriate halophyte species capable of thriving under high salinity. Promising candidates frequently employed include *Atriplex halimus*, *Salicornia europaea*, *Tamarix* spp., *Phragmites australis*, and *Suaeda maritima* (Tallou et al. 2024). A field study on mine tailing reclamation tested tilling with marble waste, pig slurry, and their combination. After 5 years, Atriplex halimus successfully colonized all plots, with MW raising pH and enhancing metal uptake, especially Pb and Cd in leaves (Acosta et al. 2018). These results highlight the plant's strong potential for effective use in saline, metal-contaminated sites due to its dual tolerance to salinity and capacity for metal removal. The core mechanisms driving this remediation involve efficient salt uptake and compartmentalization within plant vacuoles to minimize cellular toxicity (Ammu et al. 2025). Research advancements are enhancing phytodesalination efficiency. These include developing genetically modified plants with improved salt tolerance and uptake capacity and employing soil amendments like biochar alongside microbial inoculants to optimize soil conditions and boost plant health and productivity (Hussain et al. 2018). Furthermore, repeated phytodesalination cycles also lead to progressive improvements in overall soil quality contaminated with metal and salts (Lastiri-Hernândez et al. 2021).

Despite its promise, phytodesalination faces several significant challenges (Fig. 5.1). The salt-laden biomass produced during the process necessitates appropriate disposal strategies to prevent recontamination of the environment upon decomposition. Furthermore, the inherently slow growth rates of some effective halophyte species can limit the speed and scalability of remediation efforts. Another key constraint is salt accessibility, as deeply entrenched or strongly bound salts within the soil profile may be less available for plant uptake. Finally, the effectiveness of phytodesalination is highly dependent on site-specific environmental conditions, including climate, soil composition, and water availability (Ammu et al. 2025).

5.1.7 Microbial-Assisted Phytoremediation

Microorganisms serve as critical bioenhancers in phytoremediation systems, operating through three interconnected primary mechanisms that collectively transform plant-based remediation from a passive to an active process (Thijs et al. 2016). The first mechanism involves bioavailability modulation, where microbes secrete organic acids such as citrate and oxalate while facilitating redox transformations that convert metals into more plant-accessible forms (Rasool et al. 2021; Li et al. 2025). Simultaneously, these microorganisms enhance plant physiological functions through the production of phytohormones like IAA and ACC deaminase, coupled with nutrient solubilization that strengthens overall plant health and resilience (Panchal et al. 2021). The third mechanism encompasses direct detoxification pathways, where microbes employ sophisticated enzymatic processes, including reduction, methylation, and sequestration, to neutralize metal toxicity. (Chatterjee et al. 2020). These synergistic interactions create what researchers term the "rhizosphere effect," a dynamic zone of enhanced microbial activity that can dramatically increase metal uptake efficiency many-fold compared to plants functioning independently (Barra Caracciolo and Terenzi 2021).

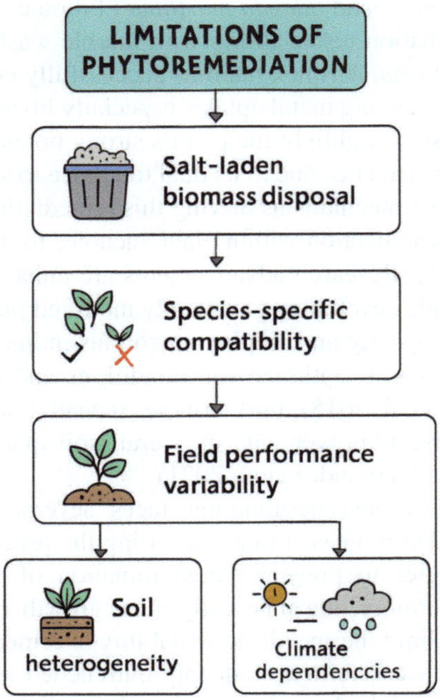

Fig. 5.1 Major limitations of phytoremediation technologies. Key constraints include: (1) Disposal challenges for salt-laden biomass generated during phytodesalination; (2) Species-specific compatibility requirements that limit plant selection; (3) Performance variability between controlled laboratory settings and field applications; (4) Soil heterogeneity affecting contaminant distribution and root penetration; (5) Climate dependencies influencing plant growth rates and treatment efficacy. These factors collectively impact the scalability and reliability of phytoremediation implementations across diverse environments

The effectiveness of this microbial enhancement stems from distinct functional groups, each contributing specialized capabilities to the remediation process. Endophytes operate within plant tissues to facilitate intra-plant metal compartmentalization, thereby reducing phytotoxicity while enhancing the translocation of metals from roots to shoots (Ekta et al. 2016). Plant growth-promoting rhizobacteria (PGPR) complement this internal activity by producing siderophores and solubilizing phosphates in the external rhizosphere environment, which simultaneously improves plant biomass production and mobilizes previously unavailable metals (Mishra et al. 2017). Mycorrhizal fungi extend this beneficial network through their extensive hyphal systems, effectively expanding the plant's rhizosphere exploration capacity and creating broader zones of metal uptake (Miransari 2011). Metal-reducing bacteria add another layer of functionality by enzymatically altering metal valence states, such as converting highly toxic Cr^{6+} to less harmful Cr^{3+}, which subsequently reduces overall toxicity while promoting metal precipitation and stabilization (Wu et al. 2016).

Studies reveal the sophisticated mechanistic insights underlying these plant-microbe partnerships across various heavy metal contamination scenarios. In arsenic-contaminated systems, the symbiotic relationship between *Betula celtiberica* and *Rhodococcus erythropolis* demonstrates remarkable efficiency, achieving 35% arsenic removal through the bacterial expression of arsenate reductase (arsC) genes that transform As(V) to the more mobile As(III) form for enhanced plant uptake (Mesa et al. 2017). This bacterial activity works in concert with *Ensifer adhaerens,* which upregulates phytochelatin synthesis in *Brassica juncea* roots, creating a comprehensive system where bacterial mobilization is coupled with enhanced plant sequestration mechanisms (Mesa et al. 2017). The chromium remediation process showcases equally sophisticated microbial contributions, where *Bacillus xiamenensis* PM14 enhances chromium uptake in *Canna indica* by 56% through the strategic production of biosurfactants, particularly surfactin, which disrupts chromium-soil complexes and increases metal bioavailability. Concurrently, this bacterial strain induces the plant's antioxidant defence systems, including superoxide dismutase (SOD) and catalase (CAT), which collectively mitigate oxidative stress during the intensive phytoextraction process (Din et al. 2020). This dual-action approach exemplifies how microbes simultaneously enhance metal uptake while protecting plant health. Copper remediation systems demonstrate the critical role of bacterial metabolites in both metal mobilization and plant stress management. *Paenibacillus mucilaginosus* increases copper bioavailability in *Medicago sativa* rhizospheres through targeted gluconic acid secretion, which chelates and solubilizes copper ions. The bacterial ACC deaminase activity provides essential stress mitigation by reducing ethylene levels, resulting in a remarkable 76% increase in root biomass that directly correlates with enhanced metal accumulation capacity (Ju et al. 2019). This integrated approach highlights how microbial partners address both the chemical and physiological challenges of metal phytoremediation.

Mercury remediation presents unique challenges that require specialized microbial capabilities, effectively demonstrated by the *Serratia marcescens* BacI56 partnership with *Juncus maritimus*. This endophytic bacterium expresses merA operons directly within plant tissues, catalyzing the critical Hg^{2+} to Hg^0 volatilization process while preventing harmful phloem retro-translocation that could redistribute mercury throughout the plant system. This sophisticated mechanism achieved 62% soil mercury reduction while maintaining plant health and preventing secondary contamination (Mello et al. 2020). Lead contamination remediation showcases the power of fungal-plant partnerships, where *Aspergillus flavus* enhances lead accumulation in *Helianthus annuus* through the production of pyoverdine siderophores that form stable lead-chelate complexes. The fungal contribution extends beyond metal mobilization, as melanin production creates protective barriers around root membranes, shielding them from metal-induced stress and enabling the plant to achieve lead uptake rates five times higher than uninoculated controls (Manzoor et al. 2021). Multi-metal contamination scenarios require particularly versatile microbial partners capable of addressing diverse metal types simultaneously. *Bacillus cereus* T1B3 exemplifies this adaptability by employing divergent strategies tailored to specific metals: for cadmium and Zn contamination, it induces metallothionein expression in *Vetiveria zizanioides*, while for chromium(VI), it facilitates enzymatic reduction to the less toxic chromium (III) form with an impressive 211% efficiency improvement. This microbial versatility proves essential for addressing the complex polymetallic contamination scenarios commonly encountered in real-world remediation projects (Nayak et al. 2018).

5.1.8 Mycoremediation

Mycoremediation—the application of fungi for environmental bioremediation—represents a sophisticated and ecologically sustainable approach to removing or transforming toxic contaminants from polluted ecosystems (Gunyar and Uztan 2021). Through synergistic enzymatic, physicochemical, and metabolic mechanisms, fungi effectively degrade hydrocarbons and immobilize heavy metals, leveraging their unique biological architecture and metabolic versatility (Dinakarkumar et al. 2024).

Fungi demonstrate exceptional bioremediation capabilities through their expansive hyphal networks, which function as highly sophisticated biological filtration systems capable of addressing complex environmental contamination scenarios (Deshmukh et al. 2016). These intricate mycelial structures operate through complementary mechanisms that maximize contaminant removal efficiency across diverse environmental conditions (Espinosa-Ortiz et al. 2022). The chitin-rich fungal cell walls contain specialized functional groups, including carboxyl, amino, and phosphate moieties, that effectively capture low-molecular-weight hydrocarbons and metallic contaminants through electrostatic and chemical binding interactions (Saravanan et al. 2021). The extensive mycelial networks provide a critical advantage by dramatically expanding rhizosphere accessibility, enabling these organisms to intercept and process contaminants across heterogeneous environmental matrices that remain inaccessible to conventional remediation approaches (Singha and Shukla 2023). Surface-level capture is complemented by active intracellular uptake mechanisms that transport contaminants into fungal cells, where sophisticated sequestration occurs through metallothionein-mediated vacuolar storage or direct enzymatic transformation pathways that neutralize toxic compounds (Robinson et al. 2021). This spatial advantage proves particularly valuable in complex contaminated sites where pollutants exist in varied concentrations and chemical forms across different soil layers and microenvironments.

Most remarkably, fungi possess the metabolic flexibility to dynamically shift between oxidative and reductive pathways in response to changing environmental conditions, making them exceptionally versatile tools for comprehensive pollutant cleanup strategies (Deshmukh et al. 2016). Under aerobic conditions, fungi deploy powerful enzymatic arsenals, including monooxygenases and dioxygenases, that efficiently oxidize complex hydrocarbons, transforming compounds such as benzene into more manageable intermediates like catechol for further metabolic processing (Rani et al. 2024). When environmental conditions shift to anaerobic states, metal-reducing fungi such as *Aspergillus niger* demonstrate remarkable adaptability by utilizing toxic metals like hexavalent Cr as alternative electron acceptors, employing specialized chromate reductases to convert highly mobile and toxic Cr(VI) into the substantially less harmful and immobile Cr(III) form (Viti et al. 2014).

Hybrid remediation systems represent the ultimate expression of fungal bioremediation sophistication. Species such as *Aspergillus*, *Penicillium*, and *Candida* integrate multiple remediation strategies by coupling biosurfactant production with enzymatic degradation processes, achieving remarkable removal of polycyclic aromatic hydrocarbons even in challenging oxygen-limited soil environments where conventional approaches typically fail (Kha Al-Hawash et al. 2018). These fungi produce biosurfactants (e.g., surfactin) that emulsify hydrophobic compounds and chelate metals via electrostatic interactions, reducing interfacial tension to enhance contaminant bioavailability for enzymatic attack and cellular uptake (Karim et al. 2025; Kaczorek et al. 2018).

White-rot fungi, including key species like *Phanerochaete chrysosporium*, *Pleurotus ostreatus*, and *Trametes versicolor*, are renowned for their exceptional bioremediation capabilities, particularly in degrading persistent and complex environmental pollutants such as those found in crude oil and contaminated soils (Al-Hawash et al. 2018; Echezonachi 2022). This unique aptitude stems from their secretion of powerful extracellular ligninolytic enzymes—specifically laccase (Lac), manganese peroxidase (MnP), and lignin peroxidase (LiP). These enzymes act synergistically to oxidatively cleave recalcitrant hydrocarbon structures, including polycyclic aromatic hydrocarbons (PAHs), alkanes, and resins, transforming them into less toxic metabolites (Pozdnyakova 2012; Al-Hawash et al. 2018). The enzymatic process generates highly reactive free radicals that destabilize the molecular architecture of pollutants, initiating biochemical pathways that can lead to their complete mineralization (Barr and Aust 1994; Kumar et al. 2024). This combination of broad substrate specificity and potent enzymatic activity underpins their diverse industrial applications in environmental cleanup."

A critical advantage of white rot fungi in bioremediation is their extensive mycelial network. This network physically penetrates contaminated soil matrices, including hard-to-reach micropores, significantly enhancing the distribution of enzymes and the overall efficiency of degradation (Al-Hawash et al. 2018). In a study white rot fungus, particularly *Agaricus bisporus*, effectively degraded petroleum hydrocarbons in contaminated soil using spent mushroom compost as inoculum. The bioremediation efficiency increased over time, with A. bisporus achieving the highest degradation rate (71.5%). Gas chromatography and ecotoxicity tests confirmed reduced hydrocarbon levels and toxicity in Esfahan Refinery soil after 3 months (Mohammadi-Sichani et al. 2019). Beyond hydrocarbon degradation, the mycelial network of *P. ostreatus* enhanced soil structure. For instance, in a study fresh spent mushroom substrate (SMS) from *P. ostreatus* significantly altered soil micromorphology in Brazilian farmlands, enhancing both topsoil and subsoil structure. The amendment increased porosity, complexity, and aggregate formation, particularly developing granular and spongy microstructures. These changes suggest that SMS directly and indirectly improves soil formation through interactions with soil fauna and fungal hyphae (Nakatsuka et al. 2016). Similarly, in another study treatment with *T. versicolor* has proven effective in hydrocarbon-contaminated environments. Its hyphal penetration is particularly beneficial in compacted areas (e.g., sandy loam soil), facilitating PAH breakdown. Additionally, this species contributes to ecological recovery by enhancing nitrogen cycling and stabilizing soil aggregates (Majcherczyk et al. 1998).

This multifaceted approach demonstrates how fungi can simultaneously address bioavailability limitations through surfactant production while providing the enzymatic machinery necessary for complete contaminant mineralization, representing a comprehensive biological solution to complex environmental contamination challenges.

5.2 Synergistic Interactions with Other Microorganisms

Synergistic interactions between microorganisms in contaminated environments represent cooperative relationships where the combined metabolic activities of multiple species significantly enhance the breakdown of complex petroleum hydrocarbons compared to individual efforts. Fungi, particularly white-rot species, frequently partner with bacteria and actinomycetes (Al-Hawash et al. 2018). In these partnerships, fungi initiate degradation by oxidizing complex hydrocarbons like PAHs, resins, and asphaltenes into smaller, more manageable intermediates. Bacteria then mineralize these intermediates to CO_2 and H_2O. (Prenafeta-Boldú et al. 2019). This division of labour is crucial because crude oil's intricate mixture of alkanes, aromatics, resins, and asphaltenes exceeds the metabolic capabilities of any single microorganism; consequently, synergistic microbial consortia achieve broader substrate coverage and more efficient overall degradation (Cao et al. 2022). Practical bioremediation strategies explicitly apply this synergy: Bioaugmentation involves inoculating contaminated sites with pre-adapted microbial consortia (e.g., fungi combined with hydrocarbon-degrading bacteria) (Mrozik and Piotrowska-Seget 2010) while Biostimulation enhances native synergistic populations by adding nutrients or surfactants. Engineered systems like constructed wetlands and bioreactors are also designed to optimize conditions for these cooperative communities (Paul et al. 2025). Examples of effective synergies include bacterial consortia like *Pseudoxanthomonas* sp. *S1-2, Bacillus* sp. *S2-A, Dietzia* sp. *CN-3, and Acinetobacter* sp. *(HC8-3S)*, (Chen et al. 2024), fungal-bacterial partnerships where fungi modify heavy compounds and provide "fungal highways" for bacterial mobility (Buda 2021), and even partnerships with cyanobacteria that supply oxygen via photosynthesis (Hadiyanto et al. 2025). The benefits of these interactions accelerated degradation rates, an expanded range of degradable hydrocarbons, increased resilience to environmental stresses (e.g., salinity, pH shifts), and reduced accumulation of toxic intermediates through more complete mineralization (Lebeau 2011). Future advancements depend on omics-based tools (metagenomics, metabolomics) to decipher intricate interspecies communication and metabolic networks, enabling the rational design of engineered consortia with predictable functions (Lea-Smith et al. 2025). However, key challenges remain, particularly in ensuring the stability and performance of synthetic consortia within the dynamic and competitive conditions of real contaminated environments.

5.3 CRISPR-Engineered Microbes for Plastic Degradation

Global plastic pollution, particularly from recalcitrant polymers like polystyrene (PS), polyethylene terephthalate (PET), and polyethylene (PE), poses a severe environmental threat (MacLeod et al. 2021). While natural microbial degradation pathways exist, their efficiency is often insufficient for remediation (Reineke and Schlömann 2023; Palit et al. 2025). Advances in genetic engineering, specifically Clustered Regularly Interspaced Short Palindromic Repeats (CRISPR)-based genome editing, now enable the precise enhancement of microbial plastic-degrading capabilities (Saini et al. 2024).

Key enzymes like Polyethylene terephthalate hydrolase (PETase) and Mono(2-hydroxyethyl) terephthalate hydrolase (MHETase), originally discovered in *Ideonella sakaiensis*, are prime targets for engineering (Ermis 2025). These enzymes have been successfully expressed heterologously in model hosts like *Escherichia coli* to facilitate PET degradation (Aytan et al. 2025). For instance, PETase secretion in *E. coli* BL21 using the pelB signal peptide enables hydrolysis of PET model compounds like bis(2-hydroxyethyl) terephthalic acid (BHET) and amorphous PET films (Shi et al. 2021). Cytoplasmic expression in specialized strains (e.g., SHuffle T7 Express) enhances yields and activity by promoting correct disulfide bond formation (Carter et al. 2024). Critically, MHETase acts synergistically with PETase, breaking down the intermediate MHET into the monomers terephthalate (TPA) and ethylene glycol (EG), a pathway now being reconstructed and optimized in engineered systems (Carr et al. 2020; Liu et al. 2025a, b).

Encoding these plastic-degrading enzymes into robust microbial chassis like *Pseudomonas putida* or *Escherichia coli* is a key strategy (Mariam et al. 2025). However, metabolic efficiency can be hampered by native regulatory mechanisms. Repressor genes often inhibit the expression of target degradation pathways, particularly under non-inducing conditions or in the presence of preferred carbon sources (catabolite repression) (Li et al. 2022; Arentshorst et al. 2021). CRISPR-Cas9 facilitates precise knockout of these repressors to enhance pathway expression. For example, in *P. putida* DLL-E4, knockout of the *crc* gene (a central repressor in carbon catabolite repression) upregulated enzymes like PnpA and accelerated the degradation of the model pollutant p-nitrophenol (PNP) (Li et al. 2022). While effects on growth were noted, this demonstrates the principle that removing repressors can derepress metabolic pathways, a strategy directly applicable to activating or enhancing plastic degradation pathways in engineered microbes.

Research on applying CRISPR to engineer plastic-degrading microbes shows significant promise, particularly for PET:

- In *Ideonella sakaiensis*: CRISPR editing allows direct optimization of the native PET-degrading system within this bacterium. This includes efforts to enhance the expression or thermostability of PETase itself, building on foundational enzyme engineering work (Tournier et al. 2020).
- In *Escherichia coli*: CRISPR has been used to introduce and optimize PET-degrading enzymes. Strains expressing both PETase and MHETase from *I. sakaiensis* can degrade PET films and utilize the resulting hydrolysates (Liu et al. 2025a, b). Further engineering includes creating synthetic fusion enzymes (e.g., PETase-MHETase) aimed at improving efficiency.
- In *Pseudomonas putida*: This robust soil bacterium is a prime chassis for degrading various pollutants. CRISPR-Cas systems have been employed to introduce heterologous pathways, such as alkane hydroxylases for potential polyethylene (PE) degradation, and to modify central metabolism (e.g., β-oxidation) to potentially funnel plastic monomers (Dharmasiddhi et al. 2025).
- In *Bacillus subtilis*: CRISPR tools like CRISPR activation (CRISPRa) are being explored to upregulate potential biodegradation genes, such as laccases, for challenging polymers like PS (Wang et al. 2024).

These advancements highlight CRISPR's potential in tailoring microbial physiology for plastic bioremediation. While challenges remain, particularly for highly recalcitrant polymers like PE and PS, the precise engineering enabled by CRISPR is paving the way for developing more effective and scalable biological solutions to plastic pollution (Fig. 5.2).

5.4 Conclusion

Advanced bioremediation strategies—phytoremediation, mycoremediation, and CRISPR-engineered microbial degradation—represent a transformative paradigm shift in environmental cleanup. These approaches harness biological systems to address complex pollutants more sustainably and cost-effectively than conventional physicochemical methods. Phytoremediation leverages hyperaccumulator plants and microbial partnerships to extract, stabilize, or volatilize heavy metals, reducing soil toxicity while minimizing ecosystem disruption. Mycoremediation exploits fungal enzymatic networks (e.g., laccases, peroxidases) to degrade recalcitrant hydrocarbons from oil spills and enhances soil structure through mycelial growth. CRISPR-engineered microbes offer groundbreaking solutions for plastic pollution, enabling targeted enhancement of enzymes like PETase and MHETase to accelerate the breakdown of PET, PE, and PS into benign monomers. Collectively, these strategies highlight the power of interdisciplinary innovation—integrating plant science,

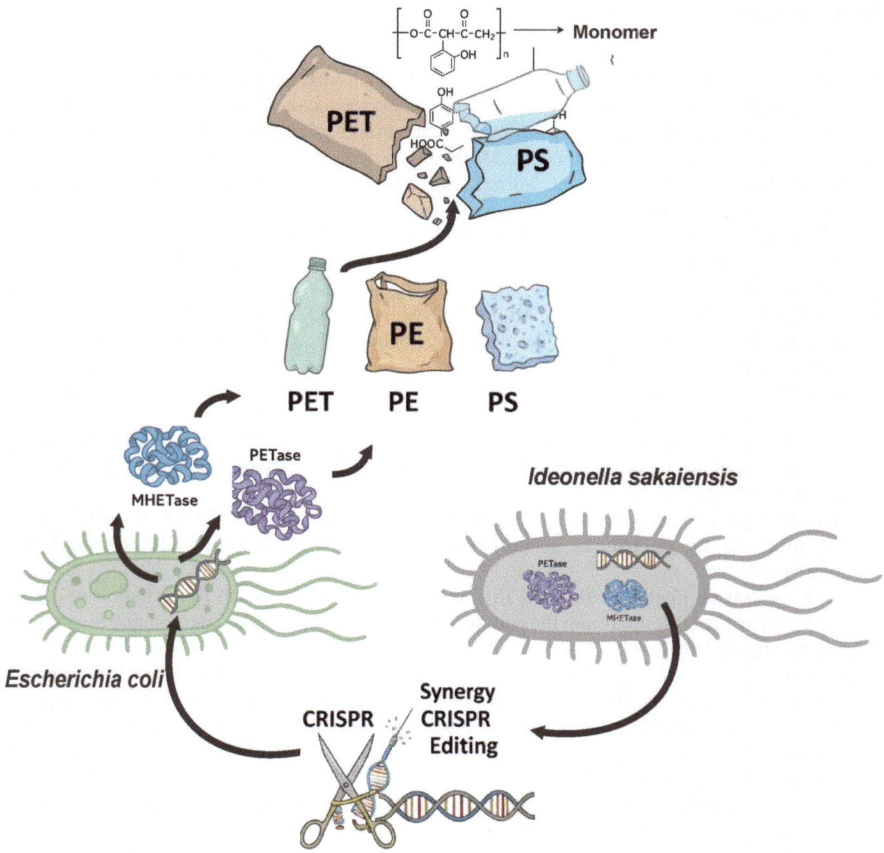

Fig. 5.2 Enzymatic degradation pathways for PET (polyethylene terephthalate). The schematic compares natural PET degradation in Ideonella sakaiensis (utilizing PETase and MHETase enzymes) with engineered PET degradation in *Escherichia coli* (achieved through CRISPR-based genetic editing to express PETase and MHETase, enabling synergistic polymer breakdown). PE (polyethylene) and PS (polystyrene) are shown as structurally distinct synthetic polymers not targeted by these specific enzymatic pathways

mycology, synthetic biology, and nanotechnology. They provide scalable, low-carbon alternatives applicable to diverse contaminants (metals, hydrocarbons, plastics) and environments (soil, water, industrial sites).

Conflict of Interest Disclosure The authors of this research article declare that there were no conflicts of interest.

Funding Acknowledgement This research received no specific grant from any funding agency in the public, commercial or not-for-profit sectors.

References

Abuzaid AS, Jahin HS, Asaad AA, Fadl ME, AbdelRahman MA, Scopa A (2021) Accumulation of potentially toxic metals in Egyptian alluvial soils, berseem clover (*Trifolium alexandrinum* L.), and groundwater after long-term wastewater irrigation. Agriculture 11(8):713

Acosta JA, Abbaspour A, Martínez GR, Martínez-Martínez S, Zornoza R, Gabarrón M, Faz A (2018) Phytoremediation of mine tailings with *Atriplex halimus* and organic/inorganic amendments: a five-year field case study. Chemosphere 204:71–78

Ahmadi F, Mohammadkhani N, Servati M (2022) Halophytes play important role in phytoremediation of salt-affected soils in the bed of Urmia Lake, Iran. Sci Rep 12:12223. https://doi.org/10.1038/s41598-022-16266-4

Akpasi SO, Anekwe IMS, Tetteh EK, Amune UO, Shoyiga HO, Mahlangu TP, Kiambi SL (2023) Mycoremediation as a potentially promising technology: current status and prospects—a review. Appl Sci 13(8):4978

Al-Hawash AB, Alkooranee JT, Zhang X, Ma F (2018) Fungal degradation of polycyclic aromatic hydrocarbons. Int J Pure Appl Biosci 6(2):8–24

Alam R, Jahan S, Hasanuzzaman M (2023) Efficiency of Helianthus annuus in rhizofiltration of arsenicfrom contaminated water. Env Technol Inn, 31, 103066. https://doi.org/10.1016/j.eti.2022.103066

Ammu KP, Malathi P, Sellamuthu KM, Jayashree R, Senthil Kumar G (2025) Phytoremediation of salt-affected soils: mechanistic insights and criteria for plant selection. Arid Land Res Manag 39(4):1–21

Anza M, Salazar O, Epelde L, Becerril JM, Alkorta I, Garbisu C (2019) Remediation of organically contaminated soil through the combination of assisted phytoremediation and bioaugmentation. Appl Sci 9(22):4757

Arentshorst M, Falco MD, Moisan M-C, Reid ID, Spaapen TOM, Van Dam J, Demirci E, Powlowski J, Punt PJ, Tsang A, Ram AFJ (2021) Identification of a conserved transcriptional activator-repressor module controlling the expression of genes involved in tannic acid degradation and gallic acid utilization in *Aspergillus niger*. Front Fungal Biol 2. https://doi.org/10.3389/ffunb.2021.681631

Awa SH, Hadibarata T (2020) Removal of heavy metals in contaminated soil by phytoremediation mechanism: a review. Water Air Soil Pollut 231(2):1–15

Aytan B, Oztug M, Ozturk HU, Kazan DİLEK, Ari S (2025) Efficient cloning and expression of IsPETase in the *E. coli* SHuffle cells; activity analysis using HR-LC/MS. Appl Biochem Microbiol 61:302–311

Azubuike CC, Chikere CB, Okpokwasili GC (2016) Bioremediation techniques—classification based on site of application: principles, advantages, limitations and prospects. World J Microbiol Biotechnol 32(11):180. https://doi.org/10.1007/s11274-016-2137-x

Balali-Mood M, Naseri K, Tahergorabi Z, Khazdair MR, Sadeghi M (2021) Toxic mechanisms of five heavy metals: mercury, lead, chromium, cadmium, and arsenic. Front Pharmacol 12:643972. https://doi.org/10.3389/fphar.2021.643972

Bañuelos GS, Dhillon KS (2011) Developing a sustainable phytomanagement strategy for excessive selenium in Western United States and India. Int J Phytoremediation 13(Suppl. 1):208–228. https://doi.org/10.1080/15226514.2011.568544

Barr DP, Aust SD (1994) Mechanisms white rot fungi use to degrade pollutants. Environ Sci Technol 28(2):78A–87A

Barra Caracciolo A, Terenzi V (2021) Rhizosphere microbial communities and heavy metals. Microorganisms 9(7):1462

Bizerea TO, Marginean O, Stroescu R, Rogobete A, Ilie C, Dezsi ȘG, Bizerea-Spiridon O (2018) The link between selenium, oxidative stress and pregnancy induced hypertensive disorders. Clin Lab 64(10):1593–1610

Buda M (2021) Soil microbial relationships in fungal-bacterial bioremediation systems: to synergy or not to be? Duluth J Adv Writ 2, 82–91

Cao Z, Yan W, Ding M, Yuan Y (2022) Construction of microbial consortia for microbial degradation of complex compounds. Front Bioeng Biotechnol 10:1051233

Carr CM, Clarke DJ, Dobson ADW (2020) Microbial polyethylene terephthalate hydrolases: current and future perspectives. Front Microbiol 11:571265. https://doi.org/10.3389/fmicb.2020.571265

Carter LM, MacFarlane CE, Karlock SP, Sen T, Kaar JL, Berberich JA, Boock JT (2024) Increased cytoplasmic expression of PETase enzymes in *E. coli*. Microb Cell Fact 23(1):319. https://doi.org/10.1186/s12934-024-02585-w

Chatterjee S, Kumari S, Rath S, Priyadarshanee M, Das S (2020) Diversity, structure and regulation of microbial metallothionein: metal resistance and possible applications in sequestration of toxic metals. Metallomics 12(11):1637–1655

Chen W, Sun J, Ji R, Min J, Wang L, Zhang J, Qiao H, Cheng S (2024) Crude oil biodegradation by a biosurfactant-producing bacterial consortium in high-salinity soil. J Mar Sci Eng 12(11):2033. https://doi.org/10.3390/jmse12112033

Chinmayee MD, Mahesh B, Pradesh S, Mini I, Swapna TS (2012) The assessment of phytoremediation potential of invasive weed *Amaranthus spinosus* L. Appl Biochem Biotechnol 167(6):1550–1559

Chollom MN, Adeyinka GC, Bakare BF (2023) Investigating the current trend of selected heavy metal pollution with possible ecological and human health effects along the uMgeni River of KwaZulu-Natal, South Africa. Int J Environ Anal Chem 104(20):9509–9527. https://doi.org/10.1080/03067319.2023.2235289

Dangi AK, Sharma B, Hill RT, Shukla P (2019) Bioremediation through microbes: systems biology and metabolic engineering approach. Crit Rev Biotechnol 39(1):79–98. https://doi.org/10.1080/07388551.2018.1500997

Deshmukh R, Khardenavis AA, Purohit HJ (2016) Diverse metabolic capacities of fungi for bioremediation. Indian J Microbiol 56(3):247–264

Dharmasiddhi IPW, Chen J, Arab B, Lan C, Euler C, Chou CP, Liu Y (2025) Engineering a cross-feeding synthetic bacterial consortium for degrading mixed PET and nylon monomers. Processes 13(2):375

Din BU, Rafique M, Javed MT, Kamran MA, Mehmood S, Khan M et al (2020) Assisted phytoremediation of chromium spiked soils by *Sesbania sesban* in association with *Bacillus xiamenensis* PM14: a biochemical analysis. Plant Physiol Biochem 146:249–258. https://doi.org/10.1016/j.plaphy.2019.11.010

Dinakarkumar Y, Ramakrishnan G, Gujjula KR, Vasu V, Balamurugan P, Murali G (2024) Fungal bioremediation: an overview of the mechanisms, applications and future perspectives. Environ Chem Ecotoxicol 6:293–302

Echezonachi SO (2022) The role of white rot fungi in bioremediation. In: Microbes and microbial biotechnology for green remediation. Elsevier, pp 305–320

Egendorf SP, Groffman P, Moore G, Cheng Z (2020) The limits of lead (Pb) phytoextraction and possibilities of phytostabilization in contaminated soil: a critical review. Int J Phytoremediation 22(9):916–930. https://doi.org/10.1080/15226514.2020.1774501

Ekta K, Pravin KU, Avidit S, Priyankar R, Javvadi P, Mona Ni, Vishal T (2016) Allelopathic Effect of Cocklebur Extract on the Fertility Status of Soil in Transplanted Rice by Controlling Weed J PureApp Microbiol, 10(1):479–483

Engwa GA, Okeke PF, Nweke Nwalo F, Unachukwu M (2019) Mechanism and health effects of heavy metal toxicity in humans. In: Karcıoğlu Ö, Arslan B (eds) Poisoning in the modern world: new tricks for an old dog? IntechOpen. https://doi.org/10.5772/intechopen.82511

Ermis H (2025) A mini-review on the role of PETase in polyethylene terephthalate degradation. Rev Environ Sci Biotechnol 24:545–555

Espinosa-Ortiz EJ, Rene ER, Gerlach R (2022) Potential use of fungal-bacterial co-cultures for the removal of organic pollutants. Crit Rev Biotechnol 42(3):361–383

Galal TM, Gharib FA, Ghazi SM, Mansour KH (2017) Phytostabilization of heavy metals by the emergent macrophyte *Vossia cuspidata* (Roxb.) Griff.: a phytoremediation approach. Int J Phytoremediation 19(11):992–999

Gardea-Torresdey JL, Gonzalez JH, Tiemann KJ, Rodriguez O, Gamez G (1998) Phytofiltration of hazardous cadmium, chromium, lead and zinc ions by biomass of *Medicago sativa* (Alfalfa). J Hazard Mater 57(1):29–39. https://doi.org/10.1016/S0304-3894(97)00072-1

Gunyar OA, Uztan AH (2021) Environmental mycobiotechnology in special reference to fungal bioremediation. In: Nanotechnology applications in health and environmental sciences. Springer International Publishing, pp 361–383

Hadiyanto H, Joelyna FA, Khoironi A et al (2025) Harnessing *Chlorella vulgaris - Aspergillus niger* interactions for effective microplastic removal in aquatic ecosystems. Waste Biomass Valoriz. https://doi.org/10.1007/s12649-025-03062-0

Hanus-Fajerska EJ, Kozminska A (2016) The possibilities of water purification using phytofiltration methods: a review of recent progress. BioTechnologia 97(4):315–322

Henkler F, Brinkmann J, Luch A (2010) The role of oxidative stress in carcinogenesis induced by metals and xenobiotics. Cancers 2(2):376–396. https://doi.org/10.3390/cancers2020376

Herzig R, Nehnevajova E, Pfistner C, Schwitzguebel J-P, Ricci A, Keller C (2014) Feasibility of labile Zn phytoextraction using enhanced tobacco and sunflower: results of five and one year field scale experiments in Switzerland. Int J Phytoremediation 16:735–754. https://doi.org/10.1080/15226514.2013.856846

Hussain F, Hussain I, Khan AHA, Muhammad YS, Iqbal M, Soja G et al (2018) Combined application of biochar, compost, and bacterial consortia with Italian ryegrass enhanced phytoremediation of petroleum hydrocarbon contaminated soil. Environ Exp Bot 153:80–88

Irshad M, Ruqia B, Hussain Z (2015) Phytoaccumulation of heavy metals in natural vegetation at the municipal wastewater site in Abbottabad, Pakistan. Int J Phytoremediation 17(12):1269–1273

Jacob JM, Karthik C, Saratale RG, Kumar SS, Prabakar D, Kadirvelu K, Pugazhendhi A (2018) Biological approaches to tackle heavy metal pollution: a survey of literature. J Environ Manage 217:56–70. https://doi.org/10.1016/j.jenvman.2018.03.077

Jomova K, Alomar SY, Nepovimova E, Kuca K, Valko M (2025) Heavy metals: toxicity and human health effects. Arch Toxicol 99(1):153–209. https://doi.org/10.1007/s00204-024-03903-2

Ju W, Liu L, Fang L, Cui Y, Duan C, Wu H (2019) Impact of co-inoculation with plant-growth-promoting rhizobacteria and rhizobium on the biochemical responses of alfalfa-soil system in copper contaminated soil. Ecotoxicol Environ Saf 167:218–226. https://doi.org/10.1016/j.ecoenv.2018.10.016

Kacprzak M, Grobelak A, Grosser A, Napora A (2014) The potential of biosolid application for the phytostabilisation of metals. Desalin Water Treat 52(19–21):3955–3964

Kaczorek E, Pacholak A, Zdarta A, Smułek W (2018) The impact of biosurfactants on microbial cell properties leading to hydrocarbon bioavailability increase. Colloid Interface 2(3):35

Karim A, Malekshah RE, Ullah N, Haribabu J, Iqbal M, Hsu SC (2025) Physiochemical interaction of surfactants with metal complexes and its biological relevance: a comprehensive overview. Transit Met Chem 50:389–405

Ke T, Guo G, Liu J, Zhang C, Tao Y, Wang P et al (2021) Improvement of the Cu and Cd phytostabilization efficiency of perennial ryegrass through the inoculation of three metal-resistant PGPR strains. Environ Pollut 271:116314

Kim HS, Kim YJ, Seo YR (2015) An overview of carcinogenic heavy metal: molecular toxicity mechanism and prevention. J Cancer Prev 20(4):232–240. https://doi.org/10.15430/JCP.2015.20.4.232

Kumar L, Ragunathan V, Chugh M et al (2021) Nanomaterials for remediation of contaminants: a review. Environ Chem Lett 19:3139–3163. https://doi.org/10.1007/s10311-021-01212-z

Kumar V, Bilal M, Ferreira LFR, Iqbal HM (2023) Genomics approach to bioremediation: principles, tools, and emerging technologies. John Wiley & Sons, Inc, Hoboken

Kumar V, Pallavi P, Sen SK, Raut S (2024) Harnessing the potential of white rot fungi and ligninolytic enzymes for efficient textile dye degradation: a comprehensive review. Water Environ Res 96(1):e10959

Lajayer BA, Moghadam NK, Maghsoodi MR, Ghorbanpour M, Kariman K (2019) Phytoextraction of heavy metals from contaminated soil, water and atmosphere using ornamental plants: mechanisms and efficiency improvement strategies. Environ Sci Pollut Res 26(9):8468–8484

Lasat MM (2002) Phytoextraction of toxic metals: a review of biological mechanisms. J Environ Qual 31(1):109–120

Lastiri-Hernández MA, Álvarez-Bernal D, Moncayo-Estrada R, Cruz-Cárdenas G, Silva García JT (2021) Adoption of phytodesalination as a sustainable agricultural practice for improving the productivity of saline soils. Environ Dev Sustain 23(6):8798–8814

Lea-Smith DJ, Hassard F, Coulon F, Partridge N, Horsfall L, Parker KD et al (2025) Engineering biology applications for environmental solutions: potential and challenges. Nat Commun 16(1):3538

Lebeau T (2011) Bioaugmentation for in situ soil remediation: how to ensure the success of such a process. In: Bioaugmentation, biostimulation and biocontrol. Springer, Berlin Heidelberg, pp 129–186

Li J, Yu H, Luan Y (2015) Meta-analysis of the copper, zinc, and cadmium absorption capacities of aquatic plants in heavy metal-polluted water. Int J Environ Res Public Health 12(12):14958–14973

Li S, Tang Y, Tang L, Yan X, Xiao J, Xiang H, Wu Q, Yu R, Jin Y, Yu J, Xu N, Wu C, Wang S, Wang C, Chen Q (2022) Preliminary study on the effect of catabolite repression gene knockout on p-nitrophenol degradation in *Pseudomonas putida* DLL-E4. PLoS One 17:e0278503. https://doi.org/10.1371/journal.pone.0278503

Li H, Zhou R, Li D, Chen XW, Mo CH, Li H (2025) Synergistic remediation of cadmium and BDE-209 co-contaminated soil using *Solanum nigrum* assisted by Arbuscular mycorrhizal fungi and citric acid. Front Microbiol 16:1624164

Liu F, Wang T, Liu XH, Xu N, Pan XL (2025a) Efficient biodegradation and upcycling of polyethylene terephthalate mediated by cell-factories. Front Microbiol 16:1599470

Liu XH, Jin JL, Sun HT, Li S, Zhang FF, Yu XH et al (2025b) Perspectives on the microorganisms with the potentials of PET-degradation. Front Microbiol 16:1541913

Luo Y, Wu Y, Qiu J, Wang H, Yang L (2019) Suitability of four woody plant species for the phytostabilization of a zinc smelting slag site after 5 years of assisted revegetation. J Soils Sediments 19(2):702–715

Macfie A, Hagan E, Zhitkovich A (2010) Mechanism of DNA–protein cross-linking by chromium. Chem Res Toxicol 23(2):341–347

MacLeod M, Arp HPH, Tekman MB, Jahnke A (2021) The global threat from plastic pollution. Science 373(6550):61–65

Majcherczyk A, Johannes C, Hüttermann A (1998) Oxidation of polycyclic aromatic hydrocarbons (PAH) by laccase of *Trametes versicolor*. Enzym Microb Technol 22(5):335–341

Makarova A, Nikulina E, Avdeenkova T, Pishaeva K (2021) The improved phytoextraction of heavy metals and the growth of *Trifolium repens* L.: the role of K_2HEDP and plant growth regulators alone and in combination. Sustainability 13(5):2432

Manić L, Wallace D, Onganer PU, Taalab YM, Farooqi AA, Antonijević B, Buha Djordjevic A (2022) Epigenetic mechanisms in metal carcinogenesis. Toxicol Rep 9:778–787. https://doi.org/10.1016/j.toxrep.2022.03.037

Manousaki E, Kalogerakis N (2011) Halophytes—an emerging trend in phytoremediation. Int J Phytoremediation 13(10):959–969. https://doi.org/10.1080/15226514.2010.532241

Manzoor M, Gul I, Manzoor A, Kallerhoff J, Arshad M (2021) Optimization of integrated phytoremediation system (IPS) for enhanced lead removal and restoration of soil microbial activities. Chemosphere 277:130243. https://doi.org/10.1016/j.chemosphere.2021.130243

Mariam I, Rova U, Christakopoulos P, Matsakas L, Patel A (2025) Data-driven synthetic microbes for sustainable future. npj Syst Biol Appl 11(1):74

Meers E, Ruttens A, Hopgood M, Lesage E, Tack FMG (2005) Potential of *Brassica rapa*, *Cannabis sativa*, *Helianthus annuus* and *Zea mays* for phytoextraction of heavy metals from calcareous dredged sediment derived soils. Chemosphere 61(4):561–572

Mello IS, Targanski S, Pietro-Souza W, Stachack FFF, Terezo AJ, Soares MA (2020) Endophytic bacteria stimulate mercury phytoremediation by modulating its bioaccumulation and volatilization. Ecotoxicol Environ Saf 202:110818. https://doi.org/10.1016/j.ecoenv.2020.110818

Mesa V, Navazas A, González-Gil R, González A, Weyens N, Lauga B et al (2017) Use of endophytic and rhizosphere bacteria to improve phytoremediation of arsenic-contaminated industrial soils by autochthonous *Betula celtiberica*. Appl Environ Microbiol 83(4):e03411–e03416. https://doi.org/10.1128/AEM.03411-16

Miransari M (2011) Hyperaccumulators, arbuscular mycorrhizal fungi and stress of heavy metals. Biotechnol Adv 29(6):645–653

Mishra J, Singh R, Arora NK (2017) Alleviation of heavy metal stress in plants and remediation of soil by rhizosphere microorganisms [Mini review]. Front Microbiol 8:1706. https://doi.org/10.3389/fmicb.2017.01706

Mohammadi-Sichani M, Mazaheri Assadi M, Farazmand A et al (2019) Ability of *Agaricus bisporus*, *Pleurotus ostreatus* and *Ganoderma lucidum* compost in biodegradation of petroleum hydrocarbon-contaminated soil. Int J Environ Sci Technol 16:2313–2320. https://doi.org/10.1007/s13762-017-1636-0

Mrozik A, Piotrowska-Seget Z (2010) Bioaugmentation as a strategy for cleaning up of soils contaminated with aromatic compounds. Microbiol Res 165(5):363–375

Nakatsuka H, Oda M, Hayashi Y, Tamura K (2016) Effects of fresh spent mushroom substrate of *Pleurotus ostreatus* on soil micromorphology in Brazil. Geoderma 269:54–60. https://doi.org/10.1016/j.geoderma.2016.01.023

Nayak AK, Panda SS, Basu A, Dhal NK (2018) Enhancement of toxic Cr (VI), Fe, and other heavy metals phytoremediation by the synergistic combination of native *Bacillus cereus* strain and *Vetiveria zizanioides* L. Int J Phytoremediation 20(7):682–691. https://doi.org/10.1080/15226514.2017.1413332

Odinga CA, Kumar A, Mthembu MS, Bux F, Swalaha FM (2019) Rhizofiltration system consisting of *Phragmites australis* and *Kyllinga nemoralis*: evaluation of efficient removal of metals and pathogenic microorganisms. Desalin Water Treat 169:120–132. https://doi.org/10.5004/dwt.2019.24428

Osredkar J, Sustar N (2011) Copper and zinc, biological role and significance of copper/zinc imbalance. J Clin Toxicol S3:001

Palit P, Minkara M, Abida M, Marwa S, Sen C, Roy A, Pasha MR, Mosae PS, Saha A, Ferdoush J (2025) PlastiCRISPR: genome editing-based plastic waste management with implications in polyethylene terephthalate (PET) degradation. Biomolecules 15(5):684. https://doi.org/10.3390/biom15050684

Panchal P, Miller AJ, Giri J (2021) Organic acids: versatile stress-response roles in plants. J Exp Bot 72(11):4038–4052

Paul S, Ingti B, Das D, Joshi SR (2025) Microbial responses to hydrocarbon contamination and its bioremediation. In: Mineral transformation and bioremediation by geo-microbes. Springer, pp 545–588

Pivetz BE (2001) Phytoremediation of contaminated soil and ground water at hazardous waste sites. US Environmental Protection Agency

Platanias LC (2009) Biological responses to arsenic compounds. J Biol Chem 284(28):18583–18587. https://doi.org/10.1074/jbc.R900003200

Pozdnyakova NN (2012) Involvement of the ligninolytic system of white-rot and litter-decomposing fungi in the degradation of polycyclic aromatic hydrocarbons. Biotechnol Res Int 2012:243217

Prenafeta-Boldú FX, De Hoog GS, Summerbell RC (2019) Fungal communities in hydrocarbon degradation. In: Microbial communities utilizing hydrocarbons and lipids: members, metagenomics and ecophysiology. Springer, Cham, pp 1–36

Radziemska M, Vaverková MD, Baryła A (2017) Phytostabilization—management strategy for stabilizing trace elements in contaminated soils. Int J Environ Res Public Health 14(9):958

Rahman MM, Mostofa MG, Keya SS, Siddiqui MN, Ansary MMU, Das AK et al (2021) Adaptive mechanisms of halophytes and their potential in improving salinity tolerance in plants. Int J Mol Sci 22(19):10733

Rajendran S, Kamaludeen SPB, Subramanian A (2024) Antioxidant defense mechanism in plants exposed to mercury toxicity: response, tolerance and remediation. In: Mercury toxicity mitigation: sustainable nexus approach. Springer, pp 225–245

Rani P, Rose PK, Kidwai MK, Meenakshi (2023) *Brassica juncea* L.: a potential crop for phytoremediation of various heavy metals. In: Heavy metal toxicity: environmental concerns, remediation and opportunities. Springer, pp 285–311

Rani MHS, Nandana RK, Khatun A, Brindha V, Midhun D, Gowtham P et al (2024) Three strategy rules of filamentous fungi in hydrocarbon remediation: an overview. Biodegradation 35(6):833–861

Rasool B, Ramzani PMA, Zubair M, Khan MA, Lewińska K, Turan V et al (2021) Impacts of oxalic acid-activated phosphate rock and root-induced changes on Pb bioavailability in the rhizosphere and its distribution in mung bean plant. Environ Pollut 280:116903

Reineke W, Schlömann M (2023) Microbial degradation of pollutants. In: Environmental microbiology. Springer, Berlin Heidelberg, pp 161–290

Rezaei Z, Moghimi H (2024) Fungal-bacterial consortia: a promising strategy for the removal of petroleum hydrocarbons. Ecotoxicol Environ Saf 280:116543

Robinson JR, Isikhuemhen OS, Anike FN (2021) Fungal–metal interactions: a review of toxicity and homeostasis. J Fungi 7(3):225

Saini N, Bharti D, Singh S, Banoo R, Bundela V, Nain P, Sahgal M (2024) Genetically engineered plastic munching microbes: recent advancements and perspectives. In: Advanced strategies for biodegradation of plastic polymers. Springer Nature Switzerland, pp 193–224

Saran A, Fernandez L, Cora F, Savio M, Thijs S, Vangronsveld J, Merini LJ (2020) Phytostabilization of Pb and Cd polluted soils using *Helianthus petiolaris* as pioneer aromatic plant species. Int J Phytoremediation 22(5):459–467

Saravanan A, Kumar PS, Govarthanan M, George CS, Vaishnavi S, Moulishwaran B et al (2021) Adsorption characteristics of magnetic nanoparticles coated mixed fungal biomass for toxic Cr(VI) ions in aquatic environment. Chemosphere 267:129226. https://doi.org/10.1016/j.chemosphere.2020.129226

Shanmugapriya S, Manivannan G, Selvakumar G, Sivakumar N (2019) Extracellular fungal peroxidases and laccases for waste treatment: recent improvement. In: Recent advancement in white biotechnology through fungi: volume 3: perspective for sustainable environments. Springer International Publishing, pp 153–187

Sharma I (2020) Bioremediation techniques for polluted environment: concept, advantages, limitations, and prospects. In: Trace metals in the environment-new approaches and recent advances. IntechOpen

Sharma S, Tiwari S, Hasan A, Saxena V, Pandey LM (2018) Recent advances in conventional and contemporary methods for remediation of heavy metal-contaminated soils. 3 Biotech 8(4):216

Sharma JK, Kumar N, Singh NP, Santal AR (2023) Phytoremediation technologies and their mechanism for removal of heavy metal from contaminated soil: an approach for a sustainable environment. Front Plant Sci 14:1076876

Shi L, Liu H, Gao S, Weng Y, Zhu L (2021) Enhanced extracellular production of IsPETase in *Escherichia coli* via engineering of the pelB signal peptide. J Agric Food Chem 69(7):2245–2252. https://doi.org/10.1021/acs.jafc.0c07469

Sigua GC, Novak JM, Watts DW, Ippolito JA, Ducey TF, Johnson MG, Spokas KA (2019) Phytostabilization of Zn and Cd in mine soil using corn in combination with biochars and manure-based compost. Environments 6(6):69

Singha LP, Shukla P (2023) Microbiome engineering for bioremediation of emerging pollutants. Bioprocess Biosyst Eng 46(3):323–339

Skuza L, Szućko-Kociuba I, Filip E, Bożek I (2022) Natural molecular mechanisms of plant hyperaccumulation and hypertolerance towards heavy metals. Int J Mol Sci 23(16):9335. https://doi.org/10.3390/ijms23169335

Sladkovska T, Wolski K, Bujak H, Radkowski A, Sobol Ł (2022) A review of research on the use of selected grass species in removal of heavy metals. Agronomy 12(10):2587

Spasojević I, Mojović M, Stević Z, Spasić SD, Jones DR, Morina A, Spasić MB (2010) Bioavailability and catalytic properties of copper and iron for Fenton chemistry in human cerebrospinal fluid. Redox Rep 15(1):29–35

Sun H, Shamy M, Costa M (2013) Nickel and epigenetic gene silencing. Genes 4(4):583–595

Tallou A, Erraji H, Al Masmoudi Y, Abbatantuono F, Ali SA, Mabrouki J et al (2024) Phytoremediation of contaminated environments using halophytes: general overview. In: Integrated solutions for smart and sustainable environmental conservation. Springer, pp 143–163

Thijs S, Sillen W, Rineau F, Weyens N, Vangronsveld J (2016) Towards an enhanced understanding of plant–microbiome interactions to improve phytoremediation: engineering the metaorganism. Front Microbiol 7:341

Tournier V, Tophill C, Allais F et al (2020) An engineered PET depolymerase to break down and recycle plastic bottles. Nature 580:216–219. https://doi.org/10.1038/s41586-020-2149-4

Ugya AY, Ahmad A, Adamu H, Giwa S, Imam T (2019) Phytoextraction of heavy metals and risk associated with vegetables grown from soil irrigated with refinery wastewater. J Appl Biol Biotechnol 7(2):14–19

Vangronsveld J, Herzig R, Weyens N, Boulet J, Adriaensen K, Ruttens A et al (2009) Phytoremediation of contaminated soils and groundwater: lessons from the field. Environ Sci Pollut Res 16(7):765–794. https://doi.org/10.1007/s11356-009-0213-6

Verbruggen N, Hermans C, Schat H (2009) Molecular mechanisms of metal hyperaccumulation in plants. New Phytol 181(4):759–776. https://doi.org/10.1111/j.1469-8137.2008.02748.x

Verma P, George KV, Singh HV, Singh SK, Juwarkar A, Singh RN (2006) Modeling rhizofiltration: heavy-metal uptake by plant roots. Environ Model Assess 11(4):387–394

Viti C, Marchi E, Decorosi F, Giovannetti L (2014) Molecular mechanisms of Cr (VI) resistance in bacteria and fungi. FEMS Microbiol Rev 38(4):633–659

Wang G, Zhao W, Yuan Y, Morel JL, Chi H, Feng W et al (2020) Mobility of metal(loid)s in Pb/Zn tailings under different revegetation strategies. J Environ Manag 263:110323

Wang T, Yang WT, Gong YM, Zhang YK, Fan XX, Wang GC et al (2024) Molecular engineering of PETase for efficient PET biodegradation. Ecotoxicol Environ Saf 280:116540

Wei X, Zhou Y, Tsang DCW, Song L, Zhang C, Yin M, Liu J, Xiao T, Zhang G, Wang J (2020) Hyperaccumulation and transport mechanism of thallium and arsenic in brake ferns (*Pteris vittata* L.): a case study from mining area. J Hazard Mater 388:121756. https://doi.org/10.1016/j.jhazmat.2019.121756

Wu S, Zhang X, Chen B, Wu Z, Li T, Hu Y et al (2016) Chromium immobilization by extraradical mycelium of arbuscular mycorrhiza contributes to plant chromium tolerance. Environ Exp Bot 122:10–18

Wu M, Zhuang J, He X, Zhu F, Lai Q, Qian X et al (2025) Synthetic biology boosts the biological depolymerization and upcycling of waste plastic resources. Synth Biol Eng 3(1):10002

Yadav M, Singh M, Nanda K, Singh SP (2022) Chapter 11 - Genetically engineered plants for phytoremediation of heavy metals. In: Sharma P, Pandey A, Tong YW, Ngo HH (eds) Current developments in biotechnology and bioengineering. Elsevier, pp 223–239. https://doi.org/10.1016/B978-0-323-99907-6.00017-7

Yan, Liang-Jun, Allen, DC (2021) Cadmium-Induced Kidney Injury: Oxidative Damage as a UnifyingMechanism. Biomolecules 11(11):1575. https://doi.org/10.3390/biom11111575

Yang H, Shu Y (2015) Cadmium transporters in the kidney and cadmium-induced nephrotoxicity. Int J Mol Sci 16(1):1484–1494. https://doi.org/10.3390/ijms16011484

Yoshida S, Hiraga K, Taniguchi I, Oda K (2021) *Ideonella sakaiensis*, PETase, and MHETase: from identification of microbial PET degradation to enzyme characterization. In: Methods in enzymology, vol 648. Academic Press, pp 187–205

Zhakypbek Y, Kossalbayev BD, Belkozhayev AM, Murat T, Tursbekov S, Abdalimov E, Pashkovskiy P, Kreslavski V, Kuznetsov V, Allakhverdiev SI (2024) Reducing heavy metal contamination in soil and water using phytoremediation. Plants 13(11):1534. https://doi.org/10.3390/plants13111534

Chapter 6
Microbial Biosurfactants and Their Role in Pollutant Cleanup

Manpreet Kaur and Reena Gupta

Abstract Microbial biosurfactants are surface-active compounds produced by bacteria, yeasts and fungi, characterized by their amphiphilic nature, environmental compatibility and ability to function under extreme conditions. Their unique chemical properties and biodegradability make them highly suitable for bioremediation applications. The present chapter explores the production, properties and environmental applications of microbial biosurfactants, with a particular focus on their role in the bioremediation of recalcitrant pollutants such as hydrocarbons and heavy metals. Biosurfactants enhance the emulsification, solubilization and bioavailability of hydrophobic contaminants, facilitating microbial degradation and detoxification processes. They also act as chelating agents for heavy metals, aiding in their removal from contaminated sites. Advances in the production of biosurfactants using alternative substrates and solid-state fermentation have improved their economic feasibility and sustainability, aligning with circular economic principles. This chapter also highlights recent research trends, including the use of biosurfactants as agricultural and biocidal agents, underscoring their potential as eco-friendly alternatives to synthetic surfactants in environmental and industrial applications.

Keywords Biosurfactants · Bioremediation · Environment · Microbes · Pollutants

6.1 Introduction

Petroleum and its derivatives remain the most widely used energy resources globally. However, their extraction, transportation, and consumption contribute to severe environmental contamination, including groundwater and subsoil pollution due to elevated oxygen depletion from intensified biological and chemical activity. Large-scale ecological disruptions, such as oil spills in marine and terrestrial

M. Kaur · R. Gupta (✉)
Department of Biotechnology, Himachal Pradesh University, Summerhill, Shimla, India

N. K. Arora, J. Mishra (eds.), *Biotechnological Solutions for a Sustainable Future*, Microorganisms for Sustainability 55,
https://doi.org/10.1007/978-981-95-3989-5_6

ecosystems, exemplify the catastrophic consequences of such pollution (Singh and Rathore 2019; Wei et al. 2020). These pollutants exhibit low biodegradability, posing long-term risks to biodiversity, soil health, and aquatic systems.

Traditional remediation strategies often rely on synthetic surfactants like sodium lauryl sulfate (SLS) and acetyl trimethyl ammonium bromide (ATAB), which act as metal ion scavengers or emulsifiers to immobilize contaminants (Rasheed et al. 2020). However, their recalcitrant nature and persistence in ecosystems result in secondary pollution, undermining remediation efforts (Farias et al. 2021; Zang et al. 2021). This has spurred interest in sustainable alternatives, particularly microbial surface-active compounds (MSACs), which combine high efficacy with environmental compatibility (Eras-Muñoz et al. 2022).

MSACs, produced by bacteria, fungi, and yeast, are amphiphilic molecules capable of solubilizing hydrophobic pollutants like petroleum hydrocarbons, polycyclic aromatic hydrocarbons (PAHs), and heavy metals. By acting as emulsifiers, foaming agents, and solubilizers, MSACs enhance the dispersion, bioavailability, and subsequent microbial degradation of these contaminants (Bolan et al. 2023). Structurally, MSACs are classified into high molecular-weight bioemulsifiers (e.g., lipoproteins) and low molecular-weight biosurfactants (e.g., glycolipids), each suited for specific remediation roles (Kashif et al. 2022). Their advantages include low toxicity, high biodegradability, and stability across extreme pH, temperature, and salinity gradients, alongside inherent antimicrobial properties (Naughton et al. 2019; Carolin et al. 2021).

Biosurfactants are particularly promising for bioremediation due to their ability to form micelles that encapsulate hydrophobic pollutants and heavy metals, enhancing their mobility and extraction (Mishra et al. 2021). Their efficiency is further emphasized by a low critical micelle concentration (CMC), enabling significant reductions in surface tension at minimal doses compared to synthetic counterparts (Jakinala et al. 2019; Malkapuram et al. 2021). Advances in fermentation technology, cost-effective substrates (e.g., agro-industrial waste), and genetic engineering have optimized biosurfactant yields, broadening their industrial applicability (Malkapuram et al. 2021). Beyond environmental cleanup, biosurfactants emerge as tools in agriculture for soil remediation and plant growth promotion, as well as in antimicrobial formulations (Sharma et al. 2021).

This chapter highlights the role of microbial biosurfactants in bioremediation, emphasizing their mechanisms, production innovations, and ecological benefits. It aims to synthesize current research on their efficacy in degrading petroleum hydrocarbons and sequestering heavy metals, while addressing challenges in scalability and commercial adoption. By bridging scientific knowledge and practical applications, this chapter stresses biosurfactants as pivotal agents in achieving sustainable environmental restoration.

6.2 Types of Microbial Biosurfactants

The MSACs are amphiphilic molecules capable of forming interfaces between fluids of differing polarities due to their dual hydrophilic (water-attracting) and hydrophobic (water-repelling) structural components (Pardhi et al. 2022). The hydrophilic moiety typically comprises amino acids, peptides, or carbohydrates, while the hydrophobic region consists of fatty acids, saturated/unsaturated peptides, or proteins. The efficacy of MSACs in reducing interfacial tension is directly influenced by their chemical composition. Based on molecular weight, MSACs are categorized into two primary classes: (1) high-molecular-weight biosurfactants (e.g., polysaccharides, lipopolysaccharides, lipoproteins), which excel at stabilizing water-oil emulsions, and (2) low-molecular-weight biosurfactants (e.g., glycolipids, lipopeptides), which are more effective at lowering surface tension (Fig. 6.1) (Soares da Silva et al. 2019). These structural variations shape their biological functions and industrial applications, including bioremediation and antimicrobial activity (Beck and Zibek 2020; Farias et al. 2021; Soberón-Chávez et al. 2021).

Additionally, microbial biosurfactants can be classified by their origin (bacterial, fungal, or yeast-derived) and chemical structure. Major categories include glycolipids (e.g., rhamnolipids), lipopeptides (e.g., surfactin), fatty acids, phospholipids, polymeric biosurfactants (e.g., emulsan), and particulate biosurfactants.

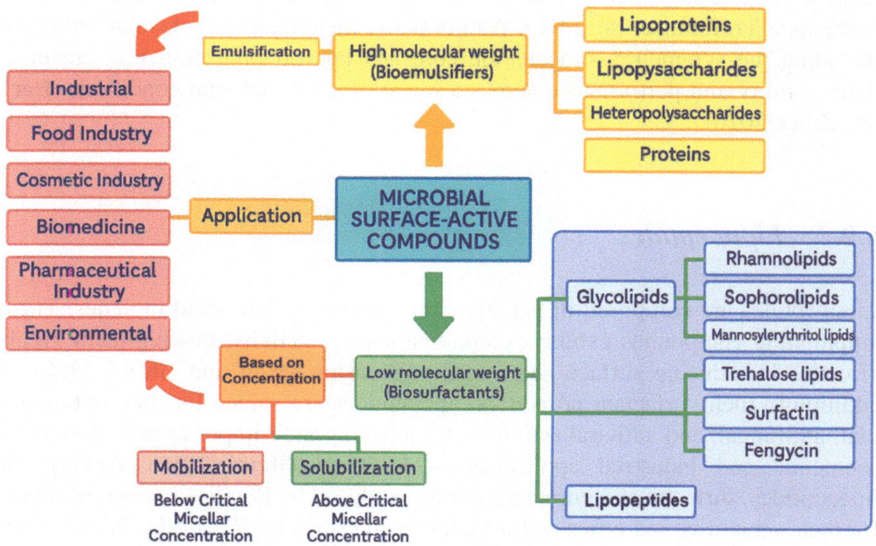

Fig. 6.1 Types and characteristics of microbial surface-active compounds. (Eras-Muñoz et al. 2022)

6.2.1 Glycolipid

Glycolipid biosurfactants are made up of both lipid and carbohydrate components. Rhamnolipids, primarily produced by *Pseudomonas aeruginosa* (Abdel-Mawgoud et al. 2010), are one of the most extensively studied glycolipids. These biosurfactants consist of one or two rhamnose sugar molecules connected to β-hydroxy fatty acid chains, which confer excellent emulsification and hydrocarbon degradation properties. Owing to these attributes, rhamnolipids play a pivotal role in bioremediation, particularly in the cleanup of oil-contaminated environments and industrial waste management, where they enhance the bioavailability of hydrophobic pollutants for microbial degradation (Mnif and Ghribi 2016).

Sophorolipids, synthesized by *Starmerella bombicola* (formerly *Candida bombicola*), are glycolipids composed of a sophorose disaccharide connected to a terminal or subterminal hydroxylated fatty acid via a β-glycosidic bond (Wang et al. 2018). These biosurfactants exhibit remarkable efficiency as emulsifiers and surface-active agents, with their structure varying between acidic (open-chain) and lactonic (closed-ring) forms, the latter showing stronger antimicrobial activity (Abhyankar et al. 2024).

Trehalolipids, produced by *Rhodococcus erythropolis*, *Mycobacterium tuberculosis*, and *Arthrobacter* species, feature trehalose sugar units linked to long-chain α-branched fatty acids (e.g., mycolic acids) through ester bonds (Shao 2010). These biosurfactants are highly effective in emulsifying hydrophobic compounds, making them invaluable for the biodegradation of hydrocarbons and other water-insoluble pollutants (Perfumo et al. 2010). Additionally, their unique molecular structure, including the symmetric disaccharide core and hydrophobic branches, enhances their stability and performance across a range of environmental conditions (Beck and Zibek 2020).

6.2.2 Lipopeptides

Lipopeptides are composed of peptide chains linked to fatty acid moieties. These amphiphilic compounds exhibit exceptional surface activity, making them highly effective in reducing surface and interfacial tension (Mnif and Ghribi 2016). In addition to their surfactant properties, lipopeptides are valued for their antimicrobial, antifungal, and antiviral activities, establishing their importance in both environmental and industrial applications (Mnif and Ghribi 2016). Among the lipopeptides, surfactin and iturins, primarily produced by *Bacillus* species, stand out for their versatility and efficacy. Surfactin, synthesized by *B. subtilis*, is one of the most potent biosurfactants known (Jacques 2010; Théatre et al. 2021). It consists of a cyclic heptapeptide linked to a beta-hydroxy fatty acid chain, forming a unique amphipathic structure. Surfactin exhibits an impressive ability to reduce surface tension to extremely low levels, making it highly effective as an emulsifying agent

(Henkel and Hausmann 2019). Additionally, it demonstrates potent antimicrobial properties, with activity against a wide range of bacteria and viruses (Das et al. 2008).

Iturins are also predominantly produced by *Bacillus* species; however, other bacterial genera such as *Paenibacillus* have also been reported to synthesize iturin-like compounds (Chung et al. 2000). These biosurfactants disrupt the integrity of microbial cell membranes, making them highly effective in the biocontrol of plant pathogens (Dunlap et al. 2019). Beyond their role in plant protection, they contribute to soil remediation efforts by enhancing the biodegradation of hydrophobic pollutants (Pacwa-Płociniczak et al. 2011). Their ability to combat fungal infections has also sparked interest in pharmaceutical applications (Carolin et al. 2021).

6.2.3 Phospholipids and Fatty Acids

Phospholipids and fatty acids, produced by diverse bacterial genera such as *Pseudomonas* and *Bacillus*, serve as biosurfactants due to their amphipathic molecular architecture (Lang 2002; Varjani and Upasani 2016). Phospholipids consist of a glycerol backbone esterified to two hydrophobic fatty acid chains and a hydrophilic phosphate headgroup, forming bilayers or micelles that effectively reduce interfacial tension (Dervichian 1964). Fatty acids, varying in chain length (C12–C20) and degree of saturation, significantly influence membrane fluidity and surfactant efficacy, thereby affecting microbial interactions with hydrophobic substances (Mnif et al. 2017). These structural features facilitate emulsification of hydrophobic substrates such as alkanes and polyaromatic hydrocarbons by solubilizing them into bioaccessible micelles, thus enhancing microbial uptake and catabolic activity. For instance, phosphatidylethanolamines produced by *Acinetobacter* species have been shown to enhance crude oil degradation by increasing the bioavailable surface area (Bhadra et al. 2022). Such mechanisms can be employed in bioremediation where biosurfactants optimize the degradation of hydrophobic compounds through enzymatic pathways, including alkane hydroxylases (Elumalai et al. 2021).

6.2.4 Polymeric Biosurfactants

Polymeric biosurfactants are high-molecular-weight amphiphiles with unique structural motifs that enable efficient emulsification, dispersion, and solubilization of hydrophobic substrates. A well-characterized example is emulsan, a lipoheteropolysaccharide produced by *Acinetobacter calcoaceticus* RAG-1. Emulsan's structure comprises a polysaccharide backbone of repeating *N*-acetylgalactosamine and *N*-acetyltalosaminuronic acid residues, esterified with hydroxylated C12–C18 fatty acid chains. This amphipathic architecture allows emulsan to adsorb at oil-water interfaces, reducing surface tension and stabilizing hydrocarbon-in-water emulsions critical for petroleum degradation (Rosenberg 1991).

Alginates, another class of polymeric biosurfactants, are linear anionic polysaccharides synthesized by *Azotobacter vinelandii* and *P. aeruginosa*. Structurally, alginates consist of β-1,4-linked D-mannuronate (M) and L-guluronate (G) residues. The ratio of M/G blocks determines their gel-forming capacity, as guluronate-rich regions crosslink with divalent cations (e.g., Ca^{2+}) to form hydrogels This property enables alginates to sequester heavy metals (e.g., Pb^{2+}, Cd^{2+}) via carboxyl group coordination, making them pivotal in heavy metal bioremediation (Sutirman et al. 2021).

Other bacterial species produce structurally distinct polymeric surfactants. For instance, *Acinetobacter calcoaceticus* A2 synthesizes biodispersan, a polysaccharide-protein complex that disperses hydrophobic aggregates in aqueous systems (Rosenberg et al. 1988). Marine bacteria like *Halomonas* spp. secrete sulfated exopolysaccharides with hydrophobic domains that enhance polycyclic aromatic hydrocarbon (PAH) bioavailability. These polymers often exhibit strain-specific modifications, such as acetyl or sulfate groups, which tailor their affinity for diverse pollutants.

6.3 Production of Biosurfactants

The production of biosurfactants begins with the selection of an appropriate microbial source, followed by cultivation under optimized conditions to promote microbial growth. These microorganisms are then induced to produce biosurfactants, which are subsequently recovered through extraction and purification processes (Fig. 6.2).

6.3.1 Microorganisms

Biosurfactants are synthesized by a diverse spectrum of microorganisms, including bacteria, yeasts, fungi, and extremophiles, each contributing unique structural and functional properties to these biomolecules (Martins and Martins 2018; Van Renterghem et al. 2018). The ecological versatility and metabolic adaptability of these microbes enable biosurfactant production under varied environmental conditions. Table 6.1 summarizes key microbial producers, their biosurfactant types, and applications.

6.3.1.1 Bacterial Producers

Bacterial biosurfactant is a major category amongst microbially derived biosurfactants. *P. aeruginosa* is a model organism for rhamnolipid production, a glycolipid class renowned for its emulsification efficiency and role in hydrocarbon degradation

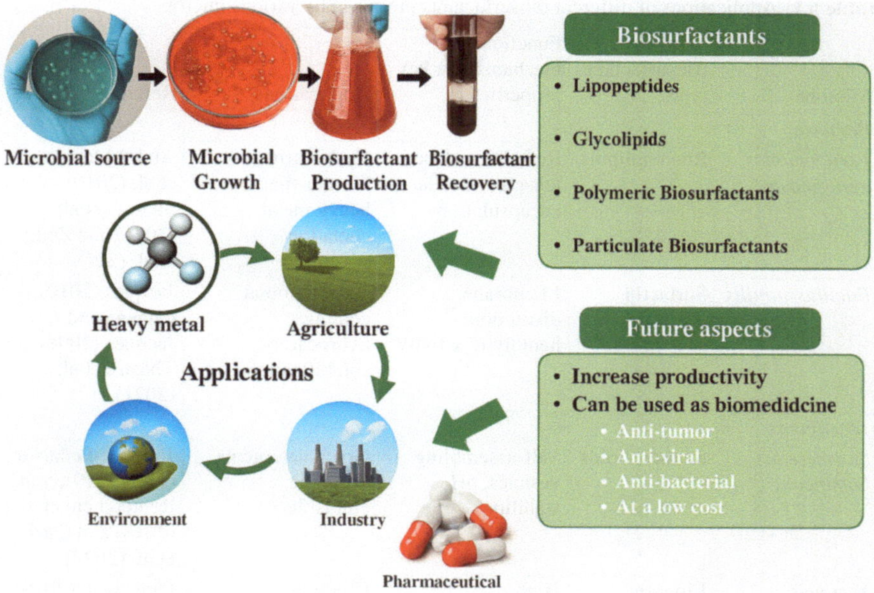

Fig. 6.2 Schematic illustration of biosurfactant production and their potential applications. (Modified and adapted from Kumar et al. 2021)

(Sharma et al. 2021). Similarly, *B. subtilis* synthesizes surfactin, a cyclic lipopeptide with exceptional surface tension reduction and antimicrobial activity, making it suitable for pharmaceuticals and oil recovery (Ongena and Jacques 2008). Other bacterial genera include *A. calcoaceticus*, which produces emulsan, a high-molecular-weight bioemulsifier used in oil spill remediation (Rosenberg et al. 1988). and *Rhodococcus erythropolis*, which generates trehalolipids, stabilizing heavy metal-contaminated soils (Shao 2010). *Lactobacillus delbrueckii*: secretes glycoprotein biosurfactants with anti-adhesive properties against pathogens and *Serratia marcescens* produces serrawettin, a lipopeptide effective in nanoparticle synthesis (Das et al. 2008). *Actinomycetes* are well known for synthesizing lipopeptides with antiviral properties (Mujumdar et al. 2019).

6.3.1.2 Yeast-Derived Biosurfactants

Yeasts are promising producers of glycolipids with applications in green chemistry and cosmetics. *S. bombicola* synthesizes sophorolipids, which reduce surface tension and are employed in biodegradable detergents (Qazi et al. 2022). *Yarrowia lipolytica* utilizes hydrophobic substrates (e.g., alkanes, waste oils) to produce mannosylerythritol lipids (MELs), which exhibit antiviral activity and skin-moisturizing properties (Pang et al. 2019). Other notable yeasts include *Pseudozyma antarctica*,

Table 6.1 Applications of different biosurfactants produced by various microbes

Microbe	Biosurfactant type	Functional mechanism & key properties	Applications	References
Bacteria				
Pseudomonas aeruginosa	Rhamnolipids	Reduces surface tension, micellar encapsulation	Hydrocarbon bioremediation, heavy metal chelation	Abdel-Mawgoud et al. (2010), Sharma et al. (2021) and Zeng et al. (2018)
Bacillus subtilis	Surfactin	Membrane disruption, hemolytic activity	Antimicrobial coatings, hydrocarbon solubilization	Jacques (2010), Ongena and Jacques (2008), and Théatre et al. (2021)
Yeast/fungi				
Starmerella bombicola	Sophorolipids	Self-assembling vesicles, pH stability	Green detergents, cosmetic emulsifiers	Jiménez-Peñalver et al. (2019), Van Renterghem et al. (2018) and Qazi et al. (2022)
Yarrowia lipolytica	Liposan	High emulsification index	Crude oil dispersion, food industry stabilizers	Pang et al. (2019) and Sarubbo et al. (2022)
Actinobacteria				
Rhodococcus erythropolis	Trehalolipids	Extreme thermostability (80 °C), metal complexation	Arctic oil spill remediation, radionuclide removal	Kuyukina and Ivshina (2019) and Shao (2010)
Specialized producers				
Acinetobacter calcoaceticus	Emulsan	High-molecular-weight bioemulsifier	Oil-in-water emulsions for EOR	Rosenberg et al. (1988) and Mujumdar et al. (2019)
Bacillus licheniformis	Lichenysin	Cyclic lipopeptide with low CMC	Antifungal seed coatings, tar sand extraction	Dunlap et al. (2019) and Pacwa-Płociniczak et al. (2011)
Pseudomonas putida	Rhamnolipids	Non-pathogenic strain, high yield	Agricultural pesticide removal	Wittgens et al. (2011) and Lamilla et al. (2021)
Serratia marcescens	Serrawettin	Low molecular weight, antimicrobial synergy	Biofilm disruption in medical devices	Bjerk et al. (2021) and Das et al. (2008)

producing MELs with high emulsification capacity for food-grade applications, and *C. albicans*, known for cell-bound biosurfactants that enhance biofilm dispersal.

6.3.1.3 Fungal Biosurfactant Producers

Fungi, though less explored, offer unique biosurfactants with structural complexity. *Aspergillus fumigatus* produces helvolic acid, a glycolipid with antifungal and anticancer properties (Chain et al. 1943). *Penicillium chrysogenum* generates penicillic acid, a surfactant effective in PAH degradation (Sobrinho et al. 2013). Others include *Ustilago maydis*, which synthesizes cellobiose lipids used for heavy metal sequestration (Beck 2020), and *Fusarium* spp., which produces fusaric acid, a biosurfactant with herbicidal activity.

6.3.1.4 Biosurfactant from Extremophiles

Recent studies have identified extremophiles as promising sources of biosurfactants with unique properties suitable for harsh industrial applications. For instance, *Geobacillus thermodenitrificans*, a thermophile, produces lipopeptides stable at temperatures up to 80 °C, making them ideal for high-temperature industrial processes (Li et al. 2023). Similarly, *Halomonas anticariensis*, a halophile, synthesizes glycolipids functional in high-salinity environments, beneficial for applications in saline conditions (Martínez-Cánovas et al. 2004). Beyond these, several other extremophiles have been identified as biosurfactant producers. *Alcanivorax borkumensis* is a marine bacterium that thrives in hydrocarbon-rich environments and produces biosurfactants that facilitate the degradation of oil pollutants. Its ability to form biofilms around oil droplets and emulsify hydrocarbons makes it instrumental in bioremediation efforts following oil spills (Yakimov et al. 1998). *Natrialba magadii is* an alkaliphilic and halophilic archaeon. *N. magadii* has adapted to survive in soda lakes with high salinity and alkaline pH. Its biosurfactants are stable under these extreme conditions (Schultz and Rosado 2019). Terahara et al. (2019) reported that *Streptomyces otsuchiensis*, isolated from marine sediments, produces biosurfactants with potential applications in biotechnological processes. Rhamnolipids produced by different species of *Pseudomonas* produce rhamnolipid biosurfactants, which show potential in environmental remediation (Devale et al. 2022).

The unique properties of biosurfactants produced by extremophiles—such as stability under high temperatures, salinity, and extreme pH—position them as game changers in the biosurfactant industry. Their robustness allows for applications in harsh industrial processes where conventional biosurfactants may fail, including enhanced oil recovery, bioremediation of polluted environments, and formulations in the food and pharmaceutical industries requiring stability under extreme conditions (Schultz and Rosado 2019).

6.3.2 *Substrate Diversity in Biosurfactant Production*

The selection of substrates is pivotal for optimizing biosurfactant yield, functionality, and commercial viability. Substrates not only influence microbial metabolic pathways but also determine the structural properties of biosurfactants, such as their emulsification capacity and CMC (critical micelle concentration) (Sarubbo et al. 2022). Recent advances emphasize the use of low-cost, renewable, and waste-derived substrates to align with circular economy principles (Jimoh et al. 2021). The substrates can be divided into four classes, supported by cutting-edge applications:

6.3.2.1 Hydrophilic Substrates

Hydrophilic substrates, which are rich in soluble sugars, are widely used due to their rapid microbial assimilation. Pure carbohydrates such as glucose and sucrose are particularly effective with engineered strains. For instance, *P. putida* KT2440 produces rhamnolipids using glucose, achieving a 65% reduction in surface tension (Wittgens et al. 2011). Agro-industrial waste also serves as a valuable substrate. Sugarcane molasses, for example, when used with *P. aeruginosa* MTCC 2297, results in 9.8 g/L of rhamnolipids, reducing production costs by 40% compared to synthetic media (Kumar et al. 2022). Molasses has demonstrated significant potential for biosurfactant production, although its processing may incur additional costs (Begum et al. 2023; Jimenez et al. 2020).

In a study cheese whey was valorized as a sustainable substrate for glycoprotein biosurfactant production (3.5 g/L yield) using *L. casei* MRTL3, demonstrating eco-friendly waste-to-resource potential. The biosurfactant exhibited anti-biofilm activity against *Salmonella enterica*, disrupting biofilm formation by 60–70%, likely due to interfacial and antimicrobial properties. (Kachrimanidou et al. 2022).

6.3.2.2 Hydrophobic Substrates: Oils and Hydrocarbons

Hydrophobic substrates induce biosurfactant synthesis by mimicking natural hydrocarbon-rich environments. Vegetable oils, such as sunflower oil and palm oil, are particularly effective. For example, *S. bombicola* produces sophorolipids using oil, reducing interfacial tension and making it ideal for green detergents (Qazi et al. 2022). Similarly, in a study eco-friendly rhamnolipids using food waste hydrolysate produced *by Burkholderia thailandensis* E264 in a bioreactor achieved a yield of 0.6–0.8 g/L over 72 h. (Kumar et al. 2022). Petroleum derivatives also play a significant role. *Rhodococcus erythropolis* DSM 44308, for instance, yields trehalolipids from crude oil, which degrade PAHs in contaminated soils (Kuyukina and Ivshina 2019). Additionally, *Acinetobacter* produces emulsifiers, emulsan, and alasan, mainly used in microbial enhanced oil recovery and biodegradation of toxic compounds (Mujumdar et al. 2019). However, due to safety and regulatory concerns, hydrocarbon-based substrates might be less favorable for biosurfactant production targeting the food, cosmetic, or pharmaceutical sectors (Qamar and Pacifico 2023).

6.3.2.3 Waste-Derived Substrates

Waste-derived substrates offer a dual benefit by reducing production costs by 30–60% while addressing waste management challenges. Waste oils, such as used cooking oil and soybean soapstock, are particularly effective. For instance, in a study, *Y. lipolytica* was genetically optimized to enhance limonene production using waste cooking oil as the substrate. Initial production rates were low, achieving only 6.1 mg/L. After optimization, which involved overexpressing key genes in the mevalonate pathway and optimizing fermentation parameters, limonene production significantly increased to 33.9 mg/L. (Pang et al. 2019). Lignocellulosic residues also serve as valuable substrates. Additionally, hydrolysates from the biodegradation of waste carbohydrate feedstocks fermented with *Pseudomonas* spp. and *Bacillus* yield biosurfactants facilitating heavy metal sequestration (Begum et al. 2023).

Glycerol, a by-product of biodiesel production, is used by *P. putida* KT2440 to produce 10.2 g/L of putisolvin, enhancing crude oil dispersion (Martins and Martins 2018). Methane is utilized by *Methylococcus capsulatus* to synthesize methane-derived glycolipids for bioremediation of chlorinated solvents. Additionally, lignin is used by *Streptomyces* spp. to produce ligninolytic enzymes and surfactants, enabling simultaneous lignocellulose degradation and emulsification.

Waste-driven biosurfactant production reduces reliance on fossil fuels, lowering carbon footprints by 25–50%. The techno-economic assessment revealed that using stearic acid as a feedstock for di-rhamnolipid production is the most cost-effective approach. This option results in the highest productivity and a relatively low production cost of $12 per kg, while maintaining a moderate carbon footprint. Stearic acid is highlighted as the best option due to its balance of productivity, cost efficiency, and environmental impact. (Noll et al. 2024). These substrates represent abundant, renewable resources, but their complex nature typically necessitates additional pretreatment steps to improve microbial accessibility, potentially increasing the cost and complexity of biosurfactant production processes (Kumar et al. 2021).

6.3.3 *Fermentation Processes*

Biosurfactants are generally produced using two methods: solid-state fermentation (SSF) and submerged fermentation (SmF), also known as liquid fermentation. While SSF is currently understudied but shows promising results, SmF is the methodology most frequently employed (Singh and Rathore 2019). The SSF mostly takes place on the surface of solid materials that have the ability to absorb or hold water, whether they contain soluble nutrients or not. Additionally, since SSF uses less water than liquid fermentation, it typically lowers the total cost of a particular bioprocess.

6.3.3.1 Solid Substrate Fermentation (SSF)

Numerous microorganisms, including fungi, bacteria, and yeasts, are employed to produce various biosurfactants. Research has shown that *Aspergillus niger* can generate biosurfactants such as emulsans when cultivated on solid substrates like wheat bran or rice husk (Asgher et al. 2020). Similarly, *Trichoderma harzianum* is known for producing biosurfactants like lipopeptides and polysaccharides on solid media, recognized for their antifungal properties (Piegza et al. 2021). Fungal species, *Penicillium citrinum,* has been reported to produce biosurfactants under solid-state fermentation (SSF) conditions, which are noted for their emulsification and antimicrobial activities (Nurhariyati et al. 2020).

Rice husk, a common agricultural waste, supports the production of emulsans and rhamnolipids in SSF (Asgher et al. 2020). *Trichoderma* species have also been explored for biosurfactant production using corn cobs as a substrate (Nunes et al. 2021). Molasses, a by-product of sugarcane, serves as an economical carbon source for biosurfactant production by both bacterial and fungal strains. Additionally, soybean cake is utilized as a source of nitrogen and carbon for lipopeptide production, while spent coffee grounds have been innovatively used for biosurfactant production by fungi and bacteria.

Among bacteria, *B. subtilis* is renowned for producing lipopeptides such as surfactin and has been successful in SSF using substrates like soybean cake and wheat bran (Jacques 2010). *P. aeruginosa* is a prominent producer of rhamnolipids and has been cultivated effectively in SSF using agro-industrial by-products (Zeng et al. 2018). *Rhodococcus* spp. are recognized for glycolipid production and have shown significant utility in SSF, particularly in bioremediation and industrial applications (Shao 2010).

Yeasts like *C. tropicalis* and *C. antarctica* are utilized in SSF for producing sophorolipids, valued for their emulsification and antimicrobial properties (Gaur et al. 2019). *Y. lipolytica* has been extensively researched for its capability to produce mannosylerythritol lipids, a type of biosurfactant, under SSF conditions (Pang et al. 2019).

6.3.3.2 Submerged Fermentation (SmF)

Submerged fermentation (SmF) is a widely adopted method for biosurfactant production due to its versatility in substrate utilization. By optimizing substrates, the production of rhamnolipids can be enhanced using carbon sources such as glucose, glycerol, and waste oils (Das and Mukherjee 2007). *B. subtilis* is extensively studied in SmF for its production of the surfactin biosurfactant (Das and Mukherjee 2007). . *Rhodococcus* spp. can synthesize glycolipids and other emulsifying agents through

SmF (Kuyukina and Ivshina 2019). *Lactobacillus rhamnosus* produces glycolipid biosurfactants in SmF, which are explored for their antimicrobial properties and safety in food and pharmaceutical applications (Surekha et al. 2016).

Submerged fermentation allows the use of a range of substrates for biosurfactant production. Both simple sugars and complex organic wastes are effective for fermentation. Glucose is a common carbon source in SmF, utilized by many biosurfactant-producing strains. Glycerol, a by-product of biodiesel production, is a cost-effective substrate that supports the production of rhamnolipids by *P. aeruginosa* and other biosurfactants. Additionally, waste products from the edible oil industry, such as waste frying oils and olive oil mill residues, serve as carbon sources in SmF for producing surfactants like rhamnolipids. Molasses, a by-product from sugarcane and beet sugar production, is another economical carbon source for biosurfactant production by various bacterial strains, including *B. subtilis* and *P. aeruginosa* (Mohanty et al. 2021). Yeast extract is commonly used as a nitrogen source in SmF, supporting microbial growth and boosting biosurfactant synthesis in bacteria such as *B. subtilis* (Das and Mukherjee 2007). Additionally, ammonium nitrate and ammonium sulfate are frequently used nitrogen sources for producing biosurfactants like surfactin and rhamnolipids.

6.3.4 Optimization of Production

Enhancing biosurfactant production requires a holistic approach that combines microbial strain engineering, medium optimization, fermentation control, and efficient downstream processing (Kumar et al. 2021). The selection of elite microbial strains (*Pseudomonas*, *Bacillus*, *Candida*) and their genetic modification through CRISPR-Cas9, recombinant DNA technology, or mutagenesis (e.g., UV/chemical treatments) can amplify biosurfactant yields by overexpressing biosynthetic genes (e.g., *rhlABC* for rhamnolipids) or silencing competing pathways, as demonstrated in engineered *B. subtilis* strains achieving five-fold higher surfactin titers (Wang et al. 2018; Jiménez-Peñalver et al. 2019; Jimoh et al. 2021). Medium composition plays an equally critical role, with carbon sources like agro-industrial waste (e.g., sunflower oil residues) and optimized C:N ratios boosting yields by several folds. Fermentation parameters—pH (6.5–7.5), temperature (30–37 °C), aeration (1–2 vvm), and agitation (150–300 rpm)—must be tightly controlled to maximize microbial activity, particularly for aerobic species like *P. aeruginosa*, where dissolved oxygen levels are pivotal (De Lima et al. 2009; Garcia-Ochoa and Gomez 2009). Post-fermentation, downstream processing begins with cell separation via centrifugation or tangential flow filtration, followed by concentration techniques such as foam fractionation or solvent extraction (e.g., ethyl acetate for glycolipids) to isolate biosurfactants. Final purification using acid precipitation, HPLC, or dialysis achieves >95% purity, essential for high-value applications (Rathankumar et al. 2023). Recent advances emphasize sustainability, with strains like *Y. lipolytica* producing 22 g/L lipopeptides from slaughterhouse waste, slashing feedstock costs by

70% (Ingham and Winterburn 2022). Future innovations, including efficient bioprocess modeling and immobilized cell reactors, promise to further streamline production, positioning biosurfactants as scalable, eco-friendly alternatives to synthetic surfactants.

6.4 Applications for Environmental Management

Hydrophobic organic pollutants, such as hydrocarbons, pose significant challenges in environmental remediation due to their low solubility, which often hinders biodegradation (Zeng et al. 2018). Biosurfactants—environmentally compatible amphiphiles produced by microorganisms—address this issue by enhancing the dispersion, emulsification, and solubilization of such contaminants (Zouari et al. 2018). Unlike synthetic counterparts, biosurfactants are biodegradable, less toxic, and stable under extreme pH, temperature, and salinity conditions. Their specificity for target pollutants minimizes collateral ecological disruption, making them a sustainable choice for large-scale remediation.

By reducing interfacial tension between fluids and oil-solid matrices, biosurfactants weaken capillary forces that trap pollutants, thereby promoting their mobilization and accessibility to degrading microbes (Malkapuram et al. 2021). Biosurfactants increase the hydrophobicity of microbial cell surfaces, enabling stronger interactions between bacteria and hydrophobic substrates. This adhesion improves microbial uptake and degradation efficiency. They mobilize trapped hydrocarbons (e.g., petroleum, polyaromatic hydrocarbons) by solubilizing them into aqueous phases, accelerating microbial degradation. Biosurfactants emulsify oils and heavy metals, enabling their removal via filtration or microbial metabolism. Introducing biosurfactant-producing microbes (e.g., *Pseudomonas*, *Bacillus* spp.) enhances contaminant degradation in polluted environments, particularly where native microbial communities lack resilience (Malkapuram et al. 2021).

While biosurfactants show immense potential, optimizing their production costs, scalability, and congener-specific efficacy remains critical. Integrating biosurfactants with nanotechnology or engineered microbial consortia could further enhance their applicability in complex environmental matrices (Gaur et al. 2022).

6.4.1 *Bioremediation of Petroleum*

While petroleum hydrocarbons are biodegradable (60–90% degradation potential), their natural breakdown is often slow, posing challenges for efficient environmental remediation (Alaidaroos 2023). Biosurfactants play a pivotal role in accelerating this process by enhancing the bioavailability of hydrophobic pollutants, enabling microbes to metabolize them more effectively (Machado et al. 2020). Research in this field typically focuses on two key substrates: soil from oil-contaminated industrial sites and artificially polluted soils designed to replicate real-world conditions (Baruah et al. 2024). Notably, petroleum sludge and hydrocarbon-rich waste have proven to be viable substrates for biosurfactant production, particularly when using microbial strains isolated from contaminated environments. For example, *Pseudomonas* and *Bacillus* species—dominant biosurfactant producers in oil-polluted ecosystems—have demonstrated remarkable efficacy in degrading hydrocarbons (Martins and Martins 2018). Marine-derived strains such as *B. subtilis* AS2, *B. licheniformis* AS3, and *B. velezensis* AS4 further showed this potential, exhibiting high emulsification capacity for low-molecular-weight hydrocarbons like diesel and crude oil fractions (Prakash et al. 2021).

Due to the complex composition of petroleum, researchers often rely on reductions in total petroleum hydrocarbons (TPH) as a standardized metric to evaluate biodegradation efficiency across studies (Steliga et al. 2020). This approach allows for consistent comparisons of biosurfactant-driven remediation strategies, which may involve bioaugmentation (introducing biosurfactant-producing microbes) or direct application of biosurfactant formulations (Kakde and Sharma 2024). Such strategies not only enhance TPH removal but also align with sustainable practices by repurposing industrial waste as feedstocks for biosurfactant synthesis.

6.4.2 Bioremediation of Polycyclic Aromatic Hydrocarbons (PAHs)

Polycyclic aromatic hydrocarbons (PAHs)—hydrophobic organic compounds characterized by fused aromatic rings—are persistent environmental pollutants linked to carcinogenic, mutagenic, and toxic effects on human health. These contaminants primarily originate from anthropogenic activities, including incomplete combustion of fossil fuels, industrial processes (e.g., coal gasification), and the use of wood preservatives like creosote, which contains PAHs. High-molecular-weight PAHs (e.g., benzo[a]pyrene) pose challenges for bioremediation due to their low bioavailability, as they strongly adsorb to soil organic matter and form recalcitrant complexes (Alotaibi et al. 2024).

Biosurfactants address this limitation by solubilizing PAHs, increasing their accessibility to degrading microorganisms. For instance, *P. aeruginosa* strains producing rhamnolipids enhanced pyrene degradation in contaminated soils by forming micelles that desorb PAHs from soil matrices (Yesankar et al. 2022). Similarly, a study using *B. subtilis* surfactin demonstrated a higher reduction in phenanthrene concentrations within 30 days, compared to untreated controls (Bhadra et al. 2022). Research often employs soil samples spiked with PAHs or collected from polluted sites (e.g., industrial zones, creosote-treated sites) to simulate real-world remediation scenarios. For example, Karlapudi et al. (2018) isolated *Acinetobacter junii* from oil-contaminated soil, which degraded 89% of fluoranthene (a 4-ring PAH) within 21 days when supplemented with biosurfactants. These findings feature the dual role of biosurfactants: enhancing PAH bioavailability while stimulating microbial metabolic activity, offering a sustainable strategy to mitigate PAH contamination in soil and groundwater.

6.4.3 Bioremediation of Environments Contaminated with Metals

Aside from altering microbial metabolism, biosurfactants help with metal solubilization, dispersion, and binding, making it easier to reclaim treated soils. Metal treatment procedures include both in situ (phytoremediation, chemical mobilization, electrokinetic extraction, soil flushing, and surface capping) and ex situ (soil washing and solidification). Soil flushing and washing are especially suggested for capturing biosurfactant-producing microorganisms (Selva Filho et al. 2023).

The efficiency of metal removal is dependent on the structure and properties of biosurfactant-metal interactions. For instance, studies show that a biosurfactant secreted by *Bacillus* sp. exhibits notable heavy metal removal efficiency, including cadmium, manganese, lead and mercury at double the critical micelle concentration (CMC), often leading to co-precipitation (Mnif et al. 2022). Concentration of biosurfactants is another critical parameter, as higher concentrations generally lead to increased removal efficiency.

When evaluating biosurfactants in contaminated environments, it is important to consider their effects on soil quality comprehensively. Therefore, assessments often include tests for plant germination and growth in treated soils. However, while studies frequently examine the capability of biosurfactants to remove metals, the comprehensive assessment of the overall worth of the remedied soil is not consistently conducted.

6.4.4 Biosurfactants: Biocidal Agents

Biosurfactants, produced by microorganisms, exhibit biocidal properties at elevated concentrations, effectively reducing microbial biodiversity. Their bactericidal effects are attributed to mechanisms such as the generation of free hydroxyl radicals, which induce oxidative stress, disrupt cellular integrity, and lead to bacterial cell death (Gaur et al. 2019; Sharma et al. 2021). At their critical micelle concentration (CMC)—the threshold at which surfactant molecules aggregate into micelles—biosurfactants interfere with microbial metabolic pathways and alter cell surface properties, ultimately causing irreversible damage (Sajna et al. 2013). Studies have demonstrated the dose-dependent biocidal effects of biosurfactants across diverse microorganisms. For instance, sophorolipids secreted by *C. albicans* (CA-B) and *C. glabrata* (CG-B) show pronounced antibacterial activity against Gram-positive pathogens. CG-B reduced populations of *B. subtilis* MTCC 441 by 65%, while CA-B achieved a 24.2% reduction (Gaur et al. 2019). In contrast, both sophorolipids showed minimal efficacy against Escherichia *coli* MTCC723 (<5% reduction), likely due to the protective lipopolysaccharide layer in Gram-negative cell walls (Sharma et al. 2021). Similarly, rhamnolipids, another class of biosurfactants, exhibit broad-spectrum antimicrobial activity. Their minimum inhibitory concentrations (MICs) range from 50 to 1600 μg mL^{-1}, with higher concentrations required for Gram-negative bacteria (Sotirova et al. 2008).

Rhamnolipids have also proven effective against plant pathogens. In trials against *Phytophthora infestans* (causing potato late blight), leaf applications of rhamnolipids reduced infected leaf area from 9.8 cm^2 (untreated controls) to 0.06 cm^2 at high concentrations (0.3%), though slight phytotoxicity was observed at this dose (Tomar et al. 2019). Lower concentrations ($\leq$0.15%) showed diminished efficacy, underscoring the importance of dose optimization. Rhamnolipids induced zoospore lysis, highlighting their potential to curb disease spread post-infection (Tomar et al. 2019). While substrate choice influenced rhamnolipid congener composition (e.g., hydrophilic variants increased with glucose), antifungal efficacy remained consistent (Rodrigues et al. 2021).

Biosurfactants also show promise in vector control. Cyclic lipopeptides demonstrated larvicidal activity against *Anopheles culicifacies,* a malaria vector, with an LC50 of 110 μg mL^{-1} over 48 h. Mortality reached 60% at higher doses, attributed to reduced water surface tension, which disrupts egg adhesion and drowns larvae (Pradhan et al. 2018).

Biosurfactants also have viricidal activity. For example, sophorolipids exhibit antiviral activity against RNA viruses, suggesting potential utility against pathogens like SARS-CoV-2 (Daverey et al. 2021). However, their efficacy is concentration-dependent, necessitating careful evaluation for target-specific optimization. Unlike chemical biocides, biosurfactants are biodegradable and less likely to accumulate in ecosystems, aligning with global demands for eco-friendly solutions.

6.4.5 Bioremediation for Pesticides

Pesticides, while critical for managing agricultural pests and phytopathogens, pose significant environmental risks due to their persistence and accumulation in ecosystems, contaminating soil, water, and food chains (Abdel-Shafy and Mansour 2018; AI-Jawhari 2018). Addressing this challenge, biosurfactants have emerged as innovative tools for mitigating pesticide pollution. These agents excel at solubilizing hydrophobic contaminants, mirroring their effectiveness in remediating heavy metals and polycyclic aromatic hydrocarbons (PAHs). (Mishra et al. 2024). By forming micellar complexes with pesticide molecules, biosurfactants disrupt hydrogen bonds between pesticides and water, enhancing solubility and bioavailability for microbial degradation (López-Prieto et al. 2020).

Key bacterial genera such as *Pseudomonas*, *Bacillus*, and *Acinetobacter* play dual roles in bioremediation. Their biosurfactants not only mobilize heavy metals like lead (Pb) and cadmium (Cd) but also accelerate the breakdown of pesticides such as lindane and organophosphorus compounds (Pacwa-Płociniczak et al. 2011; Wan et al. 2015). For example, rhamnolipids from *P. aeruginosa* degrade the organophosphorus insecticide quinalphos efficiently (Nair et al. 2014), while glycolipids from *P. rhodesiae* and glycopeptides from *P. marginalis* enhance chlorpyrifos degradation by up to 85% (Lamilla et al. 2021).

Beyond enhancing biodegradation, biosurfactants exhibit direct pesticidal activity. Glycolipid-based formulations, patented for controlling arachnids, larvae, and aphids, physically disrupt insect cuticles (Patel et al. 2023). Rhamnolipids, for instance, damage aphid cuticle membranes, causing desiccation and mortality (Mnif and Ghribi 2016).

These multifaceted applications position biosurfactants as sustainable alternatives to chemical remediation, bridging efficient contaminant removal with eco-friendly agricultural practices.

6.5 Challenges and Future Prospects

Despite their immense potential in environmental, industrial, and biomedical applications, the widespread adoption of microbial biosurfactants remains constrained by critical challenges in production, scalability, and cost-efficiency. A primary barrier is the high production cost, driven by expensive carbon sources (e.g., refined sugars, oils) and energy-intensive downstream purification processes (Karmakar et al. 2025). For instance, conventional substrates like glucose or glycerol account for ~30–50% of total production expenses, limiting economic feasibility for large-scale applications (Thakur et al. 2024). Furthermore, microbial biosurfactant yields often lag behind synthetic surfactants due to metabolic inefficiencies and strain-specific limitations. *Pseudomonas* and *Bacillus* strains, while prolific producers, typically yield 10–40 g/L of biosurfactants under optimized conditions—a fraction of the output achievable with petrochemical-based surfactants (Argentin et al. 2023; Thakur et al. 2024; Ganger et al. 2024).

To address these hurdles, research is pivoting toward sustainable substrate valorization. Agro-industrial waste (e.g., lignocellulosic biomass, molasses, oil refinery by-products) and low-cost feedstocks (e.g., waste cooking oil, dairy effluents) are being explored to slash costs while aligning with circular economy principles. (Mulligan 2023; Dalbanjan et al. 2024; Varjani et al. 2024). For example, in a few cases *Starmerella* enhanced the production of sophorolipid, which was achieved by using oil waste waste-to-biosurfactant conversion. Similarly, rhamnolipids produced from sugarcane bagasse were found to reduce substrate cost. Advances in synthetic biology and metabolic engineering are poised to revolutionize yields. CRISPR-Cas9 and omics-guided pathway optimization enable precise editing of biosynthetic gene clusters (e.g., rhlABC for rhamnolipids) to enhance precursor flux and suppress competitive pathways (Yadav et al. 2024). Engineered *E. coli* strains expressing heterologous surfactin synthetases have achieved titers exceeding a multifold increase over the wild-type producer (Jimoh et al. 2021; Xia and Wen 2022).

Bioprocessing innovations are critical to scaling production. Continuous fermentation systems, coupled with in-situ product recovery techniques (e.g., foam fractionation, membrane filtration), minimize downtime and improve productivity (Chen et al. 2021). For instance, integrated membrane bioreactors reduced rhamnolipid recovery costs significantly in a few instances (Anic et al. 2018). Additionally, immobilized cell systems and co-culture strategies enhance process stability and reduce contamination risks (Kapoore et al. 2021).

The regulatory landscape and market dynamics also play pivotal roles. Standardized toxicity and efficacy testing protocols are needed to streamline approvals for biomedical and agricultural use (Bjerk et al. 2021; Abdoli et al. 2025). Looking ahead, the global biosurfactant market is projected to grow at a CAGR of 5.6% (2023–2030), driven by demand for green alternatives in agriculture, cosmetics, and bioremediation (Roelants and Soetaert 2021). Emerging applications—such as biosurfactant-functionalized nanoparticles for targeted drug delivery or biofilm disruption in medical devices—highlight their versatility and industrial efficiency in a post-petrochemical era.

6.6 Conclusion

Biosurfactants are microbial surface-active chemicals widely used in environmental applications to improve soil fertility, protect plants in harsh environments, and function as natural biocides. Their principal environmental use is bioremediation methods that involve biostimulation and/or bioaugmentation tactics, making them potential alternatives for treating contaminants such as hydrocarbons, PAHs, and heavy metals. In agriculture, biosurfactants have the capacity to remove pesticides and remediate plants. Biosurfactants have several advantages over synthetic surfactants, including low toxicity, stability under changing conditions (temperature, pH, salinity), and high biodegradability, which is critical for effective environmental application. This review emphasizes biosurfactants' role in bioremediation as sustainable solutions, outperforming synthetic counterparts. Recognizing biosurfactants as superior alternatives requires exploring greener production methods utilizing agro-industrial wastes to reduce costs and optimize efficiency. The chapter also delves into biosurfactant applications in the petroleum industry and bioremediation of oil spills. Despite current economic competitiveness with chemical surfactants, advancing large-scale production from low-cost substrates is essential to enhance the feasibility and widespread adoption of these natural compounds.

Acknowledgments The authors are highly grateful to the Department of Biotechnology for providing all necessary facilities.

Conflict(s) of Interest The author(s) declare(s) that there is no conflict(s) of interest regarding the publication of this article.

References

Abdel-Mawgoud AM, Lépine F, Déziel E (2010) Rhamnolipids: diversity of structures, microbial origins and roles. Appl Microbiol Biotechnol 86(5):1323–1336. https://doi.org/10.1007/s00253-010-2498-2

Abdel-Shafy HI, Mansour MS (2018) Phytoremediation for the elimination of metals, pesticides, PAHs, and other pollutants from wastewater and soil. In: Kumar V, Kumar M, Prasad R (eds) Phytobiont and ecosystem restitution. Springer, Singapore, pp 101–136. https://doi.org/10.1007/978-981-13-1187-1_5

Abdoli S, Asgari Lajayer B, Bagheri Novair S, Price GW (2025) Unlocking the potential of biosurfactants in agriculture: novel applications and future directions. Sustainability 17(5):2110. https://doi.org/10.3390/su17052110

Abhyankar I, Hirlekar S, Prabhune A, Nisal A (2024) Bridging the gap: an investigation of biosurfactants-polymer systems. Curr Opin Colloid Interface Sci 72:101806. https://doi.org/10.1016/j.cocis.2024.101806

Al-Jawhari IF (2018) Heavy metals, polycyclic aromatic hydrocarbons (PAHs), radioactive materials, xenobiotic, pesticides, hazardous chemicals and dyes bioremediation. In: Kumar V, Kumar M, Prasad R (eds) Phytobiont and ecosystem restitution. Springer, Singapore, pp 215–229. https://doi.org/10.1007/978-981-13-1187-1_11

Alaidaroos BA (2023) Advancing eco-sustainable bioremediation for hydrocarbon contaminants: challenges and solutions. PRO 11(10):3036. https://doi.org/10.3390/pr11103036

Alotaibi FS, Alrajhi A, Alharbi S (2024) Bioremediation of high molecular weight polycyclic aromatic hydrocarbons. In: Microbes based approaches for the Management of Hazardous Contaminants, pp 378–389. https://doi.org/10.1002/9781119851158.ch24

Anic I, Apolonia I, Franco P, Wichmann R (2018) Production of rhamnolipids by integrated foam adsorption in a bioreactor system. AMB Express 8(1). https://doi.org/10.1186/s13568-018-0651-y

Argentin MN, Martins LF, Sousa MP, Bossolan NR (2023) Biosurfactant from a thermo-halophilic strain of *Bacillus alveayuensis* isolated from a Brazilian oil reservoir: production, chemical characterization, antimicrobial activity, and efficiency in wettability reversal and oil removal from oil-soaked sand. Geoenergy Sci Eng 231:212324. https://doi.org/10.1016/j.geoen.2023.212324

Asgher M, Arshad S, Qamar S-A. et al (2020) Improved biosurfactant production from *Aspergillus niger* through chemical mutagenesis: characterization and RSM optimization. SN Appl Sci 2:966. https://doi.org/10.1007/s42452-020-2783-3

Baruah NP, Goswami M, Sarma N, Chowdhury D, Devi A (2024) Pioneering technologies over time to rehabilitate crude oil–contaminated ecosystems: a review. Environ Sci Pollut Res 31(55):63576–63602. https://doi.org/10.1007/s11356-024-35442-9

Beck A, Zibek S (2020) Growth behavior of selected ustilaginaceae fungi used for Mannosylerythritol lipid (MEL) biosurfactant production—evaluation of a defined culture medium. Front Bioeng Biotechnol 8. https://doi.org/10.3389/fbioe.2020.555280

Begum W, Saha B, Mandal U (2023) A comprehensive review on production of bio-surfactants by bio-degradation of waste carbohydrate feedstocks: an approach towards sustainable development. RSC Adv 13(36):25599–25615. https://doi.org/10.1039/d3ra05051c

Bhadra S, Chettri D, Kumar Verma A (2022) Biosurfactants: secondary metabolites involved in the process of bioremediation and biofilm removal. Appl Biochem Biotechnol 195(9):5541–5567. https://doi.org/10.1007/s12010-022-03951-3

Bjerk TR, Severino P, Jain S, Marques C, Silva AM, Pashirova T, Souto EB (2021) Biosurfactants: properties and applications in drug delivery, biotechnology and ecotoxicology. Bioengineering 8(8):115. https://doi.org/10.3390/bioengineering8080115

Bolan S, Padhye LP, Mulligan CN, Alonso ER, Saint-Fort R, Jasemizad T, Wang C, Zhang T, Rinklebe J, Wang H, Siddique KH, Kirkham M, Bolan N (2023) Surfactant-enhanced mobilization of persistent organic pollutants: potential for soil and sediment remediation and unintended consequences. J Hazard Mater 443:130189. https://doi.org/10.1016/j.jhazmat.2022.130189

Carolin CF, Kumar PS, Ngueagni PT (2021) A review on new aspects of lipopeptide biosurfactant: types, production, properties and its application in the bioremediation process. J Hazard Mater 407:124827. https://doi.org/10.1016/j.jhazmat.2020.124827

Chain E, Florey H-W, Jennings M-A, Williams T-I (1943) Helvolic acid, an antibiotic produced by aspergillus fumigatus, mut. helvola yuill. Br J Exp Pathol 24(3):108–119

Chen C, Li D, Li R, Shen F, Xiao G, Zhou J (2021) Enhanced biosurfactant production in a continuous fermentation coupled with in situ foam separation. Chem Eng Process Process Intensif 159:108206. https://doi.org/10.1016/j.cep.2020.108206

Chung YR, Kim CH, Hwang I, Chun J (2000) *Paenibacillus koreensis* Sp. Nov., a new species that produces an iturin-like antifungal compound. Int J Syst Evol Microbiol 50(4):1495–1500. https://doi.org/10.1099/00207713-50-4-1495

Dalbanjan NP, Eelager MP, Narasagoudr SS (2024) Microbial protein sources: a comprehensive review on the potential usage of fungi and cyanobacteria in sustainable food systems. Food Humanit 3:100366. https://doi.org/10.1016/j.foohum.2024.100366

Das K, Mukherjee A-K (2007) Comparison of lipopeptide biosurfactants production by *Bacillus subtilis strains* in submerged and solid state fermentation systems using a cheap carbon source: some industrial applications of biosurfactants. Process Biochem 42(8):1191–1199

Das P, Mukherjee S, Sen R (2008) Antimicrobial potential of a lipopeptide biosurfactant derived from a marine *Bacillus circulans*. J Appl Microbiol 104(6):1675–1684. https://doi.org/10.1111/j.1365-2672.2007.03701.x

Daverey A, Dutta K, Joshi S, Daverey A (2021) Sophorolipid: a glycolipid biosurfactant as a potential therapeutic agent against COVID-19. Bioengineered 12(2):9550–9560. https://doi.org/10.1080/21655979.2021.1997261

De Lima C-J-B, Ribeiro E-J, Servulo E-F-C, Resende M-M, Cardoso V-L (2009) Biosurfactant production by Pseudomonas aeruginosa grown in residual soybean oil. App Biochem Biotech 152(1):156–168

Dervichian D (1964) The physical chemistry of phospholipids. Prog Biophys Mol Biol 14:263–IN3. https://doi.org/10.1016/s0079-6107(64)80008-0

Devale A, Sawant R, Pardesi K, Satpute S, Mujumdar S (2022) Rhamnolipids (RLs) green solution for global hydrocarbon pollution. In: Microbial surfactants. CRC Press, pp 208–225. https://doi.org/10.1201/9781003260165-12

Dunlap CA, Bowman MJ, Rooney AP (2019) Iturinic Lipopeptide diversity in the *Bacillus subtilis* species group—important antifungals for plant disease biocontrol applications. Front Microbiol 10. https://doi.org/10.3389/fmicb.2019.01794

Elumalai P, Parthipan P, Huang M, Muthukumar B, Cheng L, Govarthanan M, Rajasekar A (2021) Enhanced biodegradation of hydrophobic organic pollutants by the bacterial consortium: impact of enzymes and biosurfactants. Environ Pollut 289:117956. https://doi.org/10.1016/j.envpol.2021.117956

Eras-Muñoz E, Farré A, Sánchez A, Font X, Gea T (2022) Microbial biosurfactants: a review of recent environmental applications. Bioengineered 13(5):12365–12391. https://doi.org/10.1080/21655979.2022.2074621

Farias CB, Almeida FC, Silva IA, Souza TC, Meira HM, Soares da Silva RD, Luna JM, Santos VA, Converti A, Banat IM, Sarubbo LA (2021) Production of green surfactants: market prospects. Electron J Biotechnol 51:28–39. https://doi.org/10.1016/j.ejbt.2021.02.002

Garcia-Ochoa F, Gomez E (2009) Bioreactor scale-up and oxygen transfer rate in microbial processes: An overview. Biotechnol Adv 27(2):153–176. https://doi.org/10.1016/j.biotechadv.2008.10.006

Ganger S, Sonawane D, Mhatre N, Saiger F, Pratap A (2024) Biosurfactant production: limitations and new approaches. In: Microbial surfactants in pharmaceuticals and cosmetics. CRC Press, pp 26–57. https://doi.org/10.1201/9781003424864-2

Gaur VK, Regar RK, Dhiman N, Gautam K, Srivastava JK, Patnaik S, Kamthan M, Manickam N (2019) Biosynthesis and characterization of sophorolipid biosurfactant by *Candida* spp.:

application as food emulsifier and antibacterial agent. Bioresour Technol 285:121314. https://doi.org/10.1016/j.biortech.2019.121314

Gaur VK, Sharma P, Gupta S, Varjani S, Srivastava J, Wong JW, Ngo HH (2022) Opportunities and challenges in omics approaches for biosurfactant production and feasibility of site remediation: strategies and advancements. Environ Technol Innov 25:102132. https://doi.org/10.1016/j.eti.2021.102132

Henkel M, Hausmann R (2019) Diversity and classification of microbial surfactants. In: Biobased Surfactants, pp 41–63. https://doi.org/10.1016/b978-0-12-812705-6.00002-2

Inge NA, Van BD, Develter W, Soetaert EJ, Vandamme (2007) Cloning and characterization of the NADPH cytochrome P450 reductase gene (CPR) from *Candida bombicola.* FEMS Yeast Res 7(6):922–928. https://doi.org/10.1111/fyr.2007.7.issue-6, https://doi.org/10.1111/j.1567-1364.2007.00262.x

Ingham B, Winterburn J (2022) Developing an understanding of sophorolipid synthesis through application of a central composite design model. Microb Biotechnol 15(6):1744–1761. https://doi.org/10.1111/1751-7915.14003

Jacques P (2010) Surfactin and other Lipopeptides from *Bacillus* spp. In: Soberón-Chávez G (ed) Biosurfactants. Microbiology monographs, vol 20. Springer, Heidelberg, pp 57–91. https://doi.org/10.1007/978-3-642-14490-5_3

Jakinala P, Lingampally N, Kyama A, Hameeda B (2019) Enhancement of atrazine biodegradation by marine isolate *bacillus velezensis* MHNK1 in presence of surfactin lipopeptide. Ecotoxicol Environ Saf 182:109372. https://doi.org/10.1016/j.ecoenv.2019.109372

Jimenez A, Requiso P, Maloles J, Alcantara E, Alcantara V (2020) Biosurfactant production by *Streptomyces* Sp. CGS B11 using molasses and spent yeast medium. Philipp J Sci 150(1). https://doi.org/10.56899/150.01.01

Jiménez-Peñalver P, Rodríguez A, Daverey A, Font X, Gea T (2019) Use of wastes for sophorolipids production as a transition to circular economy: state of the art and perspectives. Rev Environ Sci Biotechnol 18(3):413–435. https://doi.org/10.1007/s11157-019-09502-3

Jimoh AA, Senbadejo TY, Adeleke R, Lin J (2021) Development and genetic engineering of hyper-producing microbial strains for improved synthesis of biosurfactants. Mol Biotechnol 63(4):267–288. https://doi.org/10.1007/s12033-021-00302-1

Kachrimanidou V, Alimpoumpa D, Papadaki A, Lappa I, Alexopoulos K, Kopsahelis N (2022) Cheese whey utilization for biosurfactant production: evaluation of bioprocessing strategies using novel *Lactobacillus* strains. Biomass Convers Biorefinery 12(10):4621–4635. https://doi.org/10.1007/s13399-022-02767-9

Kakde P, Sharma J (2024) Microbial bioremediation of petroleum contaminated soil: structural complexity, degradation dynamics and advanced remediation techniques. J Pure Appl Microbiol 18(4):2244–2261. https://doi.org/10.22207/jpam.18.4.28

Kapoore RV, Padmaperuma G, Maneein S, Vaidyanathan S (2021) Co-culturing microbial consortia: approaches for applications in biomanufacturing and bioprocessing. Crit Rev Biotechnol 42(1):46–72. https://doi.org/10.1080/07388551.2021.1921691

Karlapudi AP, Venkateswarulu T, Tammineedi J, Kanumuri L, Ravuru BK, Dirisala VR, Kodali VP (2018) Role of biosurfactants in bioremediation of oil pollution-a review. Petroleum 4(3):241–249. https://doi.org/10.1016/j.petlm.2018.03.007

Karmakar K, Sarkar R, Pal A, Rahaman SM, Acharjee A, Saha B (2025) Recent advances and emerging trends in Biosurfactants: a concise review. J Solut Chem 54(4):393–420. https://doi.org/10.1007/s10953-024-01425-0

Kashif A, Rehman R, Fuwad A, Shahid MK, Dayarathne N, Jamal A, Aftab MN, Mainali B, Choi Y (2022) Current advances in the classification, production, properties and applications of microbial biosurfactants—a critical review. Adv Colloid Interf Sci 306:102718. https://doi.org/10.1016/j.cis.2022.102718

Kumar A, Singh SK, Kant C, Verma H, Kumar D, Singh PP, Modi A, Droby S, Kesawat MS, Alavilli H, Bhatia SK, Saratale GD, Saratale RG, Chung S, Kumar M (2021) Microbial biosurfactant: a new frontier for sustainable agriculture and pharmaceutical industries. Antioxidants 10(9):1472. https://doi.org/10.3390/antiox10091472

Kumar R, Johnravindar D, Wong JW, Patria RD, Kaur G (2022) Economical Di-rhamnolipids biosynthesis by non-pathogenic *Burkholderia thailandensis* E264 using post-consumption food waste in a biorefinery approach. Sustainability 15(1):59. https://doi.org/10.3390/su15010059

Kuyukina MS, Ivshina IB (2019) Production of Trehalolipid Biosurfactants by Rhodococcus. In: Alvarez H (ed) Biology of Rhodococcus, Microbiology monographs, vol 16. Springer, Cham, pp 271–298. https://doi.org/10.1007/978-3-030-11461-9_10

Lamilla C, Schalchli H, Briceño G, Leiva B, Donoso-Piñol P, Barrientos L, Rocha VA, Freire DM, Diez MC (2021) A pesticide Biopurification system: a source of biosurfactant-producing bacteria with environmental biotechnology applications. Agronomy 11(4):624. https://doi.org/10.3390/agronomy11040624

Lang S (2002) Biological amphiphiles (microbial biosurfactants). Curr Opin Colloid Interface Sci 7(1–2):12–20. https://doi.org/10.1016/s1359-0294(02)00007-9

Li J, Liu Y, Zhou L, Gang H, Liu J, Sun G, Wang W, Yang S, Mu B (2023) Structural diversity of the lipopeptide biosurfactant produced by a newly isolated strain, *Geobacillus thermodenitrifcans* ME63. ACS Omega 8(24):22150–22158. https://doi.org/10.1021/acsomega.3c02194

López-Prieto A, Moldes AB, Cruz JM, Pérez Cid B (2020) Towards more ecofriendly pesticides: use of Biosurfactants obtained from the corn milling industry as solubilizing agent of copper Oxychloride. J Surfactant Deterg 23(6):1055–1066. https://doi.org/10.1002/jsde.12463

Machado TS, Decesaro A, Cappellaro ÂC, Machado BS, Van Schaik Reginato K, Reinehr CO, Thomé A, Colla LM (2020) Effects of homemade biosurfactant from *bacillus methylotrophicus* on bioremediation efficiency of a clay soil contaminated with diesel oil. Ecotoxicol Environ Saf 201:110798. https://doi.org/10.1016/j.ecoenv.2020.110798

Malkapuram ST, Sharma V, Gumfekar SP, Sonawane S, Sonawane S, Boczkaj G, Seepana MM (2021) A review on recent advances in the application of biosurfactants in wastewater treatment. Sustain Energy Technol Assess 48:101576. https://doi.org/10.1016/j.seta.2021.101576

Martínez-Cánovas M, Béjar V, Martínez-Checa F, Quesada E (2004) Halomonas anticariensis sp. nov., from Fuente de Piedra, a saline-wetland wildfowl reserve in Malaga, Southern Spain. Int J Syst Evol Microbiol 54:1329–1332. https://doi.org/10.1099/ijs.0.63108-0

Martins PC, Martins VG (2018) Biosurfactant production from industrial wastes with potential remove of insoluble paint. Int Biodeterior Biodegrad 127:10–16. https://doi.org/10.1016/j.ibiod.2017.11.005

Mishra S, Lin Z, Pang S, Zhang Y, Bhatt P, Chen S (2021) Biosurfactant is a powerful tool for the bioremediation of heavy metals from contaminated soils. J Hazard Mater 418:126253. https://doi.org/10.1016/j.jhazmat.2021.126253

Mishra N, Kulkarni BD, Abrar S, Shivale N, Agsar D (2024) Biosurfactants and biomass treatment technologies for heavy metals in wastewater. In: Bio-organic amendments for heavy metal remediation, pp 155–177. https://doi.org/10.1016/b978-0-443-21610-7.00038-0

Mnif I, Ghribi D (2016) Glycolipid biosurfactants: Main properties and potential applications in agriculture and food industry. J Sci Food Agric 96(13):4310–4320. https://doi.org/10.1002/jsfa.7759

Mnif I, Sahnoun R, Ellouz-Chaabouni S, Ghribi D (2017) Application of bacterial biosurfactants for enhanced removal and biodegradation of diesel oil in soil using a newly isolated consortium. Process Saf Environ Prot 109:72–81. https://doi.org/10.1016/j.psep.2017.02.002

Mohanty SS, Koul Y, Varjani S, Pandey A, Ngo HH, Chang J, Wong JW, Bui X (2021) A critical review on various feedstocks as sustainable substrates for biosurfactants production: a way towards cleaner production. Microb Cell Fact 20(1). https://doi.org/10.1186/s12934-021-01613-3

Mnif I, Bouallegue A, Bouassida M, Ghribi D (2022) Surface properties and heavy metals chelation of lipopeptides biosurfactants produced from date flour by *Bacillus subtilis* ZNI5: optimized production for application in bioremediation. Bioprocess Biosyst Eng 45(1):31–44

Mujumdar S, Joshi P, Karve N (2019) Production, characterization, and applications of bioemulsifiers (BE) and biosurfactants (BS) produced by *Acinetobacter* spp.: a review. J Basic Microbiol 59(3):277–287. https://doi.org/10.1002/jobm.201800364

Mulligan CN (2023) Sustainable production of biosurfactants using waste substrates. In: Advancements in biosurfactants research, pp 57–77. https://doi.org/10.1007/978-3-031-21682-4_3

Nair AM, Rebello S, Rishad KS, Asok AK, Jisha MS (2014) Biosurfactant facilitated biodegradation of Quinalphos at high concentrations by *Pseudomonas aeruginosa* Q10. Soil Sediment Contam Int J 24(5):542–553. https://doi.org/10.1080/15320383.2015.988205

Naughton P, Marchant R, Naughton V, Banat I (2019) Microbial biosurfactants: current trends and applications in agricultural and biomedical industries. J Appl Microbiol 127(1):12–28. https://doi.org/10.1111/jam.14243

Noll P, Solarte-Toro JC, Restrepo-Serna DL, Treinen C, Poveda-Giraldo JA, Henkel M, Cardona Alzate CA, Hausmann R (2024) Limits for sustainable biosurfactant production: techno-economic and environmental assessment of a rhamnolipid production process. Bioresour Technol Rep 25:101767. https://doi.org/10.1016/j.biteb.2024.101767

Nunes HMAR, Vieira IMM, Santos BLP, Silva DP, & Ruzene DS (2021) Biosurfactants produced from corncob: a bibliometric perspective of a renewable and promising substrate. Prepar Biochem Biotechnol, 52(2):123–134. https://doi.org/10.1080/10826068.2021.1929319

Nurhariyati T, Surtiningsih T, Yuliani H, Fatimah F, Sari S, Trikurniadewi N, Khiftiyah A, Indriyasari K, Ni'matuzahroh NM, Ibrahim S (2020) Corn Cob Hydrolyzate from Penicillium citrinum H9 as an alternative substrate for Biosurfactant production by Hydrocarbonoclastic Bacteria. Ecol Environ Conserv 26:169–175

Ongena M, Jacques P (2008) *Bacillus* lipopeptides: versatile weapons for plant disease biocontrol. Trends Microbiol 16(3):115–125. https://doi.org/10.1016/j.tim.2007.12.009

Pacwa-Płociniczak M, Płaza GA, Piotrowska-Seget Z, Cameotra SS (2011) Environmental applications of biosurfactants: recent advances. Int J Mol Sci 12(1):633–654. https://doi.org/10.3390/ijms12010633

Pang Y, Zhao Y, Li S, Zhao Y, Li J, Hu Z, Zhang C, Xiao D, Yu A (2019) Engineering the oleaginous yeast *Yarrowia lipolytica* to produce limonene from waste cooking oil. Biotechnol Biofuels 12(1). https://doi.org/10.1186/s13068-019-1580-y

Pardhi DS, Panchal RR, Raval VH, Joshi RG, Poczai P, Almalki WH, Rajput KN (2022) Microbial surfactants: a journey from fundamentals to recent advances. Front Microbiol 13. https://doi.org/10.3389/fmicb.2022.982603

Patel P, Patel R, Mukherjee A, Munshi NS (2023) Microbial biosurfactants for green agricultural technology. In: Singh N, Chattopadhyay A, Lichtfouse E (eds) Sustainable agriculture reviews, vol 60. Springer, Cham, pp 389–413. https://doi.org/10.1007/978-3-031-24181-9_15

Perfumo A, Smyth TJ, Marchant R, Banat IM (2010) Production and roles of biosurfactants and bioemulsifiers in accessing hydrophobic substrates. In: Timmis KN (ed) Handbook of hydrocarbon and lipid microbiology. Springer, Heidelberg, pp 1501–1512. https://doi.org/10.1007/978-3-540-77587-4_103

Piegza M, Joanna Pietrzykowska, Trojan-Piegza J, Łaba W (2021) Biosurfactants from *Trichoderma* Filamentous Fungi—A preliminary study. Biomolecules 11(4):519. https://doi.org/10.3390/biom11040519

Pradhan AK, Rath A, Pradhan N, Hazra RK, Nayak RR, Kanjilal S (2018) Cyclic lipopeptide biosurfactant from *Bacillus tequilensis* exhibits multifarious activity. 3 Biotech 8(6). https://doi.org/10.1007/s13205-018-1288-x

Prakash AA, Prabhu NS, Rajasekar A, Parthipan P, AlSalhi MS, Devanesan S, Govarthanan M (2021) Bio-electrokinetic remediation of crude oil contaminated soil enhanced by bacterial biosurfactant. J Hazard Mater 405:124061. https://doi.org/10.1016/j.jhazmat.2020.124061

Qamar SA, Pacifico S (2023) Cleaner production of biosurfactants via bio-waste valorization: a comprehensive review of characteristics, challenges, and opportunities in bio-sector applications. J Environ Chem Eng 11(6):111555. https://doi.org/10.1016/j.jece.2023.111555

Qazi MA, Wang Q, Dai Z (2022) Sophorolipids bioproduction in the yeast *Starmerella bombicola*: current trends and perspectives. Bioresour Technol 346:126593. https://doi.org/10.1016/j.biortech.2021.126593

Rasheed T, Shafi S, Bilal M, Hussain T, Sher F, Rizwan K (2020) Surfactants-based remediation as an effective approach for removal of environmental pollutants—a review. J Mol Liq 318:113960. https://doi.org/10.1016/j.molliq.2020.113960

Rathankumar A-K, Saikia K, Palanisamy S, Ahalliya R-M, Arasu M-V (2023) Purification assessment and assay of biosurfactant efficacy. Multifunctional Microbial Biosurfactants 25–50

Rodrigues AI, Gudiña EJ, Abrunhosa L, Malheiro AR, Fernandes R, Teixeira JA, Rodrigues LR (2021) Rhamnolipids inhibit aflatoxins production in *Aspergillus flavus* by causing structural damages in the fungal hyphae and down-regulating the expression of their biosynthetic genes. Int J Food Microbiol 348:109207. https://doi.org/10.1016/j.ijfoodmicro.2021.109207

Roelants SL, Soetaert W (2021) Industrial perspectives for (microbial) biosurfactants. Adv Biochem Eng Biotechnol 181:1–15. https://doi.org/10.1007/10_2021_175

Rosenberg M (1991) Basic and applied aspects of microbial adhesion at the hydrocarbon: water interface. Crit Rev Microbiol 18(2):159–173. https://doi.org/10.3109/10408419109113512

Rosenberg E, Rubinovitz C, Legmann R, Ron EZ (1988) Purification and chemical properties of *Acinetobacter calcoaceticus* A2 Biodispersan. Appl Environ Microbiol 54(2):323–326. https://doi.org/10.1128/aem.54.2.323-326.1988

Sajna KV, Sukumaran RK, Jayamurthy H, Reddy KK, Kanjilal S, Prasad RB, Pandey A (2013) Studies on biosurfactants from *Pseudozyma* Sp. NII 08165 and their potential application as laundry detergent additives. Biochem Eng J 78:85–92. https://doi.org/10.1016/j.bej.2012.12.014

Sarubbo LA, Silva MD, Durval IJ, Bezerra KG, Ribeiro BG, Silva IA, Twigg MS, Banat IM (2022) Biosurfactants: production, properties, applications, trends, and general perspectives. Biochem Eng J 181:108377. https://doi.org/10.1016/j.bej.2022.108377

Schultz J, Rosado AS (2019) Extreme environments: a source of biosurfactants for biotechnological applications. Extremophiles 24(2):189–206. https://doi.org/10.1007/s00792-019-01151-2

Selva Filho AA, Converti A, Soares da Silva RD, Sarubbo LA (2023) Biosurfactants as multifunctional remediation agents of environmental pollutants generated by the petroleum industry. Energies 16(3):1209. https://doi.org/10.3390/en16031209

Shao Z (2010) Trehalolipids. In: Soberón-Chávez G (ed) Biosurfactants, microbiology monographs, vol 20. Springer, Heidelberg, pp 121–143. https://doi.org/10.1007/978-3-642-14490-5_5

Sharma S, Verma R, Dhull S, Maiti SK, Pandey LM (2021) Biodegradation of waste cooking oil and simultaneous production of rhamnolipid biosurfactant by *Pseudomonas aeruginosa* P 7815 in batch and fed-batch bioreactor. Bioprocess Biosyst Eng 45(2):309–319. https://doi.org/10.1007/s00449-021-02661-0

Silva DJ and Queiroz AC (2002) Food analysis: Chemical and biological methods. UFV, Viçosa

Singh R, Rathore D (2019) Impact assessment of azulene and chromium on growth and metabolites of wheat and chilli cultivars under biosurfactant augmentation. Ecotoxicol Environ Saf 186:109789. https://doi.org/10.1016/j.ecoenv.2019.109789

Soares da SRCF, de ADG, Brasileiro PPF et al. (2019) Production, formulation and cost estimation of a commercial biosurfactant. Biodegradation 30, 191–201

Soberón-Chávez G, Hausmann R, Déziel E (2021) Editorial: biosurfactants: new insights in their biosynthesis, production and applications. Front Bioeng Biotechnol 9. https://doi.org/10.3389/fbioe.2021.769899

Sobrinho, Humberto B, et al. (2013) Biosurfactants: classification, properties and environmental applications. Recent developments in biotechnology 11.14:1–29

Sotirova A-V, Spasova D-I, Galabova D-N et al. (2008) Rhamnolipid–biosurfactant permeabilizing effects on gram-positive and gram-negative bacterial strains. Curr Microbiol 56:639–644. https://doi.org/10.1007/s00284-008-9139-3

Steliga T, Wojtowicz K, Kapusta P, Brzeszcz J (2020) Assessment of biodegradation efficiency of polychlorinated biphenyls (PCBs) and petroleum hydrocarbons (TPH) in soil using three individual bacterial strains and their mixed culture. Molecules 25(3):709. https://doi.org/10.3390/molecules25030709

Surekha K, Satpute Gauri R, Kulkarni Arun G, Banpurkar Ibrahim M, Banat Nishigandha S, Mone Rajendra H, Patil Swaranjit Singh, Cameotra (2016) Biosurfactant/s from Lactobacilli species: Properties challenges and potential biomedical applications. J Basic Microbiol 56(11):1140–1158. https://doi.org/10.1002/jobm.v56.11 https://doi.org/10.1002/jobm.201600143

Sutirman ZA, Sanagi MM, Wan Aini WI (2021) Alginate-based adsorbents for removal of metal ions and radionuclides from aqueous solutions: a review. Int J Biol Macromol 174:216–228. https://doi.org/10.1016/j.ijbiomac.2021.01.150

Terahara T, Naemura T, Nampo Y, Kobayashi T, Imada C, Hamada M, Tamura T (2019) *Streptomyces otsuchiensis* Sp. Nov., a biosurfactant-producing actinobacterium isolated from marine sediment. Int J Syst Evol Microbiol 69(12):3740–3744. https://doi.org/10.1099/ijsem.0.003638

Thakur V, Baghmare P, Verma A, Verma JS, Geed SR (2024) Recent progress in microbial biosurfactants production strategies: applications, technological bottlenecks, and future outlook. Bioresour Technol 408:131211. https://doi.org/10.1016/j.biortech.2024.131211

Théatre A, Hoste AC, Rigolet A, Benneceur I, Bechet M, Ongena M, Deleu M, Jacques P (2021) *Bacillus* Sp.: a remarkable source of bioactive lipopeptides. Adv Biochem Eng Biotechnol 181:123–179. https://doi.org/10.1007/10_2021_182

Tomar S, Lal M, Khan M, Singh B, Sharma S (2019) Characterization of glycolipid biosurfactant from *Pseudomonas aeruginosa* PA 1 and its efficacy against Phytophtora infestans. J Environ Biol 40(4):725–730. https://doi.org/10.22438/jeb/40/4/mrn-910

Van Renterghem L, Roelants SL, Baccile N, Uyttersprot K, Taelman MC, Everaert B, Mincke S, Ledegen S, Debrouwer S, Scholtens K, Stevens C, Soetaert W (2018) From lab to market: an integrated bioprocess design approach for new-to-nature biosurfactants produced by *Starmerella bombicola*. Biotechnol Bioeng 115(5):1195–1206. https://doi.org/10.1002/bit.26539

Varjani SJ, Upasani VN (2016) Carbon spectrum utilization by an indigenous strain of *Pseudomonas aeruginosa* NCIM 5514: production, characterization and surface active properties of biosurfactant. Bioresour Technol 221:510–516. https://doi.org/10.1016/j.biortech.2016.09.080

Varjani S, Haq I, Pandey A, Gupta VK, Bui X (2024) Waste management in climate change and sustainability perspectives. CRC Press, p 178. https://doi.org/10.1201/9781003386902

Wan J, Meng D, Long T, Ying R, Ye M, Zhang S, Li Q, Zhou Y, Lin Y (2015) Simultaneous removal of lindane, lead and cadmium from soils by Rhamnolipids combined with citric acid. PLoS One 10(6):e0129978. https://doi.org/10.1371/journal.pone.0129978

Wang H, Roelants SL, To MH, Patria RD, Kaur G, Lau NS, Lau CY, Van Bogaert IN, Soetaert W, Lin CS (2018) *Starmerella bombicola*: recent advances on sophorolipid production and prospects of waste stream utilization. J Chem Technol Biotechnol 94(4):999–1007. https://doi.org/10.1002/jctb.5847

Wei Z, Wang JJ, Gaston LA, Li J, Fultz LM, DeLaune RD, Dodla SK (2020) Remediation of crude oil-contaminated coastal marsh soil: integrated effect of biochar, rhamnolipid biosurfactant and nitrogen application. J Hazard Mater 396:122595. https://doi.org/10.1016/j.jhazmat.2020.122595

Wittgens A, Tiso T, Arndt TT, Wenk P, Hemmerich J, Müller C, Wichmann R, Küpper B, Zwick M, Wilhelm S, Hausmann R, Syldatk C, Rosenau F, Blank LM (2011) Growth independent rhamnolipid production from glucose using the non-pathogenic *Pseudomonas putida* KT2440. Microb Cell Fact 10(1). https://doi.org/10.1186/1475-2859-10-80

Xia L, Wen J (2022) Available strategies for improving the biosynthesis of surfactin: a review. Crit Rev Biotechnol 43(7):1111–1128. https://doi.org/10.1080/07388551.2022.2095252

Yadav T, Singh V, Sharma P, Singh SP (2024) Deciphering microbe-driven remediation of environmental pollutants: an omics perspective. In: Microbiome-based decontamination of environmental pollutants, pp 147–166. https://doi.org/10.1016/b978-0-443-21781-4.00002-5

Yakimov MM, Golyshin PN, Lang S, Moore ER, Abraham W, Lunsdorf H, Timmis KN (1998) *Alcanivorax borkumensis* Gen. Nov., Sp. Nov., a new, hydrocarbon-degrading and surfactant-producing marine bacterium. Int J Syst Bacteriol 48(2):339–348. https://doi.org/10.1099/00207713-48-2-339

Yesankar PJ, Pal M, Patil A, Qureshi A (2022) Microbial exopolymeric substances and biosurfactants as 'bioavailability enhancers' for polycyclic aromatic hydrocarbons biodegradation. Int J Environ Sci Technol 20(5):5823–5844. https://doi.org/10.1007/s13762-022-04068-0

Zang T, Wu H, Yan B, Zhang Y, Wei C (2021) Enhancement of PAHs biodegradation in biosurfactant/phenol system by increasing the bioavailability of PAHs. Chemosphere 266:128941. https://doi.org/10.1016/j.chemosphere.2020.128941

Zeng Z, Liu Y, Zhong H, Xiao R, Zeng G, Liu Z, Cheng M, Lai C, Zhang C, Liu G, Qin L (2018) Mechanisms for rhamnolipids-mediated biodegradation of hydrophobic organic compounds. Sci Total Environ 634:1–11. https://doi.org/10.1016/j.scitotenv.2018.03.349

Zouari O, Lecouturier D, Rochex A, Chataigne G, Dhulster P, Jacques P, Ghribi D (2018) Bio-emulsifying and biodegradation activities of syringafactin producing *Pseudomonas* spp. strains isolated from oil contaminated soils. Biodegradation 30(4):259–272. https://doi.org/10.1007/s10532-018-9861-x

Chapter 7
Microbial Nano-biotechnology for Wastewater Research: A Sustainable Approach

Neha Agarwal, Shachi Mishra, Vijendra Singh Solanki, and Sankara Rao Miditana

Abstract The contamination of aqueous resources with organic and inorganic pollutants poses a critical threat to human survival, ecosystems, and environmental health. Conventional wastewater treatment methods such as adsorption, membrane filtration, and chemical precipitation rely on nanomaterials (e.g., polymers, carbon-based, metallic, and metal oxide nanomaterials) but face limitations including toxicity, high costs, energy inefficiency, secondary pollution, and inadequate efficacy. Microbial-driven synthesis of nanoparticles (NPs) has emerged as a sustainable and innovative alternative, offering enhanced efficiency and environmental compatibility. This review systematically explores the transformative potential of microbial nano-biotechnology in water remediation, highlighting its superiority over traditional approaches. By synthesizing cutting-edge advancements, this chapter provides a roadmap for developing green, scalable, and cost-effective strategies to address global water contamination challenges, thereby catalyzing future research in sustainable wastewater treatment technologies.

Keywords Nano-biotechnology · Nanoparticles · Sustainable · Wastewater · Nanomaterials · Contaminants · Microorganisms

N. Agarwal (✉)
Department of Chemistry, Navyug Kanya Mahavidyalaya, University of Lucknow, Lucknow, Uttar Pradesh, India

S. Mishra
Department of Chemistry, Jai Prakash University, Chapra, Bihar, India

V. S. Solanki
Department of Chemistry, ISR, IPS Academy, Indore, Madhya Pradesh, India

S. R. Miditana
Department of Chemistry, Government Degree College,
Puttur, Tirupati, Andhra Pradesh, India

N. K. Arora, J. Mishra (eds.), *Biotechnological Solutions for a Sustainable Future*, Microorganisms for Sustainability 55,
https://doi.org/10.1007/978-981-95-3989-5_7

7.1 Introduction

Water is indispensable to life on Earth, yet rapid industrialization has unleashed a deluge of toxic effluents—pharmaceuticals, dyes, pesticides, and nonbiodegradable organic compounds into global water systems, intensifying pressures to remediate pollution sustainably. Conventional methods like advanced oxidation and electrochemical processes face limitations, including high energy demands, inefficiency, and toxic by-products (Agarwal et al. 2022a; Chahar et al. 2023). Nanotechnology, however, has emerged as a transformative force in wastewater treatment. Engineered nanoparticles (NPs), such as nanocatalysts, nanorods, and nanocomposites, leverage their high surface area-to-volume ratio and catalytic efficiency to degrade pollutants more effectively than traditional techniques (Baruah et al. 2019; Bera et al. 2022). Innovations like Au/TiO_2 photocatalysts for antibiotic-laden wastewater (Duo et al. 2022) and photocatalytic nanofiber membranes (Chabalala et al. 2021) exemplify nanotechnology's potential.

Yet synthesizing NPs via physical or chemical methods often involves toxic reagents and high costs (Agarwal et al. 2023). This has spurred interest in green synthesis using microbes like bacteria, fungi, and algae to produce biogenic NPs. These bioinspired methods yield cost-effective, eco-friendly nanomaterials with superior adsorption and catalytic properties (Razack and Duraiarasan 2020; Bhat et al. 2022). Microbial nanotechnology further extends to bioremediation, where microorganisms degrade pollutants synergistically with NPs, enhancing efficiency (Nahak et al. 2022; Yadav et al. 2021). For instance, biopolymer-coated metal NPs and nanocomposites have revolutionized heavy metal removal and micropollutant degradation (Ali et al. 2017; Corsi et al. 2018).

This chapter explores nanotechnology's dual role in wastewater remediation: eliminating toxins and advancing equitable clean water access. We highlight microbial bio-nanotechnology's promise, including cutting-edge innovations in green-synthesized NPs, hybrid approaches (e.g., nano-membranes combined with microbial systems), and challenges like scalability and environmental risk mitigation. By bridging gaps between nanotechnology, microbiology, and sustainability, this work charts a path toward safer water systems in an industrialized world (Fig. 7.1).

7.2 Nanotechnology in Wastewater Research

Escalating environmental pollution and global water scarcity have driven urgent research into innovative wastewater treatment systems. In this context, nanotechnology has emerged as a transformative, sustainable, and cost-effective solution for water purification. By leveraging nanomaterials (NMs) such as metal and metal oxide nanoparticles (NPs), carbon-based structures, metal-organic frameworks (MOFs), zeolites, and nanofibers, researchers are addressing contamination

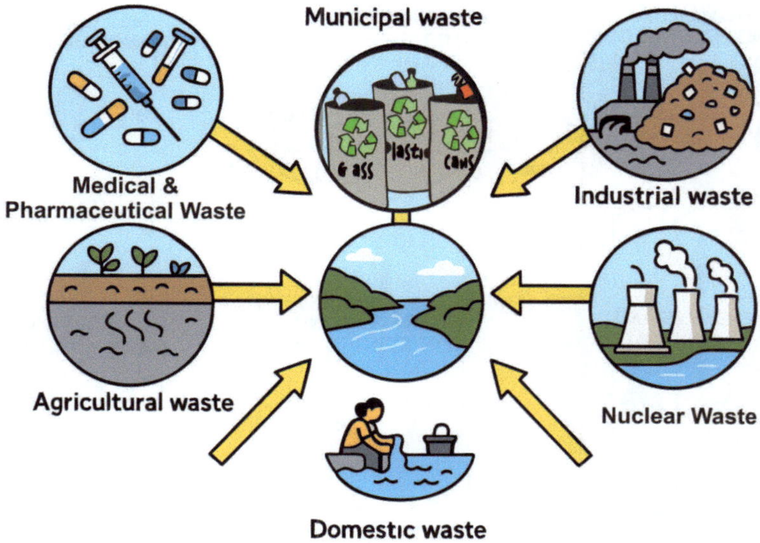

Fig. 7.1 Various sources of contamination of water

challenges with unprecedented efficiency (Qu et al. 2013a, Qu et al. 2013b; Zahmatkesh et al. 2023; Rafique et al. 2019). These materials enable advanced techniques like nano-adsorption, which exploits NPs' high surface area to capture heavy metals and organic pollutants; photocatalysis, where light-activated NPs degrade toxins; and nano-membrane filtration, which selectively removes micropollutants. Additionally, antimicrobial NPs such as silver- or copper-based variants are proving critical for pathogen elimination. Unlike conventional methods, nanotechnology minimizes energy consumption, avoids toxic by-products, and enhances scalability, positioning it as a cornerstone of next-generation wastewater remediation strategies (Fig. 7.2).

7.2.1 Nano-adsorption Technologies

Adsorption is a cornerstone of wastewater treatment due to its cost-effectiveness, versatility, and efficiency in removing diverse contaminants. Nanoparticles (NPs), particularly nano-adsorbents, have revolutionized this field owing to their high surface area-to-volume ratio, porosity, and tunable surface chemistry (Kumari et al. 2019; Pacheco et al. 2006). Metal oxide NPs such as iron, manganese, titanium, and magnesium oxides dominate adsorption applications due to their chemical stability, ease of synthesis (Kumar and Chawla 2014; Rajput et al. 2016), and regeneration potential. For example, magnetic iron oxide nanoparticles (MNPs) enable easy separation posttreatment (Dave and Chopda 2014; Shen et al. 2015), while hydrous

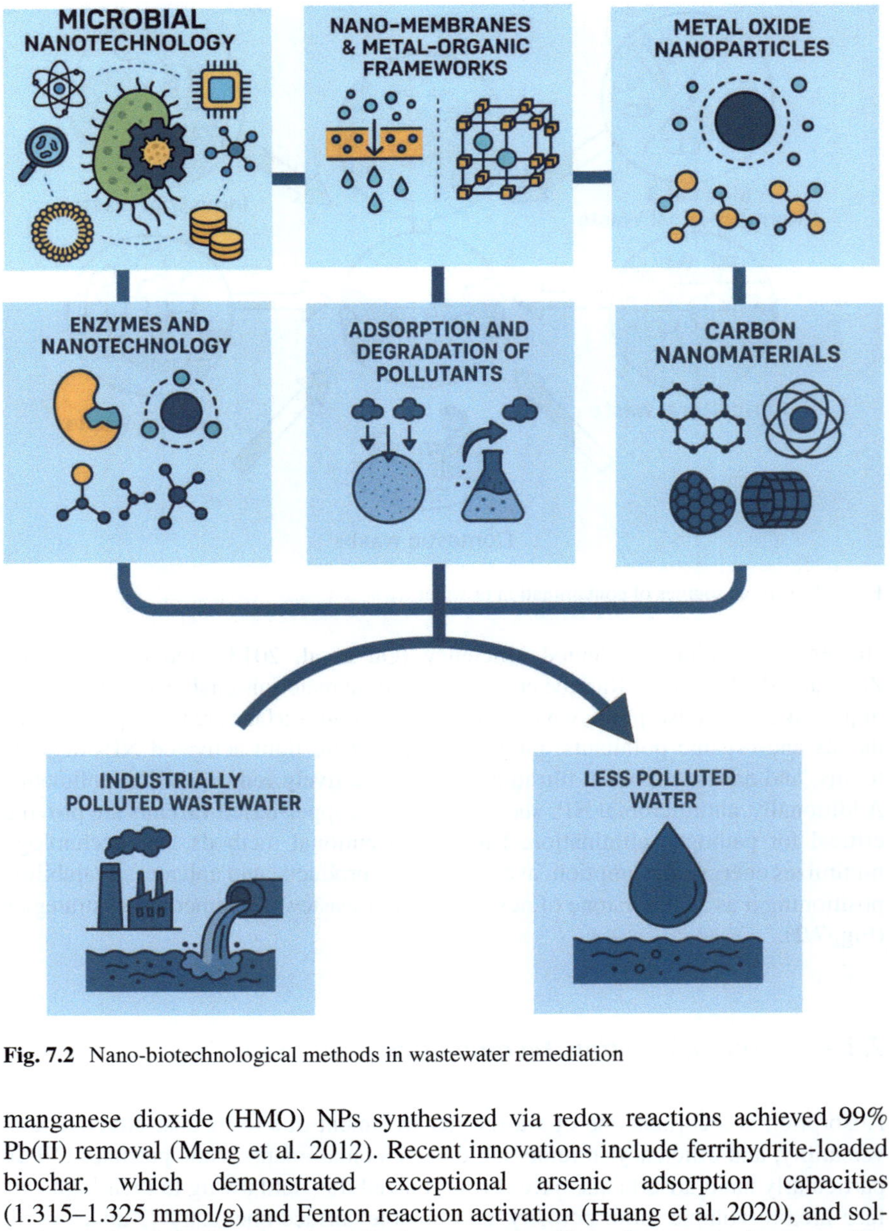

Fig. 7.2 Nano-biotechnological methods in wastewater remediation

manganese dioxide (HMO) NPs synthesized via redox reactions achieved 99% Pb(II) removal (Meng et al. 2012). Recent innovations include ferrihydrite-loaded biochar, which demonstrated exceptional arsenic adsorption capacities (1.315–1.325 mmol/g) and Fenton reaction activation (Huang et al. 2020), and solvothermally doped magnesium ferrite nanocrystallites for magnetic separation (Tang et al., 2013). Regeneration via pH adjustment ensures reusability, with studies showing minimal capacity loss even after multiple cycles (Sharma et al. 2009; Hu et al. 2006; Kyzas and Deliyanni 2013).

Carbon-based nanomaterials like graphene, carbon nanotubes (CNTs), and activated carbon further expand adsorption capabilities (Hua et al. 2012; Ray and Shipley 2015; Wang et al. 2012; Qu et al. 2013a, b). Multi-walled CNTs effectively extracted Hg(II) at pH 4.0–8.0 under agitation (Tawabini et al. 2010), while chitosan-CNT hybrids selectively adsorbed Cu(II) in mixed-metal solutions (Salam et al. 2009). Nitrogen-/bismuth-doped carbon quantum dots derived from rice husk degraded methylene blue and adsorbed Cu(II) under light exposure (Hui et al. 2021). Dendrimers, with customizable branches and hydrophobic cores, enable precision pollutant removal via hydrogen bonding, electrostatic interactions, or complexation (Crooks et al. 2001). For instance, Schiff base-modified polyamidoamine (PAMAM) dendrimers achieved near-complete mercury removal, while dendrimer-ultrafiltration systems eliminated 10 ppm Cu^{2+} with >99% efficiency (Diallo et al. 2005). These advancements underscore nano-adsorption's role in scalable, multifunctional water remediation.

7.2.2 *Nanomembranes Technology*

Nanomembranes have emerged as a revolutionary tool in water treatment, combining size-exclusion principles with nanomaterial-driven selectivity to address diverse contaminants. Their efficiency, antifouling capabilities, and cost-effectiveness make them superior to conventional filtration methods, particularly for industrial and municipal wastewater (Fan et al. 2023; Ying et al. 2017). Recent advancements in nanomaterials like carbon nanotubes (CNTs), graphene oxide (GO), and cellulose nanofibers have enhanced permeability and durability. For example, single- and multi-walled CNTs improve pollutant rejection rates through their tunable pore structures, while cellulose nanofiber membranes on polytetrafluoroethylene (PTFE) support higher microplastic removal at low operational pressures (Nain et al. 2022). Innovations extend to ceramic nanomembranes embedded with silica nanoparticles (SiO_2 NPs), which withstand extreme pH and temperature conditions in industrial effluents (Asif and Zhang 2021), and bioinspired aquaporin-based membranes that mimic natural water channels for ultrafast desalination (Tang et al. 2013).

Antimicrobial functionality is another critical advancement. In this context, silver nanoparticle (AgNP)-coated polyethersulfone (PES) membranes have shown a reduction in bacterial colonization, while MXene-integrated systems degraded organic foulants under visible light (Behboudi et al. 2023; Shaikh et al. 2024). Layer-by-layer (LbL) assemblies of polyelectrolytes and MXene nanosheets further enable tunable selectivity for pharmaceuticals and heavy metals (O'Hern et al. 2014). Despite these breakthroughs, challenges persist, including scalability, fabrication costs, and risks of nanomaterial leaching. Recent studies advocate for green synthesis methods, such as plant-derived cellulose NPs, and hybrid systems integrating AI-driven monitoring or solar photocatalysis to enhance sustainability (Krishnan et al. 2024; Fan et al. 2023). These developments position nanomembranes as pivotal to achieving scalable, next-generation water security.

7.2.3 Photocatalysis Technology

Nano-photocatalysts have revolutionized wastewater remediation by harnessing light energy to degrade pollutants into harmless by-products like water, carbon dioxide, and inorganic ions. Their high efficiency, reusability, and ability to target a broad spectrum of contaminants including pharmaceuticals, dyes, and pesticides make them economically viable for large-scale applications (Mishra and Sundaram 2023; Adeola et al. 2022). Titanium dioxide (TiO_2) and zinc oxide (ZnO) nanoparticles are widely studied for their photocatalytic activity under UV light, but recent advancements focus on visible-light-active materials. For example, iron oxide-based heterostructures, such as ferric oxide nanoplates combined with BiOCl, achieved 98% degradation of Rhodamine B within 90 min, outperforming individual components (Zheng et al. 2018). Similarly, nitrogen-doped TiO_2 nanoparticles extended light absorption into the visible spectrum, degrading 95% of ciprofloxacin in hospital wastewater (Nidheesh et al. 2024).

Despite their promise, challenges persist. Many photocatalysts, such as cadmium sulfide (CdS) quantum dots, pose toxicity risks to aquatic ecosystems, while energy-intensive UV light requirements and nanoparticle recovery issues hinder scalability (Abdullah et al. 2021). Secondary contamination is another concern. For instance, photocatalysis of triclosan, a common antimicrobial, generates chlorophenols, which are more toxic than the parent compound. Nanomembrane–photocatalyst hybrids, though innovative, face fouling and stability issues under fluctuating pH and temperature, reducing long-term efficacy (Harun-Ur-Rashid et al. 2024).

To address these limitations, research is pivoting toward green nanotechnology. Biosynthesized photocatalysts derived from plant extracts or fungi, such as *Fusarium oxysporum*-mediated ZnO NPs, offer nontoxic alternatives with comparable efficiency (Muruganandham et al. 2018). Additionally, perovskite-based materials like bismuth vanadate ($BiVO_4$) and metal-organic frameworks (MOFs) are emerging for their tunable bandgaps and high surface areas, enabling higher degradation of antibiotics under solar light (Bhuin et al. 2024; Manikandan et al. 2025). Hybrid systems integrating photocatalysis with adsorption or membrane filtration further enhance pollutant removal while mitigating by-product risks (Ejraei et al. 2019).

7.3 Microorganisms-Assisted NMs in Wastewater Research

The growing contamination of aquatic systems with hazardous pollutants has intensified the need for innovative and sustainable wastewater treatment technologies. Emerging contaminants such as heavy metals, dyes, pharmaceuticals, and pathogenic microorganisms pose significant threats to environmental and human health (Gupta et al. 2022; Jain et al. 2016). While engineered nanomaterials (NMs) like dendrimers, nanofilters, and nanomembranes offer high efficacy, their chemical

Table 7.1 Microorganism-based biogenic NPs in wastewater treatment

Biogenic NP	Microorganism used	Pollutant treated	Removal efficiency	References
Iron oxide NPs	*Aspergillus tubingensis*	Heavy metals Pb (II), Ni (II), Cu (II), and Zn (II)	More than 90%	Mahanty et al. (2020)
	Manglicolous fungi	Cr(IV)	More than 90%	Mahanty et al. (2019)
	Streptomyces thermolineatus	Cu	85%	Suganthi and Kandasamy (2017)
Iron oxide magnetic NPs	*Aspergillus niger*	Cr(VI)	More than 99%	Chatterjee et al. (2020)
Silver NPs	*Spirulina platensis*	Malachite green	–	Sidorowicz et al. (2024)
	Anabaena variabilis	Malachite green	–	Ismail et al. (2021)
	Bacillus marisflavi	Methyl red and reactive black	54.14% and 96.92%	Ahmed et al. (2020)
Manganese oxide NPs	*Marinobacter* sp.	Ag+	95%	Pei et al. (2013)
Magnesium oxide NPs	*Aspergillus terreus*	Cr(VI)	More than 97%	Saied et al. (2021)
Cadmium sulfide NPs	*Pseudomonas aeruginosa*	Cd(II)	More than 88%	Raja et al. (2022)
Palladium NPs	*Spirulina plaesis*	Pb	90%	Sayadi et al. (2018)
Platinum NPs	*Desulfovibrio vulgaris*	Ciprofloxacin and sulfamethoxazole	70% and 85%	Martins et al. (2017)
Zinc oxide NPs	*Pseudochrobactrum* sp.	Methylene blue and reactive black	Approx 100%	Siddique et al. (2021)
Silica NPs	*Actinomycetes* sp.	Decolourization of industrial effluents	80%	Mohanraj et al. (2020)

Adopted from Malik et al. (2022)
(– Not Determined)

synthesis often involves toxic reagents that risk secondary pollution (Ciambelli et al. 2020). Microorganism-assisted synthesis of NMs presents an eco-friendly alternative (Table 7.1).

Bacteria, fungi, and algae have been effectively used to biosynthesize nanoparticles (NPs) due to their metabolic diversity and environmental compatibility (Jain et al. 2021). These biogenic NPs, often capped with bioactive compounds, exhibit high stability and reactivity, particularly in aqueous environments. For instance, Mahanty et al. (2020) synthesized iron oxide nanoparticles using *Aspergillus tubingensis* from the rhizosphere of *Avicennia officinalis*, achieving significant removal of Pb(II), Cu(II), Ni(II), and Zn(II). Similarly, Nkosi et al. (2025) reported that green-synthesized iron NPs effectively reduced chemical oxygen demand (COD) and biological oxygen demand (BOD) levels in industrial effluents. Ishfaq et al.

(2023) demonstrated that manganese oxide NPs biosynthesized from microbial processes not only removed contaminants but also enhanced the growth of wheat seedlings when applied to treated wastewater. Additionally, Bernardes et al. (2025) developed silver NPs using kefir-based microbial fractions, which exhibited potent antimicrobial activity against wastewater pathogens. Mandeep and Shukla (2020) reviewed various microbial nanotechnologies, highlighting their roles in metal recovery, pollutant degradation, and biocompatible synthesis pathways.

While these results are promising, challenges such as scalability, synthesis consistency, and long-term environmental safety remain. Continued interdisciplinary research is crucial to optimizing production processes and fully harnessing the potential of microbe-assisted nanotechnologies in sustainable wastewater remediation.

7.3.1 Microbial Fuel Cells in Wastewater Remediation

Conventional wastewater treatment methods, plagued by high energy demands, greenhouse gas emissions, and toxic by-products like ammonia, are increasingly unsustainable (Nawaz et al. 2022). Microbial fuel cells (MFCs) address these challenges by harnessing electroactive bacteria to degrade organic pollutants while generating bioelectricity. These systems consist of anode and cathode chambers separated by a proton exchange membrane (PEM), where microbes such as *Geobacter sulfurreducens* oxidize organic substrates (e.g., industrial effluents, agricultural waste) at the anode, releasing electrons that travel through an external circuit to drive oxygen reduction at the cathode (Enamala et al. 2020; Butti et al. 2016). Recent innovations include graphene-coated anodes, which boosted power density by 300% in dairy wastewater treatment (Sekar et al. 2019), and biochar-cathode systems that achieved 95% chemical oxygen demand (COD) removal in textile effluents while slashing costs by 40% (Huggins et al. 2016). Hybrid MFC-algal bioreactors further enhance sustainability by sequestering CO_2 and recovering phosphorus (Tripathi et al. 2025). Despite these advances, scalability remains a hurdle: pilot-scale systems incur high costs (120–120–150/m^3) and suffer from power density drops (Nawaz et al. 2022). Emerging solutions like 3D-printed modular MFC arrays and AI-optimized biofilms aim to improve efficiency and reduce costs (You et al. 2017). Future research prioritizes low-cost PEM alternatives (e.g., chitosan membranes) and engineered microbial consortia that can improve wastewater treatment efficiency (Srivastava et al. 2015). Integrating IoT sensors for real-time monitoring further positions MFCs as viable, scalable solutions for sustainable water-energy nexus challenges (Garrido-Momparler et al., 2022).

7.3.2 Microorganisms-Based Biogenic NPs for Wastewater Remediation

The interaction between metals and microorganisms has emerged as a promising route for the development of biogenic nanoparticles (NPs), offering eco-friendly solutions for various applications including wastewater treatment, electronics, and biomedicine (Gillan 2016; Ali et al. 2017). These biogenic NPs are synthesized using microbial cells as reducing and stabilizing agents, providing a sustainable alternative to conventional chemical synthesis methods that often involve toxic reagents (Kitching et al. 2015). Recent research has emphasized the dual benefit of utilizing heavy metal-polluted wastewater as a raw material for NP biosynthesis, enabling both environmental remediation and recovery of valuable metals commonly used in industries like metallurgy, electroplating, and electronics. Microorganisms such as bacteria and fungi not only bioaccumulate these metals but also facilitate their reduction into functional nanoparticles, contributing to the efficient removal of pollutants (Malik et al. 2022).

In a recent study, *Aspergillus tubingensis* was used to biosynthesize iron oxide NPs from wastewater containing heavy metals, demonstrating excellent removal efficiencies for Pb(II), Cu(II), Ni(II), and Zn(II) (Mahanty et al. 2020). Similarly, Ahmed et al. (2022) reported green-synthesized iron NPs that achieved significant reductions in chemical oxygen demand (COD) and biological oxygen demand (BOD) in industrial wastewater. Biogenic silver nanoparticles (AgNPs), synthesized using strains like *Shewanella oneidensis* and even microbial kefir fractions, have also gained attention for their strong antimicrobial activity against waterborne pathogens and for breaking down organic pollutants. In another example, Ishfaq et al. (2023) demonstrated that manganese oxide NPs synthesized by microbes not only removed contaminants from wastewater but also promoted healthy plant growth when used for irrigation, suggesting a secondary benefit in agricultural reuse.

Further advancements have been made in the microbial synthesis of titanium dioxide (TiO_2) NPs using bacterial and fungal strains, which exhibit high photocatalytic efficiency for degrading organic pollutants in wastewater (Rathore et al. 2023). These nanoparticles are often more biocompatible and stable due to the presence of organic capping layers derived from microbial biomolecules. Additionally, research has shown that microorganisms such as *Shewanella oneidensis* and *Deinococcus radiodurans* can stabilize nanoparticles on their cell walls, improving their environmental durability and reactivity (Li et al. 2019). Zerovalent iron NPs, in particular, have emerged as highly effective for the dechlorination of organic pollutants and remediation of hazardous compounds in industrial effluents (Malik et al. 2022).

Despite their promise, challenges remain in terms of large-scale production, synthesis reproducibility, and long-term ecological impacts. Continued interdisciplinary research is essential to optimize microbial processes, explore novel microbial strains, and develop scalable production methods for biogenic NPs that are efficient, cost-effective, and environmentally safe for sustainable wastewater remediation.

7.3.3 Synergistic Effect of MFCs and Nanocomposites in Wastewater Remediation

Microbial fuel cells harness electroactive bacteria like *Geobacter sulfurreducens* to degrade organic pollutants (e.g., butyrate, acetate) into CO_2 and H_2O while generating bioelectricity, yet their commercial viability has been hindered by low power density and high costs compared to conventional fuel cells (Malik et al. 2022). To address these limitations, nanocomposites—engineered by integrating nanomaterials like carbon nanotubes (CNTs), graphene oxide (GO), and MXenes with polymers or catalytic metals are revolutionizing MFC performance. For instance, Cu-doped FeO nanoparticles boosted power output in dairy wastewater by enhancing microbial adhesion and electron transfer (Sekar et al. 2019), while GO-biohybrid anodes coated with *Pseudomonas aeruginosa* biofilms achieved 98% phenol degradation alongside doubled current density. MXene-embedded membranes further reduced oxygen crossover by 40% in textile effluent MFCs, improving proton conductivity (Shaikh et al. 2024). These nanocomposites enhance MFC efficiency through mechanisms like accelerated extracellular electron transfer, reduced cytotoxicity via polymer coatings (e.g., chitosan-coated Fe_3O_4 NPs lowering ROS by 60%), and optimized biofilm formation on nanostructured surfaces (Singh et al. 2012). Case studies highlight their potential: PANI-CNT systems achieved 92% azo dye removal in textile wastewater with 0.8 W/m^3 power output (Saroj et al. 2022), and copper ferrite ($CuFe_2O_4$) nanoparticles coated on graphene oxide ($CuFe_2O_4$/GO) system degraded 98% of ciprofloxacin in pharmaceutical effluent (Noroozi et al. 2020). Challenges such as nanomaterial leaching and scalability persist, but emerging solutions like algal-MFCs with NPs for CO_2 fixation and AI-driven designs for optimal nanocomposite–microbe pairings signal a promising future for scalable, energy-efficient wastewater treatment (Fetanat et al. 2021; Nagendranatha Reddy et al. 2019; Krishnan et al. 2024).

7.3.4 Bio-Nanofilters in Wastewater Effluent Treatment

Bio-nanofiltration represents a cutting-edge technology in wastewater treatment, focusing on the removal of toxic compounds and other pollutants. This method leverages the advancements in nanotechnology, particularly through the development of nano-webs and nanofibers produced via electrospinning techniques. These nanofibers are celebrated for their extremely high porosity and surface area, making them highly effective at trapping micropollutants (Mostafa et al. 2017). Integrating these nanofibrous structures with microbial cells, such as algae and bacteria, enhances the filtration and purification processes. For example, immobilized algal cells on electrospun chitosan nanofiber mats have shown high efficiency in removing nitrates from wastewater, offering advantages like ease of handling and cultivation in compact spaces (Eroglu et al. 2012). Recent innovations include the development of biosourced nanofiltration membranes. These are crafted from chemically recyclable aliphatic polyesters, such as P(4,5-T6GBL), which are not only biodegradable but also chemically recyclable, contributing to a sustainable life cycle (Rifan et al. 2022). Another breakthrough study introduced hybrid bio-nanofilters incorporating graphene oxide, which significantly enhanced the removal efficiency for heavy metals and persistent organic pollutants from industrial wastewater (Feng et al. 2025). Further research has led to the synthesis of dual-functional bio-nanofilters that combine bacterial removal capabilities with heavy metal adsorption, showcasing the potential for broad-spectrum water purification solutions (Yin et al. 2023). These developments underline the potential of bio-nanofilters not only as economical options but also as highly effective systems for comprehensive wastewater remediation.

7.3.5 Combined Effect of NMs with Microbial Enzymes in Remediation

Microbial enzymes (laccases, microbial oxidoreductases, peroxidases, monooxygenases, dioxygenases, lipases, proteases, etc.) combined with nanotechnology constitute a highly effective tool for wastewater remediation in comparison to other techniques because of their higher efficiency, selectivity, cost-effectiveness, and nontoxicity compared to other chemical catalysts Microbial enzymes including oxidoreductases (e.g., laccases, peroxidases, monooxygenases, dioxygenases) as well as lipases and proteases when integrated with nanotechnology, represent a highly effective strategy for wastewater remediation due to their superior catalytic efficiency and selectivity, favorable cost-effectiveness, and reduced toxicity relative to conventional chemical catalysts. The combined effect of nanotechnology with enzymes is highly useful to make NMs less harmful to the environment. Enzymes in combination with NMs create sufficient steric hindrance and minimize their cell interaction, thereby decreasing the surface energy (Dwevedi 2019). Enzymes are highly specific catalysts and eco-friendly; their presence makes NMs more efficient in green energy production and bioremediation. The large surface area of NMs improves the immobilization efficiency of enzymes, and they can be separated from the reaction mixture when magnetic NMs are used. Basically, the immobilization of enzymes leads to changes in structures, but in the presence of NMs, such modifications are not observed (Secundo 2013). Numerous studies have highlighted the synergistic integration of enzymes with nanotechnology. For example, Darwesh et al. (2019) demonstrated that immobilized peroxidase efficiently remediated wastewater by removing recalcitrant red and blue azo dyes within 4 h when treated separately. When the combination of both dyes was used, it took 6 h for the complete removal of dyes (Darwesh et al. 2019). Bacterial laccase is also widely used for the treatment of industrial effluents (Agarwal et al. 2022b).

7.4 Exploring Green Microbial Nanotechnology in Remediation of Wastewater

Green microbial nanotechnology offers a sustainable and eco-friendly approach to wastewater remediation by integrating microorganisms (bacteria, algae, fungi) with nanomaterials to enhance pollutant removal efficiency (Rath et al. 2024). This emerging approach harnesses microbes as living nano-factories capable of biosynthesizing nanoparticles in situ, which eliminates the need for toxic chemical synthesis processes and minimizes secondary pollution (Hamida et al. 2020). The biogenic nanomaterials produced via microbial pathways typically exhibit high surface area, tunable surface chemistry, and enhanced reactivity, translating to superior adsorption capacities for a broad spectrum of contaminants including heavy metals, dyes, and other recalcitrant pollutants (Mohanapragash et al. 2025). Simultaneously, photosynthetic microbes such as microalgae effectively take up excess nutrients (nitrogen and phosphorus) from wastewater during growth, achieving significant nutrient removal while generating substantial biomass (Hu et al. 2015; Mallick 2002). The resultant biomass can be harvested and valorized (for example, as biofuel feedstock or organic fertilizer), thus creating a circular treatment process that not only purifies water but also recovers valuable resources. Furthermore, incorporating nanotechnology into microbial systems for instance, using nano-engineered membranes or magnetic bio-nanocomposites has been shown to improve pollutant uptake rates and simplify the separation of biomass or catalytic particles from treated effluents (Singh et al. 2022; Yin et al. 2013). Such integrated nano–bio approaches achieve high removal efficiencies for both nutrients and toxic pollutants without generating hazardous by-products or excessive sludge, underscoring their efficiency and environmental compatibility (Singh et al. 2022). Overall, the convergence of microbial processes with nanomaterial-enabled functions yields innovative wastewater treatment strategies that are not only highly effective in contaminant removal but also sustainable in the long term, aligning with green chemistry and circular economy principles (Hamida et al. 2020; Singh et al. 2022).

7.5 Future Perspectives

Microbial-assisted nanotechnology can address escalating wastewater challenges. Green-synthesized NMs exhibit unparalleled efficiency in degrading recalcitrant pollutants like pharmaceuticals, heavy metals, and microplastics. The integration of genetically engineered bacteria with NMs can provide selective advantages including high surface area, tunable reactivity, and biocompatibility thereby minimizing secondary toxicity risks while enhancing catalytic performance, as demonstrated in prior studies by Ren et al. (2022). Emerging innovations are poised to amplify this potential of microbial-NM synergies, such as *Shewanella*-nanohybrids for uranium sequestration (Burgos et al. 2008). Waste-to-resource systems, like biochar combined with zerovalent iron nanoparticles, enhanced biohydrogen yields from grass fermentation (Yang and Wang 2019). However, scalability challenges such as NM recovery and microbial viability are being tackled through 3D-bioprinted biofilm reactors retaining higher efficiency over conventional methods (Finny 2024), while eco-toxicity concerns are mitigated by chitosan-coated ZnO NPs reducing *Sorghum bicolor* (L.) Moench plants (Rani and Hooda 2024). Standardized regulatory frameworks and interdisciplinary collaborations will be critical to translating lab-scale breakthroughs into global, sustainable solutions aligned with SDGs.

7.6 Conclusion

Microbial-assisted nanotechnology represents a paradigm shift in sustainable wastewater treatment, merging ecological safety with high efficiency. By harnessing microbes to synthesize nanomaterials, this approach minimizes chemical toxicity while enhancing pollutant degradation and resource recovery, as evidenced by biohydrogen production systems and algal-gold nanoparticle hybrids. The dual focus on remediation and energy generation positions these technologies as vital tools for achieving water security and reducing industrial carbon footprints. However, transitioning from lab-scale breakthroughs to global implementation requires scalable bioreactor designs, robust regulatory standards, and partnerships across academia, industry, and policymakers. By bridging innovation with sustainability, microbial bio-nanotechnology offers a transformative pathway to address twenty-first-century environmental challenges while advancing the vision of a cleaner, resource-efficient future.

References

Abdullah U, Ali M, Pervaiz E (2021) An inclusive review on recent advancements of cadmium sulfide nanostructures and its hybrids for photocatalytic and electrocatalytic applications. Mol Catal 508:111575. https://doi.org/10.1016/j.mcat.2021.111575

Adeola AO, Abiodun BA, Adenuga DO, Nomngongo PN (2022) Adsorptive and photocatalytic remediation of hazardous organic chemical pollutants in aqueous medium: a review. J Contam Hydrol 248:104019. https://doi.org/10.1016/j.jconhyd.2022.104019

Agarwal N, Solanki VS, Awasthi RR, Bind V et al (2022a) Release and accumulation of pharmaceuticals in the environment: critical risk assessment for environment, ecosystem and human health in book pharmaceuticals: boon or bane. In: Agarwal N (ed). Nova Science Publishers, New York, pp 1–23. https://doi.org/10.52305/GPMC4427

Agarwal N, Solanki VS, Gacem A, Hasan MA, Pare B, Srivastava A, Singh A, Yadav VK, Yadav KK, Lee C et al (2022b) Bacterial laccases as biocatalysts for the remediation of environmental toxic pollutants: a green and eco-friendly approach—a review. Water 14:4068. https://doi.org/10.3390/w14244068

Agarwal N, Solanki VS, Pare B, Singh N, Jonnalagadda SB (2023) Current trends in nanocatalysis for green chemistry and its applications—a mini-review. Curr Opin Green Sustain Chem 41:10078. https://doi.org/10.1016/j.cogsc.2023.100788

Ahmed T, Noman M, Shahid M, Niazi MBK, Hussain S, Manzoor N, Wang X, Li B (2020) Green synthesis of silver nanoparticles transformed synthetic textile dye into less toxic intermediate molecules through LC-MS analysis and treated the actual wastewater. Environ Res 191:110142. https://doi.org/10.1016/j.envres.2020.110142

Ahmed SF, Mofijur M, Ahmed B, Mehnaz T, Mehejabin F, Maliat D, Hoang AT, Shafiullah G (2022) Nanomaterials as a sustainable choice for treating wastewater. Environ Res 214:113807. https://doi.org/10.1016/j.envres.2022.113807

Ali I, Peng C, Naz I, Khan ZM, Sultan M, Islam T, Abbasi IA (2017) Phytogenic magnetic nanoparticles for wastewater treatment: a review. RSC Adv 7(64):40158–40178. https://doi.org/10.1039/C7RA04738J

Asif MB, Zhang Z (2021) Ceramic membrane technology for water and wastewater treatment: a critical review of performance, full-scale applications, membrane fouling and prospects. Chem Eng J 418:129481. https://doi.org/10.1016/j.cej.2021.129481

Baruah A, Chaudhary V, Malik R, Tomer VK (2019) Nanotechnology in water and wastewater treatment: theory and applications. In: Nanotechnology based solutions for wastewater treatment. Elsevier, Amsterdam, pp 337–368. https://doi.org/10.1016/B978-0-12-813902-8.00017-4

Behboudi A, Mohammadi T, Ulbricht M (2023) High performance antibiofouling hollow fiber polyethersulfone nanocomposite membranes incorporated with novel surface-modified silver nanoparticles suitable for membrane bioreactor application. J Ind Eng Chem 119:298–314. https://doi.org/10.1016/j.jiec.2022.11.049

Bera SP, Godhaniya M, Kothari C (2022) Emerging and advanced membrane technology for wastewater treatment: a review. J Basic Microbiol 62:245–259. https://doi.org/10.1002/jobm.202100259

Bernardes LMM, Malta SM, Santos ACC et al (2025) Green synthesis, characterization, and antimicrobial activity of silver nanoparticles from water-soluble fractions of Brazilian Kefir. Sci Rep 15:10626. https://doi.org/10.1038/s41598-025-95616-4

Bhat SA, Sher F, Hameed M, Bashir O, Kumar R, Vo D-VN, Ahmad P, Lima EC (2022) Sustainable nanotechnology-based wastewater treatment strategies: achievements, challenges and future perspectives. Chemosphere 288:132606. https://doi.org/10.1016/j.chemosphere.2021.132606

Bhuin A, Udayakumar S, Gopalarethinam J, Mukherjee D, Girigoswami K, Ponraj C, Sarkar S (2024) Photocatalytic degradation of antibiotics and antimicrobial and anticancer activities of two-dimensional ZnO nanosheets. Sci Rep 14(1):10406. https://doi.org/10.1038/s41598-024-59842-6

Burgos WD, McDonough JT, Senko JM, Zhang G, Dohnalkova AC, Kelly SD, Gorby Y, Kemner KM (2008) Characterization of uraninite nanoparticles produced by Shewanella oneidensis MR-1. Geochim Cosmochim Acta 72(20):4901–4915. https://doi.org/10.1016/j.gca.2008.07.016

Butti SK, Velvizhi G, Sulonen ML, Haavisto JM, Koroglu EO, Cetinkaya AY, Mohan SV (2016) Microbial electrochemical technologies with the perspective of harnessing bioenergy: maneuvering towards upscaling. Renew Sust Energ Rev 53:462–476. https://doi.org/10.1016/j.rser.2015.08.058

Chabalala MB, Gumbi NN, Mamba BB, Al-Abri MZ, Nxumalo EN (2021) Photocatalytic nanofiber membranes for the degradation of micropollutants and their antimicrobial activity: recent advances and future prospects. Membranes 11:678. https://doi.org/10.3390/membranes11090678

Chahar M, Khaturia S, Singh HL, Solanki VS, Agarwal N, Sahoo DK, Yadav VK, Patel A (2023) Recent advances in the effective removal of hazardous pollutants from wastewater by using nanomaterials—a review. Front Environ Sci 2023:11. https://doi.org/10.3389/fenvs.2023.1226101

Chatterjee S, Mahanty S, Das P, Chaudhuri P, Das S (2020) Biofabrication of iron oxide nanoparticles using manglicolous fungus *Aspergillus niger* BSC-1 and removal of Cr(VI) from aqueous solution. Chem Eng J 385:123790. https://doi.org/10.1016/j.cej.2019.123790

Ciambelli P, Guardia GL, Vitale L (2020) Nanotechnology for green materials and processes. In: Studies in surface science and catalysis. Elsevier, Amsterdam, pp 97–116. https://doi.org/10.1016/B978-0-444-64337-7.00007-0

Corsi I, Winther-Nielsen M, Sethi R, Punta C, Della Torre C, Libralato G et al (2018) Ecofriendly nanotechnologies and nanomaterials for environmental applications: key issue and consensus recommendations for sustainable and eco-safe nanoremediation. Ecotoxicol Environ Saf 154:237–244. https://doi.org/10.1016/j.ecoenv.2018.02.037

Crooks RM, Zhao MQ, Sun L, Chechik V, Yeung LK (2001) Dendrimer-encapsulated metal nanoparticles: synthesis, characterization, and applications to catalysis. Acc Chem Res 34(3):181–190. https://doi.org/10.1021/ar000110a

Darwesh OM, Matter IA, Eida MF (2019) Development of peroxidase enzyme immobilized magnetic nanoparticles for bioremediation of textile wastewater dye. J Environ Chem Eng 7:102805. https://doi.org/10.1016/j.jece.2018.11.049

Dave PN, Chopda LV (2014) Application of iron oxide nanomaterials for the removal of heavy metals. J Nanotechnol 2014:1–14. https://doi.org/10.1155/2014/398569

Diallo MS, Christie S, Swaminathan P, Johnson JH, Goddard WA (2005) Dendrimer enhanced ultrafiltration. Recovery of Cu (II) from aqueous solutions using PAMAM dendrimers with ethylenediamine core and terminal NH2 groups. Environ Sci Technol 39(5):1366–1377. https://doi.org/10.1021/es048961r

Duo J, Li W, Wang Y, Wang S, Wufuer R, Pan X (2022) Photothermal catalytic degradation of lomefloxacin with nano Au/TiO2. Water 14:339. https://doi.org/10.3390/w14030339

Dwevedi A (2019) Solutions to environmental problems involving nanotechnology and enzyme technology. Academic Press, Cambridge. https://doi.org/10.1016/C2016-0-04550-3

Ejraei A, Aroon MA, Ziarati Saravani A (2019) Wastewater treatment using a hybrid system combining adsorption, photocatalytic degradation and membrane filtration processes. J Water Process Eng 28:45–53. https://doi.org/10.1016/j.jwpe.2019.01.003

Enamala MK, Dixit R, Tangellapally A, Singh M, Dinakarrao SMP, Chavali M, Chandrasekhar K (2020) Photosynthetic microorganisms (algae) mediated bioelectricity generation in microbial fuel cell: concise review. Environ Technol Innov 19:100959. https://doi.org/10.1016/j.eti.2020.100959

Eroglu E, Agarwal V, Bradshaw M, Chen X, Smith S, Raston C, Iyer K (2012) Nitrate removal from liquid effluents using algae-immobilized chitosan nanofiber mats. Green Chem 14:2682–2685. https://doi.org/10.1039/C2GC35970G

Fan X, Wei G, Quan X (2023) Carbon nanomaterial-based membranes for water and wastewater treatment under electrochemical assistance. Environ Sci Nano 10(1):11–40. https://doi.org/10.1039/D2EN00545J

Feng G, Zhang H, Chen J, Fang Z, Hu X (2025) Modified graphene oxide (GO) embedded in nanofiltration membranes with high flux and anti-fouling for enhanced surface water purification. J Environ Chem Eng 13(1):115172. https://doi.org/10.1016/j.jece.2024.115172

Fetanat M, Keshtiara M, Low Z-X, Keyikoglu R, Khataee A, Orooji Y, Chen V, Leslie G, Razmjou A (2021) Machine learning for advanced design of nanocomposite ultrafiltration membranes. Ind Eng Chem Res 60(14):5236–5250. https://doi.org/10.1021/acs.iecr.0c05446

Finny AS (2024) 3D bioprinting in bioremediation: a comprehensive review of principles, applications, and future directions. PeerJ 12:e16897. https://doi.org/10.7717/peerj.16897

Garrido-Momparler V, Peris M (2022) Smart sensors in environmental/water quality monitoring using IoT and cloud services. Trends Environ Anal Chem.35 : e00173. https://doi.org/10.1016/j.teac.2022.e00173

Gillan DC (2016) Metal resistance systems in cultivated bacteria: are they found in complex communities? Curr Opin Biotechnol 38:123–130. https://doi.org/10.1016/j.copbio.2016.01.012

Gupta NM, Kirar JS, Deswal Y, Deswal L, Chandra K (2022) Pollution by pharma industries: an overview. In: Agarwal N (ed) Book pharmaceuticals: boon or bane. Nova Science Publishers, Newyork, pp 1–23. https://doi.org/10.52305/GPMC4427

Hamida RS, Ali MA, Redhwan AMO, Bin-Meferij MM (2020) Cyanobacteria—a promising platform in green nanotechnology: a review on nanoparticles fabrication and their prospective applications. Int J Nanomedicine 15:6033. https://doi.org/10.2147/ijn.s256134

Harun-Ur-Rashid M, Pal K, Imran AB (2024) Hybrid nanocomposite fabrication of Nanocatalyst with enhanced and stable photocatalytic activity. Top Catal 67:17–45. https://doi.org/10.1007/s11244-023-01809-4

Hu J, Chen GH, Lo IMC (2006) Selective removal of heavy metals from industrial wastewater using maghemite nanoparticle: performance and mechanisms. J Environ Eng-Asce 132(7):709–715. https://doi.org/10.1061/(ASCE)0733-9372(2006)132:7(709)

Hu W, Yin J, Deng B, Hu Z (2015) Application of nano TiO_2 modified hollow fiber membranes in algal membrane bioreactors for high-density algae cultivation and wastewater polishing. Bioresour Technol 193:135–141. https://doi.org/10.1016/j.biortech.2015.06.070

Hua M, Zhang S, Pan B et al (2012) Heavy metal removal from water/wastewater by nanosized metal oxides: a review. J Hazard Mater 211:317–331. https://doi.org/10.1016/j.jhazmat.2011.10.016

Huang Y, Gao M, Deng Y, Khan ZH, Liu X, Song Z, Qiu W (2020) Efficient oxidation and adsorption of As(III) and As(V) in water using a Fenton-like reagent, (ferrihydrite)-loaded biochar. Sci Total Environ:136957. https://doi.org/10.1016/j.scitotenv.2020.136957

Huggins T, Latorre A, Biffinger J, Ren Z (2016) Biochar based microbial fuel cell for enhanced wastewater treatment and nutrient recovery. Sustainability 8:169. https://doi.org/10.3390/su8020169

Hui KC, Ang WL, Sambudi NS (2021) Nitrogen and bismuth-doped rice husk-derived carbon quantum dots for dye degradation and heavy metal removal. J Photochem Photobiol A Chem 418:113411. https://doi.org/10.1016/j.jphotochem.2021.113411

Ishfaq M, Ahmad M, Hussain A, Afzal M (2023) Remediation of wastewater by biosynthesized manganese oxide nanoparticles and its effects on development of wheat seedlings. Front Plant Sci 14:1263813. https://doi.org/10.3389/fpls.2023.1263813

Ismail GA, Allam NG, El-Gemizy WM, Salem MA (2021) The role of silver nanoparticles biosynthesized by *Anabaena variabilis* and *Spirulina platensis* cyanobacteria for malachite green removal from wastewater. Environ Technol 42(28):4475–4489. https://doi.org/10.1080/09593330.2020.1766576

Jain R, Dominic D, Jordan N, Rene ER, Weiss S, van Hullebusch ED, Hübner R, Lens PNL (2016) Preferential adsorption of cu in a multi-metal mixture onto biogenic elemental selenium nanoparticles. Chem Eng J 284:917–925. https://doi.org/10.1016/j.cej.2015.08.144

Jain K, Patel A, Pardhi V, Flora S (2021) Nanotechnology in wastewater management: a new paradigm towards wastewater treatment. Molecules 26:1797. https://doi.org/10.3390/molecules26061797

Kitching M, Ramani M, Marsili E (2015) Fungal biosynthesis of gold nanoparticles: mechanism and scale up. Microb Biotechnol 8(6):904–917. https://doi.org/10.1111/1751-7915.12151

Krishnan A, Sundaram T, Nagappan B, Devarajan Y, Bhumika. (2024) Integrating artificial intelligence in nanomembrane systems for advanced water desalination. Results Eng 24:103321. https://doi.org/10.1016/j.rineng.2024.103321

Kumar R, Chawla J (2014) Removal of cadmium ion from water/wastewater by nano-metal oxides: a review. Water Qual Expo Health 5:215–226. https://doi.org/10.1007/s12403-013-0100-8

Kumari P, Alam M, Siddiqi WA (2019) Usage of nanoparticles as adsorbents for waste water treatment: an emerging trend. Sustain Mater Technol 22:00128. https://doi.org/10.1016/j.susmat.2019.e00128

Kyzas GZ, Deliyanni EA (2013) Mercury (II) removal with modified magnetic chitosan adsorbents. Molecules 18(6):6193–6214. https://doi.org/10.3390/molecules18066193

Li J, Webster TJ, Tian B (2019) Functionalized nanomaterial assembling and biosynthesis using the extremophile *Deinococcus radiodurans* for multifunctional applications. Small 15:1900600. https://doi.org/10.1002/smll.201900600

Mahanty S, Bakshi M, Ghosh S, Gaine T, Chatterjee S, Bhattacharyya S, Das S, Das P, Chaudhari P (2019) Mycosynthesis of iron oxide nanoparticles using manglicolous fungi isolated from Indian sundarbans and its application for the treatment of chromium containing solution: synthesis, adsorption isotherm, kinetics and thermodynamics study. Environ Nanotechnol Monit Manag 12:100276. https://doi.org/10.1016/j.enmm.2019.100276

Mahanty S, Chatterjee S, Ghosh S, Tudu P, Gaine T, Bakshi M et al (2020) Synergistic approach towards the sustainable management of heavy metals in wastewater using mycosynthesized iron oxide nanoparticles: biofabrication, adsorptive dynamics and chemometric modeling study. J Water Process Eng 37:101426. https://doi.org/10.1016/j.jwpe.2020.101426

Malik S, Dhasmana A, Preetam S, Mishra YK, Chaudhary V, Bera SP, Ranjan A, Bora J, Kaushik A, Minkina T, Jatav HS, Singh RK, Rajput VD (2022) Exploring microbial-based green nanobiotechnology for wastewater remediation: a sustainable strategy. Nanomaterials (Basel) 12(23):4187. https://doi.org/10.3390/nano12234187

Mallick N (2002) Biotechnological potential of immobilized algae for wastewater N, P and metal removal: a review. Biometals 15:377–390. https://doi.org/10.1023/a:1020238520948

Manikandan V, Vinoth Kumar J, Elango D, Subash V, Jayanthi P, Dixit S, Singh S (2025) Metal-organic frameworks (MOFs): multifunctional platforms for environmental sustainability. Chem Record n/a(n/a):e202400257. https://doi.org/10.1002/tcr.202400257

Martins M, Mourato C, Sanches S, Noronha JP, Crespo MB, Pereira IA (2017) Biogenic platinum and palladium nanoparticles as new catalysts for the removal of pharmaceutical compounds. Water Res 108:160–168. https://doi.org/10.1016/j.watres.2016.10.071

Meng X, Hongjie W, Di L et al (2012) Removal of Pb (II) from aqueous solution by hydrous manganese dioxide: adsorption behaviour and mechanism. J Environ Sci 25(3):479–486. https://doi.org/10.1016/S1001-0742(12)60100-4

Mishra S, Sundaram B (2023) A review of the photocatalysis process used for wastewater treatment. Mater Today Proc 7:147. https://doi.org/10.1016/j.matpr.2023.07.147

Mohanapragash AG, Kaleeswari RK, Meena S, Baskar M, Umamaheswari T, Selvamurugan M, Ramesh T (2025) Innovative nanoremediation techniques for soil contamination: exploring metal based nanomaterial approaches—a review. Soil Sediment Contam Int J:1–69. https://doi.org/10.1080/15320383.2025.2477818

Mohanraj R, Gnanamangai BM, Poornima S, Oviyaa V, Ramesh K, Vijayalakshmi G et al (2020) Decolourisation efficiency of immobilized silica nanoparticles synthesized by actinomycetes. Mater Today Proc 48:129–135. https://doi.org/10.1016/j.matpr.2020.04.139

Mostafa MK, Mahmoud AS, Sary El-deen RA, Peters RW (2017) Application of entrapped nano zero valent iron into cellulose acetate membranes for domestic wastewater treatment. In: Proceedings of the environmental aspects, applications and implications of Nanomaterials and Nanotechnology 2017–topical conference at the 2017 AIChE Annual Meeting, Minneapolis, MN, USA

Muruganandham H, Vishnukumar P, Vivekanandhan S (2018) Biogenic approaches for SiO_2 nanostructures: exploring the sustainable platform of nanofabrication: processing and characterization. In: Green and sustainable advanced materials: processing and characterization. Wiley, Newark, pp 107–141. https://doi.org/10.1002/9781119407089.ch5

Nagendranatha Reddy C, Nguyen HTH, Noori MT, Min B (2019) Potential applications of algae in the cathode of microbial fuel cells for enhanced electricity generation with simultaneous nutrient removal and algae biorefinery: current status and future perspectives. Bioresour Technol 292:122010. https://doi.org/10.1016/j.biortech.2019.122010

Nahak BK, Preetam S, Sharma D, Shukla S, Syväjärvi M, Toncu DC, Tiwari A (2022) Advancements in net-zero pertinency of lignocellulosic biomass for climate neutral energy production. Renew Sust Energ Rev 161:112393. https://doi.org/10.1016/j.rser.2022.112393

Nain A, Sangili A, Hu S-R, Chen C-H, Chen Y-L, Chang H-T (2022) Recent progress in nanomaterial-functionalized membranes for removal of pollutants. iScience 25(7):104616. https://doi.org/10.1016/j.isci.2022.104616

Nawaz A, Haq I, Qaisar K, Gunes B, Raja SI, Mohyuddin K, Amin H (2022) Microbial fuel cells: insight into simultaneous wastewater treatment and bioelectricity generation. Process Saf Environ Prot 161:357–373. https://doi.org/10.1016/j.psep.2022.03.039

Nidheesh PV et al (2024) Doped TiO_2 for antibiotic removal. Water Res 250:121055. https://doi.org/10.1201/9781003208662

Nkosi NC, Basson AK, Ntombela ZG, Dlamini NG, Pullabhotla RVSR (2025) Green synthesis and characterization of iron nanoparticles synthesized from bioflocculant for wastewater treatment: a review. Biotechnol Notes 6:10–31. https://doi.org/10.1016/j.biotno.2024.12.001

Noroozi R, Gholami M, Farzadkia M, Jonidi Jafari A (2020) Degradation of ciprofloxacin by $CuFe_2O_4$/GO activated PMS process in aqueous solution: performance, mechanism and degradation pathway. Int J Environ Anal Chem 102(1):174–195. https://doi.org/10.1080/03067319.2020.1718669

O'Hern SC, Boutilier MSH, Idrobo J-C, Song Y, Kong J, Laoui T, Atieh M, Karnik R (2014) Selective ionic transport through tunable subnanometer pores in single-layer graphene membranes. Nano Lett 14(3):1234–1241. https://doi.org/10.1021/nl404118f

Pacheco S, Medina M, Valencia F, Tapia J (2006) Removal of inorganic mercury from polluted water using structured nanoparticles. J Environ Eng 132:342–349. https://doi.org/10.1061/(ASCE)0733-9372(2006)132:3(342)

Pei Y, Chen X, Xiong D, Liao S, Wang G (2013) Removal and recovery of toxic silver ion using deep-sea bacterial generated biogenic manganese oxides. PLoS One 8(12):e81627. https://doi.org/10.1371/journal.pone.0081627

Qu X, Alvarez PJ, Li Q et al (2013a) Applications of nanotechnology in water and wastewater treatment. Water Res 47(12):3931–3946. https://doi.org/10.1016/j.watres.2012.09.058

Qu XL, Brame J, Li Q, JJP A (2013b) Nanotechnology for a safe and sustainable water supply: enabling integrated water treatment and reuse. Acc Chem Res 46(3):834–843. https://doi.org/10.1021/ar300029v

Rafique M, Tahir MB, Sadaf I (2019) Advanced research in nanosciences for water technology. Springer, pp 95–131. https://doi.org/10.1007/978-3-030-02381-2

Raja RK, Hazir S, Balasubramani G, Sivaprakash G, Obeth ESJ, Boobalan T, Pugazhendhi A, Raj RHK, Arun A (2022) Green nanotechnology for the environment. In: Handbook of microbial nanotechnology. Elsevier, Amsterdam, pp 461–478. https://doi.org/10.1016/B978-0-12-823426-6.00006-1

Rajput S, Pittman CU Jr, Mohan D (2016) Magnetic magnetite (Fe3O4) nanoparticle synthesis and applications for lead (Pb^{2+}) and chromium (Cr^{6+}) removal from water. J Colloid Interface Sci 468:334–346. https://doi.org/10.1016/j.jcis.2015.12.008

Rani N, Kusum, Hooda V (2024) Chitosan/ZnO nanocomposites for improving the growth and reducing the toxicity of Zn in *Sorghum bicolor* (L.) Moench plants. Acta Physiol Plant 46:67. https://doi.org/10.1007/s11738-024-03693-1

Rath P, Bhardwaj LK, Yadav P, Bhardwaj AK (2024) A synthesis of biogenic nanoparticles (NPs) for the treatment of wastewater and its application: a review. In: Bhardwaj AK, Srivastav AL, Rai S (eds) Biogenic wastes-enabled nanomaterial synthesis. Springer, Cham. https://doi.org/10.1007/978-3-031-59083-2_5

Rathore C, Yadav VK, Gacem A, AbdelRahim SK, Verma RK, Chundawat RS, Gnanamoorthy G, Yadav KK, Choudhary N, Sahoo DK, Patel A (2023) Microbial synthesis of titanium dioxide nanoparticles and their importance in wastewater treatment and antimicrobial activities: a review [review]. Front Microbiol 14. https://doi.org/10.3389/fmicb.2023.1270245

Ray PZ, Shipley HJ (2015) Inorganic nano-adsorbents for the removal of heavy metals and arsenic: a review. RSC Adv 5(38):29885–29907. https://doi.org/10.1039/C5RA02714D

Razack SA, Duraiarasan S (2020) A way to create a sustainable environment: green nanotechnology–with an emphasis on noble metals. In: Hussain CM (ed) The ELSI handbook of nano-

technology: risk, safety, ELSI and commercialization. Wiley-Scrivener, Austin, pp 359–425. https://doi.org/10.1002/9781119592990.ch14
Ren B, Jin Y, zhao L, Cui C, Song X (2022) Enhanced Cr(VI) adsorption using chemically modified dormant *Aspergillus niger* spores: process and mechanisms. J Environ Chem Eng 10(1):106955. https://doi.org/10.1016/j.jece.2021.106955
Rifan H, Robin M, Cywar, Eugene YX, Chen GS (2022) Sustainable nanofiltration membranes based on biosourced fully recyclable polyesters and green solvents. J Membr Sci Lett 2(1):100016. https://doi.org/10.1016/j.memlet.2022.100016
Saied E, Eid AM, Hassan E-D, Salem S, Radwan AA, Halawa M, Saleh FM, Saad HA, Saied EM, Fouda A (2021) The catalytic activity of biosynthesized magnesium oxide nanoparticles (MgO-NPs) for inhibiting the growth of pathogenic microbes, tanning effluent treatment, and chromium ion removal. Catalysts 11(7):821. https://doi.org/10.3390/catal11070821
Salam MA, Makki MSI, Abdelaal MYA (2009) Preparation and characterization of multi-walled carbon nanotubes/chitosan nanocomposite and its application for the removal of heavy metals from aqueous solution. J Alloy Compd 509(5):2582–2587. https://doi.org/10.1016/j.jallcom.2010.11.094
Saroj AL, Rawat P, Sharma AL (2022) Polymer composites for dye-sensitized solar cells. In: Polymer electrolytes and their composites for energy storage/conversion devices. CRC Press, pp 231–259. https://doi.org/10.1201/9781003208662-11
Sayadi MH, Salmani N, Heidari A, Rezaei MR (2018) Bio-synthesis of palladium nanoparticle using *Spirulina platensis* alga extract and its application as adsorbent. Surf Interfaces 10:136–143. https://doi.org/10.1016/j.surfin.2018.01.002
Secundo F (2013) Conformational changes of enzymes upon immobilisation. Chem Soc Rev 42:6250–6261. https://doi.org/10.1039/C3CS35495D
Sekar AD, Jayabalan T, Muthukumar H, Chandrasekaran NI, Mohamed SN, Matheswaran M (2019) Enhancing power generation and treatment of dairy waste water in microbial fuel cell using Cu-doped iron oxide nanoparticles decorated anode. Energy 172:173–180. https://doi.org/10.1016/j.energy.2019.01.102
Shaikh RA, Al-Othman A, Tawalbeh M, Shamayleh A, Nancarrow P (2024) Development of MXene incorporated PVDF based membranes for an enhanced performance in higher temperature PEM fuel cells. Process Saf Environ Prot 189:985–994. https://doi.org/10.1016/j.psep.2024.06.126
Sharma YC, Srivastava V, Singh VK, Kaul SN, Weng CH (2009) Nano-adsorbents for the removal of metallic pollutants from water and wastewater. Environ Technol 30(6):583–609. https://doi.org/10.1080/09593330902838080
Shen YF, Tang J, Nie ZH et al (2015) Preparation and application of magnetic Fe_3O_4 nanoparticles for wastewater purification. Sep Purif Technol 68(3):312–319. https://doi.org/10.1016/j.seppur.2009.05.020
Mandeep, Shukla P (2020) Microbial nanotechnology for bioremediation of industrial wastewater [mini review]. Front Microbiol 11:590631. https://doi.org/10.3389/fmicb.2020.590631
Siddique K, Shahid M, Shahzad T, Faisal Mahmood F, Nadeem H, Saif ur Rehman M, Hussain S, Sadak O, Gunasekaran S, Kamal T, Ahmad I (2021) Comparative efficacy of biogenic zinc oxide nanoparticles synthesized by *Pseudochrobactrum* sp. C5 and chemically synthesized zinc oxide nanoparticles for catalytic degradation of dyes and wastewater treatment. Environ Sci Pollut Res 28:28307–28318. https://doi.org/10.1007/s11356-021-12575-9
Sidorowicz A, Fais G, Desogus F, Loy F, Licheri R, Lai N, Cao G, Concas A (2024) Eco-friendly photocatalytic treatment of dyes with Ag nanoparticles obtained through sustainable process involving *Spirulina platensis*. Sustainability 16:8758. https://doi.org/10.3390/su16208758
Singh AV, Vyas V, Salve TS, Cortelli D, Dellasega D, Podestà A, Milani P, Gade WN (2012) Biofilm formation on nanostructured titanium oxide surfaces and a micro/nanofabrication-based preventive strategy using colloidal lithography. Biofabrication 4(2):025001. https://doi.org/10.1088/1758-5082/4/2/025001

Singh S, Shankar S, Shikha (2022) Microbial synthesis of nanoparticles for wastewater remediation. In: Rai JPN, Saraswat S (eds) Nano-biotechnology for waste water treatment, vol 111. Water Science and Technology Library. https://doi.org/10.1007/978-3-031-00812-2_5

Sivaraj R, Salam HA, Rajiv P, Rajendran V (2015) Green nanotechnology: the solution to sustainable development of environment. In: Environmental Sustainability. Springer, Berlin/Heidelberg, pp 311–324. https://doi.org/10.1007/978-81-322-2056-5_18

Srivastava P, Yadav AK, Mishra BK (2015) The effects of microbial fuel cell integration into constructed wetland on the performance of constructed wetland. Bioresour Technol 195:223–230. https://doi.org/10.1016/j.biortech.2015.05.072

Suganthi SH, Kandasamy R (2017) A novel single step synthesis and surface functionalization of iron oxide magnetic nanoparticles and thereof for the copper removal from pigment industry effluent. Sep Purif Technol 188:458–467. https://doi.org/10.1016/j.seppur.2017.07.059

Tang CY, Zhao Y, Wang R, Hélix-Nielsen C, Fane AG (2013) Desalination by biomimetic aquaporin membranes: review of status and prospects. Desalination 308:34–40. https://doi.org/10.1016/j.desal.2012.07.007

Tawabini B, Al-Khaldi S, Atieh M et al (2010) Removal of mercury from water by multi-walled carbon nanotubes. Water Sci Technol 61(3):591–598. https://doi.org/10.2166/wst.2010.897

Tripathi A, Das S, Ghangrekar MM, Dubey BK (2025) Improving algae-assisted bioelectrochemical system with the integration of advanced cell disruption process for lipid recovery: A mini-review. Bioresour Technol Rep 29:102041. https://doi.org/10.1016/j.biteb.2025.102041

Wang X, Guo Y, Yang L et al (2012) Nanomaterials as sorbents to remove heavy metal ions in wastewater treatment. J Environ Anal Toxicol 2(7):1–7. https://doi.org/10.4172/2161-0525.1000154

Yadav V, Ali J, Garg MC (2021) Biosorption of methylene blue dye from textile-industry wastewater onto sugarcane bagasse: response surface modeling, isotherms, kinetic and thermodynamic modeling. J Hazard Toxic Radioact Waste 25:04020067. https://doi.org/10.1061/(asce)hz.2153-5515.0000572

Yang G, Wang J (2019) Synergistic enhancement of biohydrogen production from grass fermentation using biochar combined with zero-valent iron nanoparticles. Fuel 251:420–427. https://doi.org/10.1016/j.fuel.2019.04.059

Yin Z, Sadok A, Sailem H, McCarthy A, Xia X, Li F, Garcia MA, Evans L, Barr AR, Perrimon N, Marshall CJ, Wong ST, Bakal C (2013) A screen for morphological complexity identifies regulators of switch-like transitions between discrete cell shapes. Nat Cell Biol 15(7):860–871. https://doi.org/10.1038/ncb2764

Yin J, Huang G, Xiao H, Chen N, An C, Chao T-C, Feng R, Read S (2023) Bioinspired and dual-functional nanocellulose aerogels for water disinfection and heavy metal removal. Nano Today 51:101918. https://doi.org/10.1016/j.nantod.2023.101918

Ying Y, Ying W, Li Q, Meng D, Ren G, Yan R, Peng X (2017) Recent advances of nanomaterial-based membrane for water purification. Appl Mater Today 7:144–158

You J, Preen RJ, Bull L, Greenman J, Ieropoulos I (2017) 3D printed components of microbial fuel cells: towards monolithic microbial fuel cell fabrication using additive layer manufacturing. Sustain Energy Technol Assess 19:94–101. https://doi.org/10.1016/j.seta.2016.11.006

Zahmatkesh S, Hajiaghaei-Keshteli M, Bokhari A, Sundaramurthy S, Panneerselvam B, Rezakhani Y (2023) Wastewater treatment with nanomaterials for the future: a state-of-the-art review. Environ Res 216:114652. https://doi.org/10.1016/j.envres.2022.114652

Zheng Y, Zhang X, Zhao J, Yang P (2018) Assembled fabrication of α-Fe_2O_3/BiOCl hetero junctions with enhanced photocatalytic performance. Appl Surf Sci 430:585–594. https://doi.org/10.1016/j.apsusc.2017.06.097

Chapter 8
Advancing Radwaste Remediation: Innovations, Strategies, and Future Perspectives

Sachchit Patel, Yogita Pandey, Siddraj Thakur, Hitanshu Kapadiya, Ronakkumar Bhadaniya, and Arpit Shukla

Abstract Human activities typically produce trash, particularly due to urbanization and the expansion of big urban areas, resulting in inadequate garbage management. Consequently, it exerts a significant influence on both human health and the natural environment. Conventional approaches to handling radioactive waste involve strategies for safely managing, storing, and disposing of radioactive materials produced by various human activities. Efficient management of radioactive waste is crucial to protect human health and save the environment. "Innovative Radwaste Remediation Technologies" refers to recently created and advanced methods or technologies specifically designed to address the challenges associated with the management and purification of radioactive waste, commonly referred to as radwaste. Biological remediation of radioactive waste entails the use of living organisms, such as bacteria and plants, to mitigate or decrease the impact of radioactive contaminants on the environment. Physicochemical procedures, such as vitrification, solidification and stabilization, and ion exchange, can efficiently recover contaminated soil that contains radionuclides. Ensuring the appropriate disposal of radioactive waste is of utmost importance due to its capacity to inflict enduring harm on ecosystems. Nevertheless, a potential resolution is emerging in the field of bioremediation, which employs microorganisms to counteract the detrimental consequences of radioactive waste. This chapter provides an analysis of an effective cleanup plan implemented at the Hanford Nuclear Reservation, a very polluted location in the United States. This case study demonstrates the capacity for creative and enduring solutions to practical problems, emphasizing the significance of comprehending the intricate interplay among various elements within a system, offering an ecologically sound and enduring alternative for tackling this urgent matter.

Keywords Radwaste · Nuclear waste · Radiation · Phytoremediation · Vitrification

---

S. Patel · Y. Pandey · S. Thakur · H. Kapadiya · R. Bhadaniya · A. Shukla (✉)
School of Biotechnology and Bioengineering, Institute of Advanced Research, Gandhinagar, Gujarat, India
e-mail: arpit.shukla@iar.ac.in

N. K. Arora, J. Mishra (eds.), *Biotechnological Solutions for a Sustainable Future*, Microorganisms for Sustainability 55,
https://doi.org/10.1007/978-981-95-3989-5_8

8.1 Introduction

Radioactive waste or radwaste is referred to as material that contains or is polluted with radionuclides at levels that exceed prescribed clearance criteria, and for which there is no intended future use (International Atomic Energy Agency [IAEA] n.d.). Radioactive waste consists of radioactive materials that are typically produced as by-products of nuclear power generation and other uses of nuclear fission or nuclear technology, such as research and medicine (Jeswani and Khelurkar 2015). Radioactive waste is harmful to many living forms and the environment, and is subject to government regulations to safeguard human health and the environment.

The classification of radioactive waste is a systematic process that categorizes waste based on its characteristics, origin, and level of radioactivity. The International Atomic Energy Agency (IAEA) provides a widely accepted classification system for radioactive waste. Radwaste can be classified as follows:

- Low-level waste (LLW): Low-level waste consists of radioactive materials with relatively low levels of radioactivity. Items in this category encompass contaminated tools, protective clothes, and reactor components (International Atomic Energy Agency [IAEA] n.d.).
- Intermediate-level waste (ILW): Intermediate-level waste exhibits greater amounts of radioactivity in comparison to low-level waste. The assortment comprises resins, filters, and reactor components (International Atomic Energy Agency [IAEA] n.d.).
- High-level waste (HLW): High-level waste consists of very radioactive substances, usually produced from the depleted fuel of nuclear reactors. It contains substances that have radioisotopes with a long half-life (International Atomic Energy Agency [IAEA] n.d.).
- Very low-level waste (VLLW): Certain nations classify low-level waste into even lower categories, such as very low-level waste, which has extremely low quantities of radioactivity and is regarded as less dangerous (International Atomic Energy Agency [IAEA] n.d.).
- Very short-lived waste (VSLW): This refers to waste that can be held for a specific duration, often a few years, in order for its radioactivity to reduce through the process of radioactive decay. It can later be exempted from regulatory oversight based on authorized protocols by the regulatory agency for disposal as regular trash, for utilization, or for controlled release. This class includes trash that contains radionuclides with extremely brief half-lives, typically employed for research and medicinal applications (International Atomic Energy Agency [IAEA] n.d.).
- Exempt waste (EW): This waste has a negligible level of radioactivity and no longer requires oversight from the regulatory authorities. Once the regulatory body approves the substance, it ceases to be classified as radioactive waste (International Atomic Energy Agency [IAEA] n.d.).

The importance of radioactive waste is in its potential to pose risks to human health and the environment due to the ionizing radiation emitted by the radioactive materials it contains. Effective control and proper elimination of radioactive waste are essential in order to limit these potential hazards. Some key aspects of the significance of radioactive waste are as follows:

- Radiation exposure and health risks: Inadequate management or improper disposal of radioactive waste can result in radiation exposure, which can have detrimental health consequences, including the development of cancer and genetic abnormalities (International Atomic Energy Agency [IAEA] n.d.).
- Environmental impact: Discharging radioactive compounds into the environment can pollute the air, water, and soil, thereby impacting ecosystems and species (International Atomic Energy Agency [IAEA] n.d.)
- Long-term storage challenges: Secure and durable storage solutions are necessary to prevent potential leakage and contamination of high-level radioactive waste, such as spent nuclear fuel.
- Regulatory framework and standards: Adherence to international standards and regulations is crucial to ensure the safe management and disposal of radioactive waste (International Atomic Energy Agency [IAEA] n.d.).
- Public perception and trust: The adoption and sustainability of nuclear technology rely heavily on the crucial factors of public perception and confidence in the handling of radioactive waste.

The remediation of radioactive waste comprises the use of numerous approaches to minimize the environmental and health hazards linked to radioactive pollutants. The selection of the remediation approach is contingent upon several aspects, including the kind and magnitude of the contamination, the properties of the radioactive substances, and the particular environmental circumstances. Techniques or strategies used for remediation of radwaste are as follows: traditional radwaste management approaches, biological remediation strategies, physicochemical remediation methods, and innovative remediation technologies.

Microorganisms have continually demonstrated their resilience by living in harsh conditions, including those with high levels of ionizing radiation (Fredrickson et al. 2004). Their capacity to tolerate and undergo consistent genetic alterations has resulted in the creation of recombinant mutants that are being utilized for the mitigation of diverse contaminants, including heavy metals, hydrocarbons, and even radioactive waste (radwaste) (Brim et al. 2000). The processes that have been widely studied in microbes for radwaste remediation are bioaccumulation, biotransformation, biosorption, biosolubilization, and bioprecipitation (Shukla et al. 2017).

Researchers are now investigating novel techniques to improve the effectiveness, sustainability, and practicality of using biological methods to manage radioactive waste, which present promising prospects for the future of radwaste bioremediation. This chapter aims to provide a comprehensive overview of radioactive waste, its classification, hazards, and regulatory frameworks, while highlighting the significance of safe management and disposal practices. It further emphasizes recent advances in bioremediation techniques, particularly microbial strategies, as sustainable and innovative solutions for mitigating radioactive contamination.

8.2 Conventional Radwaste Disposal Methods Practiced Globally

Significant expertise has been gained in the field of radioactive waste management since the initial disposal of radioactive waste in Oak Ridge, Tennessee, in 1944. The initial disposal site, designed for the disposal of "actively contaminated broken glassware or materials that were not sufficiently clean for other purposes," consisted of a basic trench filled with unprocessed garbage, situated in the Oak Ridge facility (Rule 2011). Other nuclear sites and waste producers in the United States and other nations also followed similar tactics during the initial stages of nuclear power's development (Won et al. 1997).

Currently, there are many types of disposal sites for low- and intermediate-level radioactive wastes, which include both near-surface facilities and designed geological repositories. Over 100 low- and intermediate-level waste disposal sites are either currently in operation or have been used in the past (Canton 2021). Additionally, more than 42 repositories are currently being developed in the Member States of the International Atomic Energy Agency (IAEA) (Won et al. 1997).

Low-level radioactive waste is produced by a range of sources, such as medical facilities, research organizations, and nuclear power plants. Regulatory frameworks developed by specific nations govern the storage and handling of this radwaste. Several nations employ near-surface disposal facilities for LLRW (Canton 2021). These facilities are specifically intended to segregate waste from the surrounding environment for a predetermined. Engineered near-surface facilities utilize engineered barriers, such as liners and coverings, to effectively separate the waste. These facilities are specifically engineered to inhibit the migration of radionuclides into the surrounding environment (Kennedy Jr 1993). Certain nations choose to utilize deep geological repositories, which involve storing LLRW in subterranean formations at significant depths in order to enhance isolation from the biosphere. Following closure, LLRW disposal sites require extended monitoring and upkeep to guarantee ongoing safety and environmental preservation. LLRW management strategies differ among countries. The IAEA offers guidelines and suggestions to promote the standardization of procedures on a worldwide scale (Canton 2021).

The interment of intermediate and high-level radioactive waste is a multifaceted matter, with disposal approaches differing across various nations. A major strategy involves the utilization of deep geological repositories (DGRs), such as the Onkalo repository in Finland, which employs a multi-barrier system to effectively separate the waste from the biosphere (Posiva Oy n.d.). The construction is presently in progress at Onkalo, and comprehensive information on the project can be found on Posiva's website (Posiva Oy n.d.). The Yucca Mountain Project in the United States aimed to function as a repository for HLW but encountered obstacles of a political and regulatory nature. The Yucca Mountain repository's present condition is ambiguous, and several resolutions are being contemplated (US Nuclear Regulatory Commission n.d.).

Sweden is now engaged in the active implementation of the KBS-3 technique, which entails the containment of waste within copper canisters and its subsequent disposal in a deep geological repository designed for storing spent nuclear fuel. The initiative is being spearheaded by the Swedish Nuclear Fuel and Waste Management Company (SKB) (Swedish Nuclear Fuel and Waste Management Company n.d.).

8.3 Limitations and Challenges of Traditional Techniques

- Storage limitations: Conventional radwaste management typically entails the storage of radioactive waste in specially designed facilities or repositories. A primary obstacle is the restricted capacity of these storage facilities (International Atomic Energy Agency [IAEA] n.d.).
- Long-term containment: Securing the enduring confinement of radioactive waste is a substantial difficulty. Radioactive substances can retain their dangerous properties for an extended period of time, necessitating the use of strong methods for confinement (Board and Council 2006). Environmental impact: Conventional means of disposal, such as burial or shallow land burial, might potentially lead to environmental pollution. Mitigating the potential adverse effects of these approaches on ecosystems and water supplies is a significant challenge.
- Public perception and acceptance: The important factor in radwaste management systems is the acceptability of the public. The absence of public trust and comprehension might hinder the deployment of certain methodologies or the construction of novel disposal sites (Boice Jr et al. 2014).
- Cost and economics: The process of developing and implementing efficient radioactive waste management procedures might incur significant costs. The issue lies in achieving equilibrium between the economic dimensions of waste management and the imperative of maintaining safety, and adhering to regulatory requirements.
- Regulatory compliance: Complying with regulations for the disposal and management of radioactive waste is an intricate and demanding undertaking. Complying with ever-changing global and domestic regulations presents an additional level of complexity.

8.4 Environmental Impact and Long-Term Concerns

The substantial considerations in nuclear energy and waste management revolve around the environmental effect and long-term concerns related to radioactive waste. They can include the following:

- Radioactive waste has the potential to contaminate the air, water, and soil, which can pose significant dangers to ecosystems and the health of humans (US Environmental Protection Agency [EPA] n.d.).
- The possible migration of radionuclides into groundwater poses a long-term danger, which might have an impact on the availability of drinking water (US Environmental Protection Agency [EPA] 2018).
- Radioactive emissions can have an influence on plant and animal life, resulting in alterations to the variety of species and the functioning of ecosystems (Beir 2005).
- Radioactive waste disposal facilities may necessitate enduring land-use restrictions, constraining the use of impacted regions for agriculture, housing, or other activities.
- Long-term concerns include the possibility of human exposure to radioactive substances, which might result in heightened risks of cancer and other health complications (United Nations Scientific Committee on the Effects of Atomic Radiation [UNSCEAR] n.d.)
- Long-term factors to consider are the possibility of deliberate acts of terrorism or sabotage at radioactive waste storage or disposal sites (Agency 2011).
- To guarantee the secure confinement and separation of radioactive waste for extended periods of time, it is necessary to establish enduring management strategies that account for evolving institutions, societies, and technologies.
- The decommissioning of nuclear reactors and the handling of ensuing waste present enduring difficulties in terms of safety, cost, and environmental consequences (Agency 1999).

8.5 Innovative Remediation Technologies

Recently, there has been a major focus on the development of cutting-edge technologies for radwaste remediation. These technologies are designed to tackle the difficulties that come with older procedures. These innovative methods utilize plants, microbes, and advanced biomaterials as effective, sustainable, and eco-friendly solutions for radwaste cleanup.

8.5.1 Phytoremediation

Phytoremediation is the process of employing plants to reduce or eliminate pollutants from the environment. Some plant species possess the capacity to gather and accumulate radionuclides, providing a natural and economical method of cleanup (Meagher 2000). This approach holds great potential for effectively eliminating radioactive pollutants from soil (Salt et al. 1995). Hyperaccumulators, a specific type of plant species, possess a unique capability to absorb and accumulate significant levels of radionuclides from the soil (Chaney and Baklanov 2017). An exemplary instance is the sunflower (*Helianthus annuus*), which has undergone thorough examination due to its potential in the remediation of radioactive waste (Ali et al. 2013). Sunflowers possess extensive root systems that enable them to take radionuclides from the soil and deposit them in their tissues (Dushenkov et al. 1997). The collected plant biomass, which contains increased levels of radionuclides, can thereafter be carefully handled and disposed of, reducing the dissemination of pollutants. Phytoremediation is a cost-effective and affordable method for cleaning up polluted areas, especially suitable for extensive locations with minimal levels of pollution. Phytoremediation also has the benefit of producing plant residue that is rich in radioactive metals (Malhotra et al. 2014). However, there are drawbacks associated with this method, including its time-consuming and burdensome nature, the quantity of biomass generated, the age of the plant, and the influence of toxins on flora (Malhotra et al. 2014). Additionally, it is constrained by the rate at which plants may grow. Additional time may be necessary in comparison to traditional clean-up methods (Mwegoha 2008). Another issue to consider is the time-consuming nature of excavating, disposing, or incinerating the plants employed. In contrast, phytoextraction and degrading methods may take several years to complete (Pulford and Watson 2003). Furthermore, the efficacy of phytoremediation might be constrained by other parameters, including the duration of growth, climatic conditions, root penetration, soil composition, and the extent of pollution (Salido et al. 2003) (Fig. 8.1).

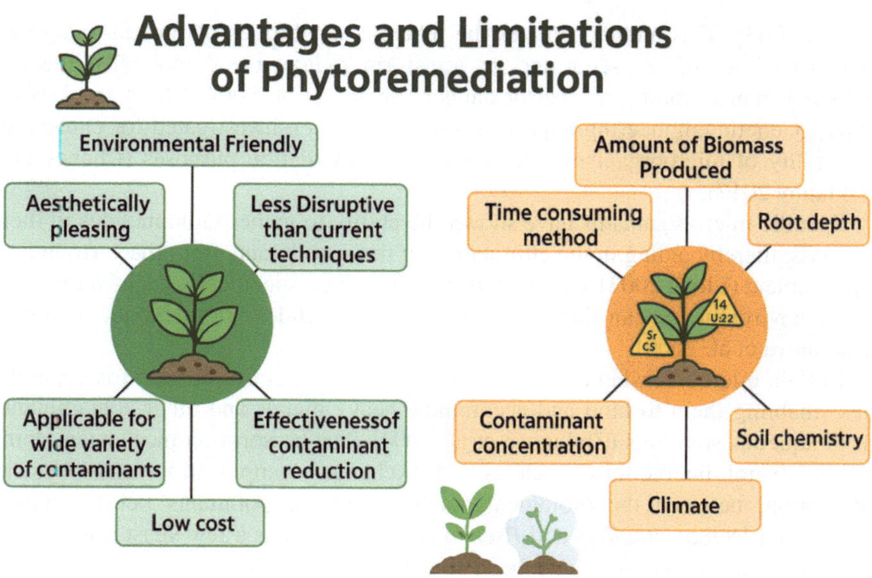

Fig. 8.1 Advantages and limitations of phytoremediation. (Adapted from Malhotra et al. (2014))

8.5.2 Nanotechnology

Nanotechnology encompasses the deliberate manipulation of materials at the nanoscale, often employing particles with diameters ranging from 1 to 100 nm (Roco 2003). Regarding the cleanup of radioactive waste, nanoparticles possess distinctive characteristics that render them very efficient in collecting, immobilizing, or altering radionuclides. The adsorption properties of iron oxide nanoparticles and their potential for removing radioactive nuclides from aqueous solutions have been extensively researched (Fan et al. 2012; O'Hara et al. 2016). Carbon nanotubes (CNTs) are a distinct category of nanomaterials that show promise in the field of radioactive waste (radwaste) cleanup (Yu et al. 2014). Due to their elevated surface area and distinctive structure, they are well suited for the process of adsorbing and isolating radionuclides. Titanium dioxide nanoparticles have photocatalytic properties that can be utilized to break down organic constituents in radioactive waste. Research has been conducted on nanomaterials, including nanoparticles, nanotubes, and nanocomposites, to explore their ability to absorb radionuclides from polluted environments (Liu et al. 2022b).

8.5.3 Microbial-Assisted Remediation

Microbial bioremediation is the process of employing microorganisms, which can be either naturally present or genetically modified, to convert or break down pollutants. This method can improve the breakdown of radionuclides in soil and water (Lloyd 2003). This method utilizes the inherent capabilities of specific microbial agents to immobilize, sequester, or transform radioactive compounds, thereby reducing their mobility or level of danger (Suzuki et al. 2004; Lloyd and Lovley 2001). Genetic engineering approaches have also been investigated for enhancing the ability of microorganisms for targeted bioremediation purposes (Chaney and Baklanov 2017).

Specific microorganisms have shown the ability to gather radionuclides in their biomass, thereby aiding in the elimination of these pollutants from the surrounding environment (Lloyd 2003). Certain radiotrophic fungi, specifically adapted to radioactive environments, can thrive and collect radionuclides via metabolic activities (Zhdanova et al. 2000).

Certain microbes, whether living or dead, exhibit effective biosorption capabilities, enabling them to bind and accumulate heavy metals and other contaminants from aqueous solutions (Zhdanova et al. 2004). The biosorption mechanism comprises several interactions, such as ion exchange, complexation, and physical adsorption, between the biological material and the pollutants being targeted (Lagergren 1898a). *Bacillus* and *Pseudomonas,* as well as a few yeast species like *Saccharomyces cerevisiae*, have been studied for their biosorption capabilities for radionuclides like uranium and cesium (Kapoor and Viraraghavan 1995).

Additionally, complexation may also occur where metal ions form stable complexes with specific functional groups on the biosorbent surface, such as sulfhydryl, amino, or carboxyl groups (Volesky and Holan 1995). Adsorption involves the physical adherence of metal ions to the surface of the biosorbent. Van der Waals forces, hydrogen bonding, and electrostatic interactions also play a significant role in this process (Lagergren 1898b).

Another process of bio-solubilization of radioactive waste entails employing microorganisms to increase the solubility and mobility of radionuclides, thereby enabling their extraction from solid matrices such as soil or waste materials (Lovley and Phillips 1992). This process is frequently linked to microbial activities that facilitate the breakdown of mineral phases and increase the accessibility of radionuclides. The mechanisms involved in bio-solubilization are as follows:

1. Microbial dissolution of minerals: In this process, microorganisms generate organic acids, such as citric acid and oxalic acid, that have the capacity to dissolve mineral phases and liberate bound radionuclides, thus enhancing their bioavailability (Gadd 2010).
2. Chelation and complexation: This involves the release of chelating agents or ligands that can form stable complexes with radionuclides. This process enhances the solubility of radionuclides and makes their transportation easier (Merroun and Selenska-Pobell 2008).
3. Biosurfactant production: Certain microorganisms secrete biosurfactants that have the ability to increase the solubility of radionuclides by modifying the surface tension of the solution (Das et al. 2008).

Studies indicate bioprecipitation of radioactive waste employing microorganisms. For instance, a few bacteria are known to facilitate the formation of less soluble uranium minerals by enzymatically reducing uranium, resulting in its precipitation (Kulkarni et al. 2013). Sulfate-reducing bacteria have shown the ability to solubilize radionuclides and to form solid particles by changing their chemical properties (Gu et al. 2003). Bioprecipitation of solid particles of rare earth elements, including those that have radioactive isotopes, is also reported (Banerjee et al. 2021).

These remediation methods provide improved efficacy in the extraction and immobilization of radionuclides, resulting in expedited and more efficient cleanup procedures.

8.6 Physicochemical Remediation Methods

Contaminated soil containing radionuclides can be effectively restored using physicochemical techniques such as soil washing, soil flushing, and electrokinetic remediation. Alternatively, some techniques, such as vitrification, solidification/stabilization, and the deployment of passive or reactive barriers to restrict the movement of pollutants and prevent the spread of radionuclides into the surrounding environment, have been reported (Cameselle and Gouveia 2019).

8.6.1 Vitrification

Vitrification is a remediation technique that involves converting contaminated soil into a substance that resembles glass (Yuan et al. 2017). An electric current is applied directly to the contaminated soil using a series of electrodes surrounding the polluted spot in order to heat it (Fig. 8.2). Temperatures over 1000 °C are required to achieve the soil components' melting point. The glass-like materials effectively capture metals and other inorganic impurities, while the organic contaminants are burned or vaporized and then collected in a hood located above the treatment site. The vitrification of low-activity waste was investigated by Taylor-Pashow et al. (2018). The predominant radionuclide present in the trash is technetium, namely the isotope 99Tc. The authors examined the removal of Tc from a condensate stream of off-gas operations using reductive precipitation with stannous chloride. The initial findings indicated successful segregation and fixation of Tc; however, the scientists emphasized the need for further investigation before implementing this technique on a larger scale. In contrast, Yuan et al. (2017) utilized microwave heating at a temperature of 1200 °C to imitate the vitrification process of soil contaminated with radioactive substances. The study of the finished material revealed an amorphous structure when microwave sintering was used, whereas traditional sintering technology discovered a crystalline silica structure (Majling et al. 1995). It has been noted that microwave technology is more effective than the traditional sintering approach for vitrifying radioactive contaminated soil (Liu et al. 2022a).

The primary benefits are the on-site application, the brief duration of treatment, the effective stabilization of inorganic pollutants, and the elimination of organic substances. The primary limitation is the substantial energy demand.

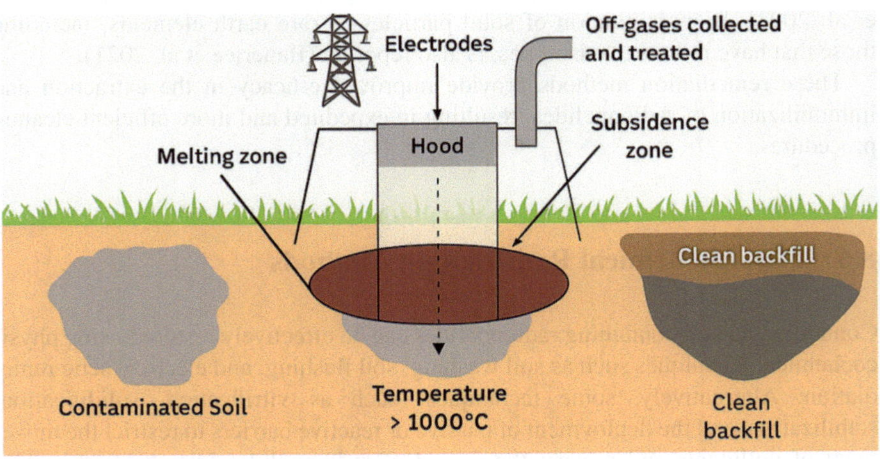

Fig. 8.2 Vitrification. (Adapted from Cameselle and Gouveia (2019))

8.6.2 Solidification and Stabilization

The process of solidification and stabilization is used to immobilize pollutants in contaminated soils, rather than for their degradation, destruction, or removal (Shen et al. 2019). Despite the pollutants remaining in the same location, their detrimental impact on the environment is not evident because of their confinement in the solid phase, with limited breakdown in interstitial water. Consequently, the pollutants are trapped in solid states and are not capable of being absorbed by living organisms (Kogbara 2014). For instance, Qi et al. (2015) documented the synthesis of a mesoporous calcium-silicate substance that effectively traps and immobilizes several metals and radionuclides. The authors documented the successful elimination of cobalt and radionuclides from a liquid waste stream using calcium-silicate material, which effectively prevents the spread of radioactive pollutants. The materials that have absorbed radionuclides can be disposed of by solidifying them into a ceramic or glass-ceramic composition. This technique ensures that the radionuclides are immobilized before being transferred to a landfill (Guo et al. 2017). The process of stabilization/solidification (S/S) using cement or cement-like materials is a fascinating method for treating soils contaminated with radionuclides (Falciglia et al. 2013). Thorium oxide (ThO_2) was added to the soil, and then the soil was stabilized and solidified using various combinations of Portland cement and barite aggregates. Different ratios of soil to binder were used. The outcome demonstrated effective immobilization of thorium and substantial containment of γ-radiation in comparison to the untreated soil. Martins de Freitas and Al-Tabbaa (2013) suggested the use of low pH cements to solidify ILRW through a process called cementation. They assert that the presence of hyperalkaline pore fluids undermines the efficacy of the solidification/stabilization process when using Portland cement. Furthermore, the production of Portland cement is responsible for substantial CO_2 emissions. Low pH cements are a more environmentally viable option and have emerged as a promising alternative for the solidification and stabilization of nuclear waste. The authors have stated that reactive magnesium oxide and magnesium phosphate cements possess significant benefits in terms of decreased permeability, increased density, and the ability to regulate acid–base interactions.

Falciglia et al. (2014) examined the improved bonding of thorium oxide-contaminated soil in Portland cement using barite aggregates. The cement material effectively contained the thorium-contaminated soil, yielding optimal outcomes when using increasing proportions of the binding ingredient. The primary advantages of this study are the use of barite aggregates, which resulted in substantial confinement of γ radiation and a remarkable decrease in the leaching of contaminants. The sole disadvantage of barite aggregates is a minor decrease in the mechanical strength of the resulting cement mixture. The workers concluded that this technology is a viable and effective method for remediating soil contaminated with low levels of radionuclides (Falciglia et al. 2015).

8.6.3 Ion Exchange

Ion exchange is a commonly employed technique for the treatment of radioactive waste (radwaste). This method entails the transfer of ions between a solid medium (often a resin) and a liquid medium that contains impurities. Ion exchange is frequently used in the field of radioactive waste management to selectively eliminate certain radionuclides from liquid waste streams. Ion exchange is a process where ions are exchanged between a solid phase, such as an ion exchange resin, and a liquid phase, such as wastewater, in a reversible manner. The resin is specifically engineered to selectively adsorb some ions while desorbing others (Adamson and Gast 1967). Typical examples of ion exchange resins are strong acid cation exchange resins, strong base anion exchange resins, and chelating resins. The choice of resin is contingent upon the particular radionuclides found in the waste stream (Perrin et al. 1981). Ion exchange is a very efficient method for eliminating several types of radionuclides, including cesium-137, strontium-90, and certain actinides, from liquid radioactive waste. Ion exchange is frequently employed as a component of a multi-stage treatment procedure for radioactive waste, enhancing other methods like precipitation or membrane filtration (Wang et al. 2005). Once the ion exchange resin has reached its saturation threshold, it might undergo regeneration in order to be reused. Regeneration generally entails the passage of a regenerant solution through the resin in order to replace the trapped ions (Helfferich 1995). Ion exchange has been effectively utilized by several nuclear plants and research organizations for the remediation of radioactive waste. Case studies and operational experiences offer significant perspectives. Table 8.1 presents a comparative overview of various physicochemical techniques employed for the remediation of radioactive waste.

8.7 Case Study

The Hanford Nuclear Reservation in Washington State, USA, a former plutonium production site, epitomizes the enduring environmental challenges of nuclear legacy waste. Between 1943 and 1987, uranium-laden wastewater discharges contaminated groundwater with soluble uranium (VI) (UO_2^{2+}), a mobile and ecotoxic radionuclide threatening ecosystems and human health. Traditional remediation methods proved inadequate for such large-scale contamination, prompting researchers to explore bioremediation—a strategy that utilizes the metabolic prowess of dissimilatory metal-reducing bacteria (DMRB) to immobilize uranium.

Central to this approach was the exploitation of Geobacteraceae, a family of anaerobic bacteria capable of enzymatically reducing soluble uranium (VI) to insoluble uranium (IV) (UO_2), which precipitates as uraninite nanoparticles. This reductive immobilization hinges on extracellular electron transfer (EET) mechanisms, where *Geobacter* species transfer electrons from organic substrates like acetate to uranium (VI) via membrane-bound c-type cytochromes and conductive pili, often

Table 8.1 Comparative overview of various physicochemical techniques employed for the remediation of radioactive waste

Method	Process description	Key features/ applications	Limitations	References
Ion exchange	Radionuclides are removed by replacing ions in the liquid with ions on a solid resin.	Highly efficient for cations/anions (e.g., cesium, strontium) in liquid waste.	Resins can saturate quickly; regeneration produces secondary waste. Limited to low-to-intermediate activity waste.	Ojovan et al. (2019) and IAEA (2020)
Precipitation and coprecipitation	Insoluble compounds form to remove radionuclides; coprecipitation involves nonradioactive substances.	Effective for liquid-phase radionuclide elimination.	Generates sludge requiring disposal; efficiency depends on pH and competing ions. Limited to specific radionuclides.	Ewing and Jercinovic (1986), Nash (2017)
Chemical Extraction	Complexing agents selectively isolate radionuclides (e.g., actinides) from waste.	Efficient for isolating specific radionuclides like actinides.	High cost of reagents; potential secondary organic waste. Requires precise control of chemical conditions.	Nash (2017)
Membrane filtration	Uses ultrafiltration/ reverse osmosis to separate radionuclides by molecular size.	Reduces liquid waste volume; effective for low-to-moderate radioactivity.	Membrane fouling; high energy consumption. Unsuitable for highly radioactive solutions.	Drioli et al. (2015) and IAEA (2020)
Sorption processes	Radionuclides adsorb onto solid surfaces (e.g., activated carbon, zeolites).	Effective for liquid-phase separation.	Limited capacity; sensitivity to pH/ temperature. Requires disposal of spent sorbents.	Yuna (2016) and Nash (2017)
Vitrification	Radioactive waste is immobilized in a glass matrix at high temperatures.	Ideal for high-level waste; reduces leachability and ensures long-term stability.	High energy input; volatilization of radionuclides (e.g., cesium) during melting. Limited to compatible waste types.	Ojovan et al. (2019) and IAEA (2020)

termed "nanowires" (Holmes et al. 2004; Lovley et al. 2011). Acetate played a dual role: as an electron donor fueling microbial respiration and as a carbon source stimulating Geobacter biomass growth, thereby amplifying bioreduction capacity (Anderson et al. 2003; Holmes et al. 2004).

Field implementation involved injecting acetate into contaminated aquifers to engineer subsurface redox dynamics. The microbial consumption of acetate lowered groundwater redox potential (Eh), shifting the environment from aerobic to anaerobic conditions favorable for Geobacter dominance over competing microorganisms (Nevin et al. 2008). Monitoring via geochemical profiling and spectroscopic techniques, such as X-ray absorption near-edge structure (XANES), revealed rapid uranium (VI) concentration declines (>90%) and confirmed uranium (IV) as the dominant speciation in sediments (Lovley 2006). This demonstrated the scalability of in situ bioremediation, contrasting sharply with energy-intensive methods like pump-and-treat.

Success at Hanford underscored bioremediation's potential as a sustainable, low-cost solution for radionuclide contamination. Beyond uranium, the principles apply to other redox-sensitive contaminants, such as technetium and plutonium, broadening their environmental relevance (Lovley et al. 2011). However, challenges remain, including the long-term stability of uranium (IV) under fluctuating redox conditions and competition from alternative electron acceptors like Fe^{3+}. Ongoing research focuses on optimizing microbial consortia and genetic engineering to enhance resilience, ensuring this biotechnology evolves to meet complex remediation demands. By integrating microbiology, geochemistry, and environmental engineering, the Hanford case study exemplifies how nature-inspired strategies can address anthropogenic environmental crises, offering a blueprint for managing radioactive contamination in the Anthropocene.

8.8 Current Challenges in Radwaste Remediation Technologies

Radioactive waste remediation presents a multidimensional scientific and engineering challenge, driven by the heterogeneous nature of radionuclides, which span diverse physical and chemical forms—including spent nuclear fuel, contaminated aqueous effluents, and solid matrices—each demanding tailored management strategies. These waste streams vary in radioactivity levels (e.g., high-level vs. low-level waste), decay pathways (α, β, or γ emitters), and environmental mobility, necessitating precise risk assessments and remediation protocols to prevent ecological dispersion through groundwater infiltration, atmospheric transport, or biotic uptake (International Atomic Energy Agency 2018). For instance, long-lived isotopes like plutonium-239 (half-life: 24,100 years) and uranium-238 (half-life: 4.5 billion years) persist in ecosystems for millennia, complicating containment due to their radiotoxicity and potential integration into biogeochemical cycles (Lutze and Ewing 1988). Traditional methods such as cementation or vitrification often fail to address the speciation-dependent mobility of these radionuclides, underscoring the need for advanced materials (e.g., metal-organic frameworks) or biogeochemical interventions (e.g., microbial redox transformations) to achieve long-term sequestration (Lopez-Fernandez et al. 2021).

Compounding this complexity is the absence of harmonized global regulatory frameworks, leading to inconsistent waste classification, disposal criteria, and safety thresholds. Disparities in national policies—such as variations in permissible uranium concentrations in groundwater—hinder transnational collaboration, delay site remediation, and elevate risks of cross-border contamination (OECD-NEA 2020). Furthermore, the exorbitant costs of deploying cutting-edge technologies (e.g., electrokinetic extraction, phytoremediation, or robotic waste retrieval) create inequitable burdens, particularly for low- and middle-income nations, where financial constraints often force reliance on suboptimal solutions like near-surface repositories lacking engineered barriers (World Nuclear Association 2025). Equally significant is the socio-technical challenge of public acceptance: misconceptions about radiation risks, distrust in institutional transparency, and ethical concerns over intergenerational equity frequently stymie project approvals, even for scientifically sound strategies. Effective communication, participatory decision-making, and community engagement are thus critical to securing the "social license" for remediation initiatives, as evidenced by the contentious debates surrounding Fukushima's water discharge plans (Cui 2021). Addressing these intertwined challenges demands interdisciplinary innovation, international policy alignment, and a commitment to balancing technical efficacy with societal trust—a task as urgent as it is formidable in the era of expanding nuclear energy reliance.

8.9 Conclusion

The remediation of radioactive waste demands a multifaceted approach, integrating innovative biological, physical, and engineered strategies to address its complex environmental and health risks. The Hanford Nuclear Reservation case study exemplifies the potential of microbial bioremediation, where Geobacteraceae species effectively immobilize uranium (VI) via reductive precipitation, leveraging extracellular electron transfer mechanisms and acetate as a dual electron donor and carbon source. Despite advances, challenges persist, including the long-term stability of sequestered radionuclides, regulatory fragmentation, and prohibitive costs, particularly for resource-limited nations. Public skepticism and ethical concerns further complicate implementation, necessitating transparent communication and interdisciplinary collaboration. Emerging technologies—such as nanotechnology, phytoremediation, and advanced oxidation processes—offer promising avenues but require optimization for scalability and sustainability. Ultimately, harmonizing global standards, fostering technological innovation, and prioritizing socio-environmental equity are critical to achieving safe, enduring solutions for radioactive waste management in an era of expanding nuclear reliance.

References

Adamson AW, Gast AP (1967) Physical chemistry of surfaces, vol 150. Interscience Publishers, New York

Agency, I. A. E (1999) Decommissioning of nuclear power plants and research reactors, pp 1–41. https://www.iaea.org/publications/5778/decommissioning-of-nuclear-power-plants-and-research-reactors

Agency, I. A. E (2011) Nuclear security recommendations on radioactive material and associated facilities, no. 14, pp 1–27. https://www.iaea.org/publications/8616/nuclear-security-recommendations-on-radioactive-material-and-associated-facilities

Ali H, Khan E, Sajad MA (2013) Phytoremediation of heavy metals—concepts and applications. Chemosphere 91(7):869–881

Anderson RT, et al. (2003) Stimulating the in situ activity of *Geobacter species*. Applied and Environmental Microbiology, 69(10):5575–5582

Banerjee A, Sarkar S, Govil T, González-Faune P, Cabrera-Barjas G, Bandopadhyay R, Salem DR, Sani RK (2021) Extremophilic exopolysaccharides: biotechnologies and wastewater remediation. Front Microbiol 12:721365

Beir VII (2005) Health risks from exposure to low levels of ionizing radiation. The National Academies Report in Brief. National Academies Press, Washington DC

Board, T. R, Council, N. R (2006) Going the distance?: the safe transport of spent nuclear fuel and high-level radioactive waste in the United States. The National Academies Press. https://doi.org/10.17226/11538

Boice JJ, Da Cool U, Harrison UJ, Mettler JF, Sinclair UW (2014) Aims and scope subscriptions. https://doi.org/10.1177/ANIB_43_3

Brim H, McFarlan SC, Fredrickson JK, Minton KW, Zhai M, Wackett LP, Daly MJ (2000) Engineering *Deinococcus radiodurans* for metal remediation in radioactive mixed waste environments. Nat Biotechnol 18(1):85–90

Cameselle C, Gouveia S (2019) Physicochemical methods for the remediation of radionuclide contaminated sites. In: Remediation measures for radioactively contaminated areas, pp 31–49. https://doi.org/10.1007/978-3-319-73398-2_2

Cameselle, C., & Gouveia, S. (2019) Physicochemical methods for the remediation of radionuclide contaminated sites. In: Remediation Measures for Radioactively Contaminated Areas, pp. 31–49. https://doi.org/10.1007/978-3-319-73398-2_2

Canton H (2021) International Atomic Energy Agency—IAEA. In: The Europa directory of international organizations 2021. Routledge, Abingdon, Oxon, pp 305–314

Chaney RL, Baklanov IA (2017) Chapter five – phytoremediation and phytomining: status and promise. In: Cuypers A, Vangronsveld J (eds) Advances in botanical research, vol 83. Academic Press, pp 189–221. https://doi.org/10.1016/bs.abr.2016.12.006

Cui L (2021) Risk communication in the post-Fukushima era. Radiat Med Prot 2(2):79–82. https://doi.org/10.1016/j.radmp.2021.04.004

Das P, Mukherjee S, Sen R (2008) Antimicrobial potential of a lipopeptide biosurfactant derived from a marine *Bacillus circulans*. J Appl Microbiol 104(6):1675–1684

Drioli E, et al. (2015) Membrane engineering for sustainable development. Desalination, 356, 56–84

Dushenkov S, Kapulnik Y, Blaylock M, Sorochisky B, Raskin I, Ensley B (1997) Phytoremediation: a novel approach to an old problem. In: Studies in environmental science, vol 66. Elsevier, Amsterdam. pp 563–572

Environmental Radiation Protection Standards for Nuclear Power Operations (40 CFR Part 190) | US EPA (n.d.) Retrieved 22 Jan 2024, from https://www.epa.gov/radiation/environmental-radiation-protection-standards-nuclear-power-operations-40-cfr-part-190

Ewing RC, Jercinovic MJ (1986) Mineralogy of uranium and thorium. Science, 234:200–206

Falciglia PP, Salvatore C, Pace F, Stefano R, Vagliasindi FGA (2013) Stabilisation/solidification of radionuclides polluted soils: a novel analytical approach for the assessment of the γ-radiation shielding capacity. Chem Eng Trans 32:223–228

Falciglia PP, Cannata S, Romano S, Vagliasindi FGA (2014) Stabilisation/solidification of radionuclide polluted soils—part I: assessment of setting time, mechanical resistance, γ-radiation shielding and leachate γ-radiation. J Geochem Explor 142:104–111

Falciglia PP, Puccio V, Romano S, Vagliasindi FGA (2015) Performance study and influence of radiation emission energy and soil contamination level on γ-radiation shielding of stabilised/solidified radionuclide-polluted soils. J Environ Radioact 143:20–28

Fan T, et al. (2012) Bioremediation of uranium-contaminated soils: A review. Environmental Science & Technology, 46(22):12473–12484

Fredrickson JK, Zachara JM, Balkwill DL, Kennedy D, Li SW, Kostandarithes HM, Daly MJ, Romine MF, Brockman FJ (2004) Geomicrobiology of high-level nuclear waste-contaminated vadose sediments at the Hanford Site, Washington State. Appl Environ Microbiol 70(7):4230–4241

Gadd GM (2010) Metals, minerals and microbes: geomicrobiology and bioremediation. Microbiology 156(3):609–643

Gu B, Brooks SC, Roh Y, Jardine PM (2003) Geochemical reactions and dynamics during titration of a contaminated groundwater with high uranium, aluminum, and calcium. Geochim Cosmochim Acta 67(15):2749–2761

Guo B, Sasaki K, Hirajima T (2017) Solidification of ettringite after uptaking selenate as a surrogate of radionuclide in glass-ceramics by using industrial by-products. J Mater Sci 52:12999–13011

Helfferich FG (1995) Ion exchange. Courier Corporation, New York

Holmes, D., Nevin, K., & Lovley, D. (2004). Comparison of 16S rRNA, nifD, recA, gyrB, rpoB and fusA genes within the family Geobacteraceae fam. nov. International Journal of Systematic and Evolutionary Microbiology, 54, 1591–1599. https://doi.org/10.1099/ijs.0.02958-0

IAEA. (2018) Classification of Radioactive Waste. Safety Standards Series No. GSG-1. Vienna: IAEA

IAEA (n.d.). TRS No. 11087: Best Practices for Radioactive Waste Disposal. Vienna: IAEA

IAEA (n.d.). TRS No. 7118: Decontamination of Nuclear Facilities. Vienna: IAEA

IAEA (n.d.). TRS No. 7441: Radioactive Waste Management Approaches. Vienna: IAEA

IAEA (n.d.). TRS No. 8154: Environmental Remediation of Radioactive Sites. Vienna: IAEA

IAEA (n.d.). TRS No. 8420: Advances in Nuclear Waste Immobilisation. Vienna: IAEA

International Atomic Energy Agency (2009) Classification of radioactive waste, IAEA safety standards series no. GSG-1. IAEA, Vienna

International Atomic Energy Agency (IAEA). (n.d.). Radioactive Waste Management: Safety Fundamentals. Vienna: IAEA.

Jeswani H, Khelurkar N (2015) A review of radioactive waste management. https://doi.org/10.1109/ICTSD.2015.7095849

Kapoor A, Viraraghavan T (1995) Fungal biosorption—an alternative treatment option for heavy metal bearing wastewaters: a review. Bioresour Technol 53(3):195–206

Kennedy Jr, WE (1993) Low-Level Radioactive Waste Disposal in the United States: An Overview of Current Commercial Regulations and Concepts. Pacific Northwest National Laboratory, Richland, WA

Kogbara RB (2014) A review of the mechanical and leaching performance of stabilized/solidified contaminated soils. Environ Rev 22(1):66–86

Kulkarni S, Ballal A, Apte SK (2013) Bioprecipitation of uranium from alkaline waste solutions using recombinant *Deinococcus radiodurans*. J Hazard Mater 262:853–861

Lagergren, S. (1898a). About the theory of so-called adsorption of soluble substances. Kungliga Svenska Vetenskapsakademiens Handlingar, Stockholm

Lagergren S (1898b) Zur theorie der sogenannten adsorption geloster stoffe. Kungl Svenska Vetenskapsakad Handl 24:1–39

Liu X, Miao Y, Luo F, Tang H, Yuan B, Zhao Y, Xie Y, Shao D, Lu X (2022a) Microwave vitrification of Sr-contaminated soil: microstructure, mechanical properties and chemical durability. J Radioanal Nucl Chem 331(1):511–522. https://doi.org/10.1007/s10967-021-08111-1

Liu Z, Xu Z, Xu L, Buyong F, Chay TC, Li Z, Cai Y, Hu B, Zhu Y, Wang X (2022b) Modified biochar: synthesis and mechanism for removal of environmental heavy metals. Carbon Res 1(1):8

Lloyd JR (2003) Microbial reduction of metals and radionuclides. FEMS Microbiol Rev 27(2–3):411–425

Lloyd JR, Lovley DR (2001) Microbial detoxification of metals and radionuclides. Curr Opin Biotechnol 12(3):248–253

Lovley, D.R. (2006) Microbial Fe(III) reduction: Physiology and ecology. Microbiology, 152, 1231–1243

Lovley, D.R., et al. (2011) Electricity production with microorganisms. Nature Reviews Microbiology, 9(1), 15–28

Lopez-Fernandez M, Jroundi F, Ruiz-Fresneda MA, Merroun ML (2021) Microbial interaction with and tolerance of radionuclides: underlying mechanisms and biotechnological applications. Microb Biotechnol 14(3):810–828. https://doi.org/10.1111/1751-7915.13718

Lovley DR, Phillips EJ (1992) Reduction of uranium by *Desulfovibrio desulfuricans*. Appl Environ Microbiol 58(3):850–856. https://doi.org/10.1128/aem.58.3.850-856.1992

Lutze W, Ewing RC (1988) Radioactive Waste Forms for the Future. Amsterdam: North Holland. https://doi.org/10.1016/C2009-0-03078-1

Majling J, Znasik P, Agrawal D, Cheng J, Roy R (1995) Conventional and microwave sintering of condensed silica fume. J Mater Res 10(10):2411–2414. https://doi.org/10.1557/JMR.1995.2411

Malhotra R, Agarwal S, Gupta P (2014) Phytoremediation of radioactive metals. J Civil Eng Environ Technol 1:75–79

Martins de Freitas R, Al-Tabbaa A (2013) Development of low-pH cements for immobilisation of intermediate level radioactive waste: achievements and challenges. In: International conference on radioactive waste management and environmental remediation, vol 56017, p V001T01A012. https://doi.org/10.1115/ICEM2013-96059

Meagher RB (2000) Phytoremediation of toxic elemental and organic pollutants. Curr Opin Plant Biol 3(2):153–162

Merroun ML, Selenska-Pobell S (2008) Bacterial interactions with uranium: an environmental perspective. J Contam Hydrol 102(3–4):285–295

Mwegoha WJS (2008) The use of phytoremediation technology for abatement soil and groundwater pollution in Tanzania: opportunities and challenges. J Sustain Dev Africa 10(1):140–156

Nash KL (2017) Separation chemistry of transuranium elements. Chemical Reviews, 117(10):9238–9272

Nevin KP, et al. (2008) Power generation with microbial fuel cells. Nature Reviews Microbiology, 6(7):599–604

O'Hara MJ, Carter JC, Warner CL, Warner MG, Addleman RS (2016) Magnetic iron oxide and manganese-doped iron oxide nanoparticles for the collection of alpha-emitting radionuclides from aqueous solutions. RSC Adv 6(107):105239–105251

OECD/NEA (2006) Radioactive waste management programmes in OECD/NEA member countries, radioactive waste management. OECD Publishing, Paris. https://doi.org/10.1787/9789264012110-en

OECD Nuclear Energy Agency (OECD-NEA) (2020) Radioactive Waste Management Review. Paris: OECD Publishing

Ojovan MI, et al. (2019) An Introduction to Nuclear Waste Immobilisation. Elsevier

Perrin DD, Dempsey B, Serjeant EP (1981) pKa prediction for organic acids and bases, vol 1. Springer, Berlin

Posiva – Repository in ONKALO (n.d.) Retrieved 22 Jan 2024, from https://www.posiva.fi/en/index/finaldisposal/researchandfinaldisposalfacilitiesatonkalo.html

Pulford ID, Watson C (2003) Phytoremediation of heavy metal-contaminated land by trees—a review. Environ Int 29(4):529–540

Qi G, Lei X, Li L, Yuan C, Sun Y, Chen J, Chen J, Wang Y, Hao J (2015) Preparation and evaluation of a mesoporous calcium-silicate material (MCSM) from coal fly ash for removal of Co (II) from wastewater. Chem Eng J 279:777–787

Roco MC (2003) Broader societal issues of nanotechnology. J Nanopart Res 5:181–189

Rule P (2011) Nuclear regulatory commission. Fed Regist 76(80):23208–23218

Salido AL, Hasty KL, Lim J-M, Butcher DJ (2003) Phytoremediation of arsenic and lead in contaminated soil using Chinese brake ferns (*Pteris vittata*) and Indian mustard (*Brassica juncea*). Int J Phytoremediation 5(2):89–103

Salt DE, Blaylock M, Kumar NPBA, Dushenkov V, Ensley BD, Chet I, Raskin I (1995) Phytoremediation: a novel strategy for the removal of toxic metals from the environment using plants. Bio/Technology 13(5):468–474

Shen Z, Jin F, O'Connor D, Hou D (2019) Solidification/stabilization for soil remediation: an old technology with new vitality. Environ Sci Technol 53(20):11615–11617. https://doi.org/10.1021/acs.est.9b04990

Shukla A, Parmar P, Saraf M (2017) Radiation, radionuclides and bacteria: an in-perspective review. J Environ Radioact 180:27–35. https://doi.org/10.1016/j.jenvrad.2017.09.013

SKB (n.d.) We take care of the Swedish radioactive waste. Retrieved 22 Jan 2024, from https://skb.com/

Suzuki Y, Kelly SD, Kemner KM, Banfield JF (2004) Enzymatic U (VI) reduction by *Desulfosporosinus* species. Radiochim Acta 92(1):11–16

Taylor-Pashow KML, McCabe DJ, Nash CA (2018) Tc removal from the waste treatment and immobilization plant low-activity waste vitrification off-gas recycle. Sep Sci Technol 53(12):1925–1934

United Nations Environment Programme, International Atomic Energy Agency (2018) Radiation protection of the public and the environment, pp 1–51. https://www.iaea.org/publications/11183/radiation-protection-of-the-public-and-the-environment

UNSCEAR. (n.d.) Sources and Effects of Ionizing Radiation. New York: United Nations

US Environmental Protection Agency (EPA). (2018) Radiation Protection Guidance for Diagnostic and Interventional X-Ray Procedures. Washington, DC: EPA

US Nuclear Regulatory Commission (n.d.) Radioactive Waste Management Regulations. https://www.nrc.gov/

Volesky B, Holan ZR (1995) Biosorption of heavy metals. Biotechnol Prog 11(3):235–250

Wang LK, Hung Y-T, Shammas NK (2005) Physicochemical treatment processes, vol 3. Springer, New York

Won K, Han J, Bonne A (1997) Radioactive waste disposal: global experience and challenges. IAEA Bull 39(1):33–41

World Nuclear Association (2025) Radioactive waste – myths and realities. https://world-nuclear.org/information-library/nuclear-fuel-cycle/nuclear-waste/radioactive-wastes-myths-and-realities:contentReference[oaicite:0]{index=0}

Yu J-G, Zhao X-H, Yu L-Y, Jiao F-P, Jiang J-H, Chen X-Q (2014) Removal, recovery and enrichment of metals from aqueous solutions using carbon nanotubes. J Radioanal Nucl Chem 299:1155–1163

Yuan X, Qing Q, Zhang S, Lu X (2017) Vitrification of radioactive contaminated soil by means of microwave energy. AIP Conf Proc 1820(1):040032

Yuna, L. (2016) Waste management strategies for nuclear facilities. Journal of Environmental Radioactivity, 162–163, 34–41

Zhdanova NN, Zakharchenko VA, Vember VV, Nakonechnaya LT (2000) Fungi from Chernobyl: mycobiota of the inner regions of the containment structures of the damaged nuclear reactor. Mycol Res 104(12):1421–1426

Zhdanova NN, Tugay T, Dighton J, Zheltonozhsky V, Mcdermott P (2004) Ionizing radiation attracts soil fungi. Mycol Res 108(9):1089–1096

Chapter 9
Applied Biotechnology in the Restoration of Cultural Heritage

Mamta Kodarkar and Srikanth Mutnuri

Abstract Cultural heritage assets, including monuments, manuscripts, murals, and sculptures, are increasingly threatened by environmental pollutants, climate-induced degradation, and microbial colonization. Traditional conservation techniques, often reliant on chemical solvents and synthetic polymers, pose risks to both the integrity of heritage materials and the health of conservators. In response, biotechnology has emerged as a sustainable and targeted alternative for cultural heritage restoration. This chapter explores the principles and applications of biorestoration through microbial and enzymatic interventions. It highlights the mechanisms of material deterioration and the limitations of conventional methods, followed by an in-depth discussion on biocleaning, bio-consolidation, and bio-protection techniques. Case studies, including sulfate removal from the St. Augustine Tower in Goa using *Pseudomonas* and *Arcobacter* strains, and the enzymatic treatment of paper artifacts using crude chymopapain from papaya latex, illustrate the real-world efficacy of these methods.

Keywords Cultural heritage · *Pseudomonas* · *Arcobacter* · Biocleaning · Bio-consolidation

9.1 Introduction

Cultural heritage encompassing historic sites, artifacts, paintings, and architectural monuments faces unprecedented threats from environmental pollutants, climatic fluctuations, and biological degradation (Artesani et al. 2020). Once-stable materials such as marble, limestone, paper, and textiles are increasingly vulnerable to chemical reactions (e.g., sulfate and nitrate deposition), microbial colonization, and

M. Kodarkar · S. Mutnuri (✉)
Applied Environmental Biotechnology Laboratory, Department of Biological Sciences, Birla Institute of Technology and Science, Sancoale, India
e-mail: srikanth@goa.bits-pilani.ac.in

N. K. Arora, J. Mishra (eds.), *Biotechnological Solutions for a Sustainable Future*, Microorganisms for Sustainability 55,
https://doi.org/10.1007/978-981-95-3989-5_9

physical weathering (Ludewig 2024). Traditional conservation methods, including chemical solvents, mechanical cleaning, and synthetic polymers, often fall short (Weththimuni and Licchelli 2023).

Many conventional treatments are invasive, environmentally harmful, or lack specificity, risking irreversible damage to delicate surfaces (Giacomini 2024). For instance, alkoxysilanes and polymers used to stabilize clay-rich stones frequently fail to prevent swelling-induced damage, while harsh solvents like glacial acetic acid or formamide pose health hazards to conservators (Goins 1995). The urgency to address these challenges is compounded by the alarming rate at which cultural heritage is vanishing, a crisis demanding innovative, sustainable solutions.

In response, biotechnology has emerged as a transformative approach, offering targeted and eco-friendly alternatives. Enzymes, microbial consortia, and bio-based materials enable the precise removal of pollutants (e.g., black sulfated crusts, protein-based adhesives) without harming underlying substrates (Mazzoli et al. 2018). Unlike broad-spectrum chemical treatments, biological methods exploit natural metabolic processes, such as sulfate reduction by bacteria or enzymatic hydrolysis of organic deposits (Bauer et al. 2021; Geweely 2023). These strategies align with the principles of "green conservation," minimizing ecological impact while preserving material integrity (Paolino et al. 2024).

Recent advances in microbial community analysis and omics technologies further enhance diagnostic accuracy, enabling tailored interventions for biodeterioration control. These tools enable precise identification of both active and dormant microbial populations, which is essential for designing targeted and effective conservation strategies. For instance, a comparative analysis using DNA and RNA sequencing at the Preah Vihear temple (Cambodia) demonstrated that RNA-based profiling could identify metabolically active microorganisms, which differed substantially from the total microbial DNA pool (Del Frari et al. 2019). This distinction is critical in identifying actual biodeteriogens versus incidental microbial presence. Moreover, metabolomics approaches have provided insights into the production of organic acids and other metabolites by microbes under different conditions, helping to assess their real-time impact on substrates (Del Frari and Ferreira 2020). By integrating these omics tools, conservators can make more informed decisions regarding treatment and prevention of microbial colonization, ensuring minimal intervention and maximum preservation of cultural artifacts.

This chapter explores the principles, applications, and advancements of applied biotechnology in cultural heritage restoration. It begins by examining the mechanisms of material degradation and the limitations of traditional conservation practices. Subsequent sections detail biological strategies, including microbial biocleaning, enzymatic treatments, and bio-consolidation, highlighting their scientific foundations and practical implementation. Case studies illustrate successes in removing sulfates from marble monuments, stabilizing clay-rich tuff stones, and restoring paintings with enzyme-loaded gels. Finally, this chapter discusses emerging trends, such as salt-tolerant enzymes and nanotechnology, while addressing challenges in scaling these methods for global heritage preservation.

9.2 Cultural Heritage

Cultural heritage encompasses the tangible and intangible creations of human ingenuity from architectural monuments adorned with masonry and murals to portable artworks like paintings and sculptures that embody historical narratives, aesthetic values, and collective identity (Bouchenaki 2003; Dillenberger 2004). These artifacts and traditions form an irreplaceable bridge between past and future generations, and their loss represents an irreversible rupture in cultural continuity (Parulekar-Berde et al. 2020a).

Despite its profound significance, safeguarding cultural heritage remains fraught with challenges. The field of conservation, still in its developmental stages compared to mature scientific disciplines (Otero 2022), grapples with fragmented methodologies. Approaches tailored to specific materials (e.g., stone, paper) or cultural contexts often lack interoperability, compounded by the absence of universally standardized protocols for assessment, intervention, and monitoring (Skibo and Schiffer 2008; Parulekar-Berde et al. 2020b).

The vulnerability of heritage materials arises from their dynamic interaction with environmental and biological stressors (Moncmanová 2007). Abiotic factors such as humidity fluctuations causing swelling and shrinking, extreme temperatures that accelerate chemical reactions, atmospheric pollutants like sulfur dioxide that induce corrosion, and prolonged light exposure leading to fading or embrittlement, collectively contribute to the relentless degradation of materials (Fig. 9.1) (Pinzari and Gutarowska 2021). These processes are exacerbated by biological agents: microorganisms, including bacteria, fungi, algae, and lichens, colonize surfaces under favorable conditions (Fig. 9.1), metabolizing organic substrates inherent to artifacts (e.g., pigments, binders) or introduced through pollutants and prior treatments (Parulekar-Berde et al. 2020a).

The synergy of abiotic and biotic degradation manifests in both aesthetic and structural damage (Fig. 9.1). Microbial activity produces stains and discoloration through pigment synthesis or breakdown (Strzelczyk 2004), while biofilms secrete corrosive acids and enzymes that etch mineral surfaces (e.g., limestone) or degrade organic components (e.g., paint binders, textiles) (Rivera et al. 2018; Pinzari and Gutarowska 2021). Over time, this leads to structural weakening, delamination, and, ultimately, the irreversible loss of cultural artifacts (Hamilton 1996). Understanding these decay pathways is critical for developing targeted conservation strategies that address both environmental and biological threats.

9.3 Microorganisms in Biorestoration of Cultural Heritage

Traditional methods for restoring cultural heritage often involve chemical solvents, mechanical abrasion, or synthetic polymers (Baglioni et al. 2021). While sometimes effective, these approaches can carry significant drawbacks, including risks to

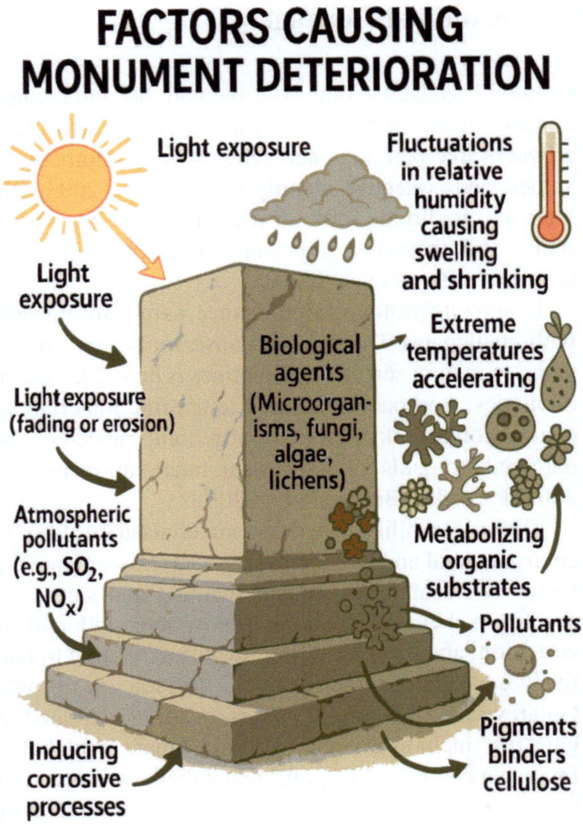

Fig. 9.1 Factors affecting the deterioration of cultural heritage

human health for conservators, potential negative impacts on the surrounding environment, and the possibility of causing irreversible damage or alteration to the delicate and often unique materials of the heritage object itself (Strlič et al. 2013). In the ongoing search for alternative restoration techniques that are safer, more sustainable, and potentially more targeted, the innovative use of living microorganisms as "restorative agents" has emerged as a significant and promising field within conservation science (Soffritti et al. 2019). This biological approach harnesses the specific metabolic capabilities of certain microbes. These organisms can be carefully isolated from various environmental matrices, sometimes even from the deteriorated artifact itself, then meticulously screened in the laboratory and selected based on desirable traits: either their capacity to selectively eliminate harmful or disfiguring deposits (both organic and inorganic substances) or their ability to induce biomineralization (bio-calcify), making them valuable tools for consolidation (Meng et al. 2023). The application of microorganisms to restore artworks and monuments generally operates through three distinct primary mechanisms, each addressing specific conservation challenges.

9.3.1 Biocleaning

This strategy employs microorganisms to remove unwanted surface accretions (Fig. 9.2), such as sulfate crusts (e.g., gypsum formed from pollution on stone), altered organic materials (like aged varnishes or protein-based glues), nitrate salt efflorescence, or biological patinas without damaging the underlying original substrate (Martino et al. 2022). Specific groups of bacteria, such as sulfate-reducing bacteria (SRBs) and nitrate-reducing bacteria (NRBs), are key players in this process (Balloi et al. 2017). They metabolize the target inorganic compounds (sulfates and nitrates, respectively), converting them into soluble or gaseous forms that can be gently removed (Soffritti et al. 2019).

Pioneering studies demonstrated the efficacy of SRB, such as *Desulfovibrio vulgaris* and *Desulfovibrio desulfuricans*, in dissolving black gypsum crusts on marble. These anaerobes reduce sulfates to hydrogen sulfide, converting calcium sulfate into soluble ions and reprecipitating harmless calcite (Gauri et al. 1989; Cappitelli and Sorlini 2005). A study by Ranalli et al. (2006) demonstrated that *D. vulgaris* effectively removed black crusts (mainly gypsum) from marble surfaces without damaging the substrate. Within 2 h, up to 85% of the sulfate content was reduced,

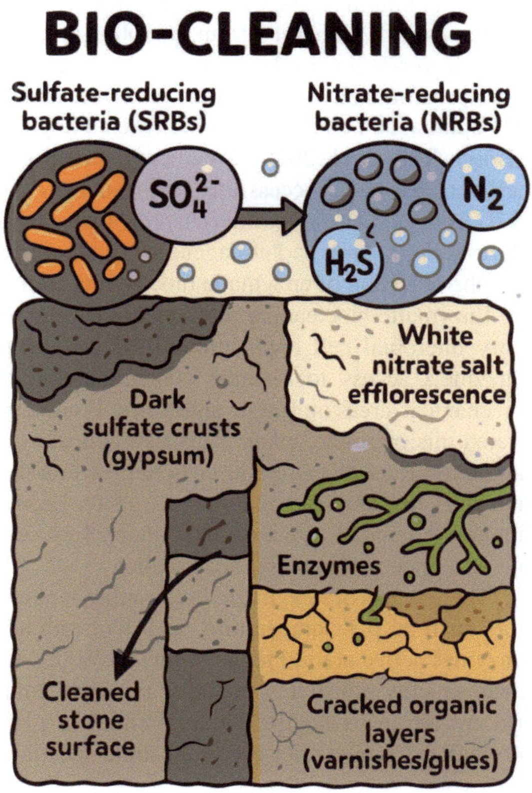

Fig. 9.2 Biocleaning action by microbes

as confirmed by XRD and ion chromatography. Additionally, SEM analysis showed intact crystal morphology in treated zones, indicating a noninvasive and selective biocleaning process.

In a study, *Pseudomonas stutzeri* was selectively applied for the removal of nitrate efflorescence from wall paintings affected by salt contamination, and the treatment showed a significant reduction in surface nitrate levels, often up to 80–90% nitrate removal within a short treatment period (90 min to 2 h) (Roig et al. 2013). Furthermore, the biocleaning was highly selective, preserving underlying pigments and materials, and was confirmed through ion chromatography and colorimetric measurements showing minimal aesthetic alteration and no chemical residues. Similarly, another study demonstrated that nitrate-reducing bacteria help mitigate salt efflorescence on tuff stones, as observed in the Matera Cathedral. (Alfano et al. 2011). Strain-specific microbial consortia enable tailored biocleaning treatments. On Florence Cathedral's polychrome walls, *D. vulgaris* combined with Carbogel outperformed laser and chemical poultices in cleaning red marlstone and green serpentine (Gioventù and Lorenzi 2013). Similarly, *Serratia ficaria* SH7 and *Pseudomonas stutzeri* CONC11 removed calcium oxalate and protein deposits from marble within 48 h (Alisi et al. 2022).

Enzymes derived from bacteria and fungi play a pivotal role in biocleaning strategies for cultural heritage preservation (Cremonesi and Casoli 2021). Among these, proteases widely sourced from microbial systems are economically significant due to their versatility in degrading protein-based contaminants like aged adhesives, glues, and organic residues. These enzymes are classified by their functional groups, pH optima, and cleavage specificity, enabling tailored applications in conservation (Bosch-Roig et al. 2015). Weathered animal glue, a common binder in historical artifacts, has been successfully removed using purified enzymes such as lipase type VII (from *Candida cylindracea*), protease type XIV (from *Streptomyces griseus* and *Aspergillus sojae*), and collagenases (from *Clostridium histolyticum*), demonstrating the precision of enzymatic biocleaning (Ranalli et al. 2005).

Enzymatic biocleaning protocols often pair enzymatic treatments with preliminary microbial cell applications to degrade bulk organic matter, enhancing the enzyme's access to residual contaminants. This two-step approach minimizes mechanical stress on fragile substrates. Pioneering work by Wendelbo (1976) illustrated this principle, using trypsin to delicately separate book pages adhered with animal glue: a milestone in enzymatic cultural heritage conservation. Since then, enzymes like amylases and proteases have been adapted to clean diverse materials, including painted canvases, parchment, frescoes, sculptures, paper, ceramics, and metals, underscoring their adaptability.

The success of enzymatic biocleaning hinges on optimizing environmental factors such as pH, temperature, and application methods. Commercially available enzymes active at mild temperatures (<30 °C) are particularly valuable for treating heat-sensitive substrates, ensuring minimal damage to delicate pigments or fibers. Recent advances highlight the potential of plant-derived proteases (e.g., papain, bromelain), which offer broad substrate specificity and stability across variable pH and temperature ranges, expanding the toolkit for sustainable conservation (Samrot et al. 2022).

While microbial methods reduce collateral damage, challenges persist, including prolonged treatment times and salinity tolerance. Recent breakthroughs include salt-resistant proteases and esterases for wall paintings (Ruginescu et al. 2022) and *Streptomyces griseus*-derived antimicrobials to inhibit biodeteriogens (Elsayed et al. 2023). Efforts to discover novel enzymes from extreme ecosystems (e.g., marine environments, alkaline lakes) and engineered microbial strains aim to enhance catalytic efficiency and substrate range. Such innovations promise to address persistent challenges in cultural heritage conservation, including the removal of complex, layered contaminants without altering the artifact's original patina or structural integrity.

9.3.2 Bio-consolidation

Bio-consolidation is a sustainable technique used to address the structural weakening of porous heritage materials such as stone, mortars, and frescoes caused by prolonged weathering and material decay (Perito et al. 2020). This method employs bio-calcifying bacteria capable of inducing the precipitation of calcium carbonate ($CaCO_3$), typically in the form of calcite, within the pore networks of the deteriorated substrates (Benedetti et al. 2023) (Fig. 9.3). The core process behind this is microbiologically induced calcium carbonate precipitation (MICP), a biological mechanism that promotes in situ mineralization to restore cohesion and structural integrity (Castro-Alonso et al. 2019).

Bio-consolidation can be achieved either biologically (MICP) or chemically through chemically induced calcite precipitation (CICP). In CICP, alkaline agents such as sodium hydroxide or potassium hydroxide are introduced to maintain high pH levels favorable for calcite formation (Park et al. 2020). While this method ensures rapid precipitation, it often results in tightly clustered, narrowly spaced crystals due to the abrupt and uncontrolled nature of the process. To address this, delaying agents like magnesium chloride or pH buffers are sometimes added. However, such additives can alter the precipitate's chemical composition and physical properties, potentially reducing compatibility with the original substrate (Putra et al. 2016). In contrast, MICP relies on microbial metabolic activity to create the optimal chemical environment for calcite formation. Microorganisms, particularly ureolytic bacteria, enzymatically regulate local pH and carbonate ion concentrations, enabling a more gradual, controlled, and uniform deposition of calcite crystals (Konstantinou and Wang 2023).

A well-characterized pathway for MICP is ureolysis, prominently demonstrated by the bacterium *Sporosarcina pasteurii*. This organism produces the enzyme urease, which hydrolyzes urea into ammonia and carbon dioxide. The resulting release of diffusible ammonia elevates the pH, facilitating the chemical conditions needed for calcium ions to react with carbonate ions and form calcium carbonate (Sridhar

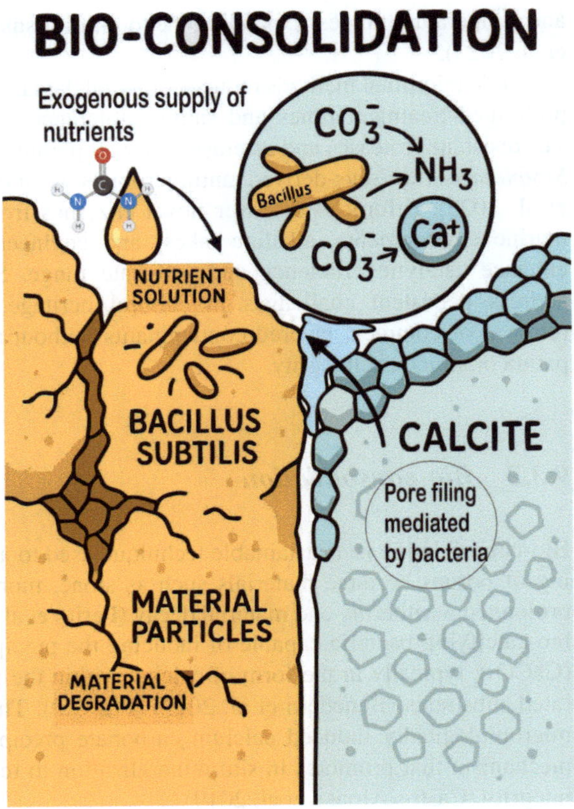

Fig. 9.3 Microbe-mediated bio-consolidation

et al. 2021). Other microbial pathways that contribute to MICP include denitrification and sulfate reduction, each playing a role in altering the geochemical environment to favor calcite precipitation (Lin et al. 2021; Murugan et al. 2021).

The common microbial strains used in bio-consolidation include *Bacillus subtilis*, *Acinetobacter* spp., and various ureolytic bacteria (Soffritti et al. 2019). When provided with suitable nutrients such as calcium sources and urea or organic acids, these microbes modify their microenvironment by increasing pH levels and generating carbonate ions. This leads to the formation of new mineral phases that act as a natural cement. The resulting calcite deposits effectively bind loose grains, reduce porosity, and significantly improve the mechanical strength and durability of the treated materials. Importantly, this biological method is highly compatible with the original material composition, offering a noninvasive and eco-friendly alternative to traditional conservation techniques. Sridhar et al. (2021) explore the innovative use of waste materials in MICP processes to promote sustainability in concrete construction. They report that industrial and agricultural waste products can serve as alternative sources of calcium and nutrients essential for bacterial activity in MICP, thereby reducing reliance on conventional chemical inputs. This approach not only mitigates waste disposal issues but also lowers the carbon footprint associated with concrete production.

Despite its advantages, MICP faces challenges in microbial viability, nutrient delivery, and long-term stability. However, its ability to produce homogenous, biogenic calcite that aligns with conservation ethics positions it as a sustainable alternative to conventional methods. Ongoing research into strain optimization and metabolic pathway engineering holds promise for refining MICP's precision and scalability, solidifying its role in heritage preservation.

9.3.3 Bio-prevention and Bio-protection

Beyond cleaning and consolidation, microorganisms can also be used proactively to prevent or inhibit future biodeterioration. This involves applying beneficial or benign microorganisms (antagonists) or their metabolic products to heritage surfaces to hinder the colonization and growth of known biodeteriogens (e.g., harmful fungi, algae, or bacteria). The mechanisms can include competition for space and nutrients, or the production of specific bioactive substances that inhibit the growth of undesirable organisms (Fig. 9.4). For example, various compounds derived from *Bacillus* species, acting as bio-fungicides, have demonstrated potential in preventing contamination and subsequent damage caused by specific detrimental microbes on heritage materials (Silva et al. 2015).

Favero-Longo and Viles (2020) highlight the emerging role of microbial-origin biocides, mainly from *Bacillus* and *Streptomyces* spp., in managing biodeterioration

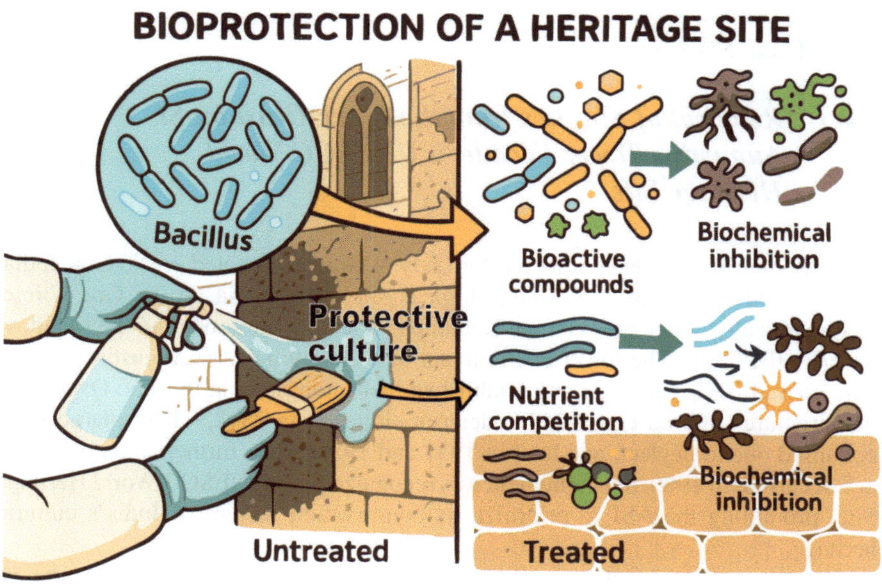

Fig. 9.4 Bio-protection of heritage sites by using protective culture

on stone cultural heritage surfaces. These biocides include metabolites inhibiting the growth of deleterious lithobionts. Earlier studies have also documented the potential of certain microorganisms to form protective biofilms or induce mineral precipitation on stone and other building materials, effectively slowing weathering processes and mitigating microbial-induced deterioration. For instance, Tiano et al. (2006) reported that applying specific bacterial strains (e.g., those capable of producing calcium carbonate) onto ornamental marble reduced microcracking by nearly 55% compared to untreated surfaces. These bio-based treatments create a consolidated layer that can enhance the substrate's resistance to environmental stressors such as acid rain, salt crystallization, and extreme temperature fluctuations (Gerbersdorf and Wieprecht 2015). Moreover, targeted microbial applications have shown promise in outcompeting harmful microbes, preventing their colonization on monuments and heritage structures.

Bioengineered microbial rock varnishes, particularly those mediated by fungi and lichens, offer a sustainable bio-protection strategy for built environments and historic structures (Gadd and Dyer 2017). These systems provide physical shielding (via fungal hyphae or lichen thalli) and biomineralization (producing carbonates, oxalates, or phosphates), which protect surfaces from weathering while avoiding the incompatibility issues of synthetic coatings (Gilbert and Stephens 2018). Their removal often accelerates deterioration, underscoring their natural protective role (Unković et al. 2023). This approach harnesses microbial metabolic processes to create durable, mineral-rich coatings that align with conservation goals for cultural heritage.

9.4 Case Studies

9.4.1 Removal of Sulfates from the Black Crust on St. Augustine Tower Facade, a UNESCO World Heritage Site

St. Augustine Tower, also known as Our Lady of Grace, is located in Old Goa, India, and is administered by the Archaeological Survey of India (ASI), Goa Circle. Geographically, it lies between the Kadamba Plateau to the south and the Mandovi River to the north. The tower is a remnant of a church built by Augustinian friars who arrived in Goa in 1572 and established their monastery at the site. Over time, the structure suffered considerable deterioration first in 1842, when a large vault collapsed due to neglect, and later in 1931 and 1938, when further structural damage occurred. In 1986, the monument was recognized as a UNESCO World Heritage Site, prompting the ASI to prioritize its conservation as part of India's cultural heritage.

The success of biocleaning hinges on selecting optimal microbial strains. Through culture-dependent and culture-independent analyses, *Pseudomonas* spp.

and *Arcobacter* spp. were identified as ideal candidates due to their ubiquity in natural environments, metabolic versatility, and capacity to utilize sulfates and diverse carbon sources. A biocleaning gel was formulated by embedding these bacterial strains in Carbogel, a biocompatible delivery matrix. This jelly-like substance was applied directly to the degraded facade surfaces under ambient temperatures (28–35 °C). Posttreatment, the Carbogel was carefully removed using sterile swabs and hydrophilic cotton pads. To maximize efficacy, fresh bacterial suspensions in Carbogel were reapplied every 15 h, with three iterative cycles conducted (Kodarkar et al. 2023).

This study underscores the potential of targeted microbial treatments in mitigating environmental degradation on heritage structures. By integrating precise architectural documentation with innovative biocleaning protocols, the approach offers a sustainable model for preserving Goa's cultural legacy while addressing gaps in existing conservation literature (Fig. 9.5).

9.4.2 Biorestoration of Paper Heritage Using Papaya Latex Enzymes

The invention of paper marked a turning point in the history of human communication, enabling the preservation of written memory and the documentation of human expeditions over time. This breakthrough facilitated the survival of historical texts that continue to offer valuable insights into the eras from which they originated. Today, a substantial portion of our cultural heritage, including paintings, maps, manuscripts, books, and other artworks, is preserved on paper (Roman et al. 2013).

One of the primary threats to paper-based heritage is cellulose degradation. This process involves the oxidation of the glucopyranose ring and hydrolysis of glycosidic bonds, which leads to acid production. These acidic by-products accelerate catalytic degradation pathways, resulting in weakened cellulose fibers and rendering the paper fragile and chemically unstable (Fig. 9.6). Over time, such

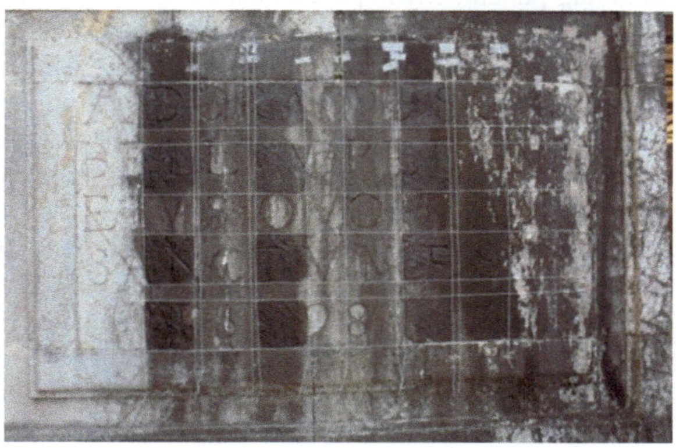

Fig. 9.5 Photograph of the stone facade of the UNESCO World Heritage Site, St. Augustine Tower, Old Goa, India, covered with black crust composed of sulfates and other residual organic matter

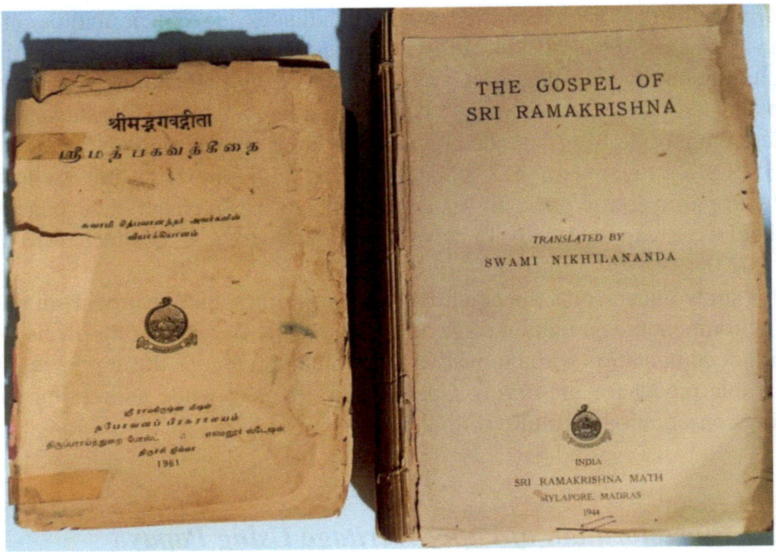

Fig. 9.6 Examples of twentieth-century books that have become brittle and brown during storage. (Source: picture taken by author; Books were obtained from Mr. Dhirajlal D Shah Library, shop no 7, Paras Niketan Cooperative Society, Chheda Nagar, Chembur, Mumbai)

degradation manifests in physical symptoms ranging from crumbling to patch formation (Yadav et al. 2020). Traditional methods of residue removal and paper restoration often involve chemical treatments, which may pose risks to both the conservator and the material. Enzyme-based approaches have emerged as a safer and more selective alternative. Enzymes offer the advantage of working under mild conditions while preserving the structural integrity of heritage materials (Vokic and Berovic 2005). In particular, proteolytic enzymes derived from plant latex are gaining attention due to their broad substrate specificity and functional stability across wide temperature and pH ranges (Samrot et al. 2022).

Papaya (*Carica papaya*) latex contains a rich assortment of proteolytic enzymes, including papain, chymopapain, caricain, and glycyl endopeptidase. Although less studied, papaya peptidase A constitutes approximately 14% of the total latex protein, in comparison to 5% for papain and 27% for chymopapain (Yu and Zhang 2020). Both papain and chymopapain exhibit strong hydrolytic activity toward proteinaceous residues. Notably, chymopapain is a cysteine protease comprising 218 amino acids with potent proteolytic and hydrolytic capabilities, making it particularly promising for biorestoration applications.

The field of biotechnology-based restoration, or *biorestoration*, is gaining momentum as conservators seek sustainable, efficient, and noninvasive solutions aligned with environmentally responsible practices. In this context, enzymes from papaya latex represent a promising tool for the safe restoration of delicate cultural materials such as historic manuscripts, murals, paintings, and postage stamps.

9.5 Conclusion

The preservation of cultural heritage in the face of escalating environmental and biological threats demands innovative, sustainable, and material-sensitive interventions. This chapter has illustrated how biotechnology, through microbial biocleaning, bio-consolidation, and enzyme-based restoration, offers a promising alternative to traditional methods that often compromise artifact integrity or pose environmental hazards. Case studies, such as the microbial treatment of sulfate-encrusted St. Augustine Tower and the enzymatic restoration of fragile paper manuscripts using papaya latex, demonstrate the precision and efficacy of these biologically driven approaches. Moreover, advances in microbial strain selection, delivery systems, and the application of eco-friendly enzymes broaden the applicability of these techniques across diverse substrates from stone to paper. As heritage scientists move toward greener conservation frameworks, the convergence of biotechnology and cultural preservation not only enhances the longevity of artifacts but also aligns with global calls for environmental responsibility.

Acknowledgments This work was supported by the Department of Science and Technology, Government of India, project number DST/TDT/SHRI/31/2018 (G). We thank Central Sophisticated Instrument Facilities, Birla Institute of Technology and Science Pilani, KK Birla Goa Campus, for allowing us to use the Scanning Electron Microscope. We are very grateful to the Archaeological Survey of India, Goa Circle (ASI), for supporting and assisting with the diagnostic work on the St. Augustine Church.

References

Alfano G, Lustrato G, Belli C, Zanardini E, Cappitelli F, Mello E, Sorlini C, Ranalli G (2011) The bioremoval of nitrate and sulfate alterations on artistic stonework: the case-study of Matera Cathedral after six years from the treatment. Int Biodeterior Biodegrad 65(7):1004–1011. https://doi.org/10.1016/j.ibiod.2011.07.010

Alisi C, Magrini D, Vettori S, Salvadori B, Vincenti M, Manna D, Bietti M, Sprocati AR (2022) Sustainable restoration guided by scientific and archival investigations: the bio-cleaning of Lorenzo Duke of Urbino's Sarcophagus, a Michelangelo's Masterpiece in the Medici Chapels. Heritage 5(4):3359–3373. https://doi.org/10.3390/heritage5040172

Artesani A, Di Turo F, Zucchelli M, Traviglia A (2020) Recent advances in protective coatings for cultural. Coatings 10:217

Baglioni M, Poggi G, Chelazzi D, Baglioni P (2021) Advanced materials in cultural heritage conservation. Molecules 26(13):3967

Balloi A, Lombardi E, Troiano F, Polo A, Capitelli F, Gulotta D, Toniolo L, Lucchini A, Daffonchio D (2017) Sulfate reducing bacteria as bio-cleaning agents: development of new methodologies and study cases. Conserv Sci Cult Herit 15(2):109–119. (2015): Special issue for second conference on biology and archaeobiology – from knowledge to preventive conservation

Bauer MA, Kainz K, Ruckenstuhl C, Madeo F, Carmona-Gutierrez D (2021) Murals meet microbes: at the crossroads of microbiology and cultural heritage. Microbial Cell 8(12):276–279. https://doi.org/10.15698/MIC2021.12.765

Benedetti F, Kratter M, Atanasio P, Mura F, Beccaccioli M, Scifo J, di Sarcina I, Tomassetti MC, Schneider K, Rossi M, Cemmi A, Nigro L, Rinaldi T (2023) Isolation of carbonatogenic bacteria for biorestoration. J Cult Herit 64:282–289. https://doi.org/10.1016/j.culher.2023.10.014
Bosch-Roig P, Lustrato G, Zanardini E, Ranalli G (2015) Biocleaning of cultural heritage stone surfaces and frescoes: which delivery system can be the most appropriate? Ann Microbiol 65(3):1227–1241. https://doi.org/10.1007/s13213-014-0938-4
Bouchenaki M (2003) The interdependency of the tangible and intangible cultural heritage. ICOMOS 14th General Assembly and International Symposium: "Place, Memory, Meaning: Preserving Intangible Values in Monuments and Sites", Victoria Falls, Zimbabwe. International Council on Monuments and Sites (ICOMOS), Paris, France
Cappitelli F, Sorlini C (2005) From papyrus to compact disc: the microbial deterioration of documentary heritage. Crit Rev Microbiol 31(1):1–10. https://doi.org/10.1080/10408410490884766
Castro-Alonso MJ, Montañez-Hernandez LE, Sanchez-Muñoz MA, Macias Franco MR, Narayanasamy R, Balagurusamy N (2019) Microbially induced calcium carbonate precipitation (MICP) and its potential in bioconcrete: microbiological and molecular concepts. Front Mater 6:126
Cremonesi P, Casoli A (2021) Enzymes as tools for conservation of works of art. J Cult Herit 50:73–87. https://doi.org/10.1016/j.culher.2021.06.005
Del Frari G, Ferreira RB (2020) Metabolomic analysis of fungi on cultural heritage substrates reveals substrate-specific metabolic activity. Int Biodeterior Biodegrad 148:104900. https://doi.org/10.1016/j.ibiod.2020.104900
Del Frari G, Trovão J, Gil F, Ferreira RB (2019) RNA-based molecular analysis of fungal communities on cultural heritage: case study of the Preah Vihear temple, Cambodia. Int Biodeterior Biodegrad 145:104816
Dillenberger J (2004) A theology of artistic sensibilities: the visual arts and the church. Wipf and Stock Publishers. https://doi.org/10.1016/j.ecoleng.2009.02.006
Elert K, Rodriguez-Navarro C (2022) Degradation and conservation of clay-containing stone: a review. Constr Build Mater 330:127226. https://doi.org/10.1016/j.conbuildmat.2022.127226
Elsayed Y, El-Kadi S, El-Rian M, Mabrouk N (2023) The efficiency of microbial culture extracts as green antimicrobial products against some microorganisms colonizing the historic oil paintings. Sci Cult 9(2):127–143. https://doi.org/10.5281/zenodo.8283084
Favero-Longo SE, Viles HA (2020) A review of the nature, role and control of lithobionts on stone cultural heritage: weighing-up and managing biodeterioration and bioprotection. World J Microbiol Biotechnol 36(7):100. https://doi.org/10.1007/s11274-020-02878-3
Gadd GM, Dyer TD (2017) Bioprotection of the built environment and cultural heritage. Microb Biotechnol 10(5):1152–1156. https://doi.org/10.1111/1751-7915.12750
Gauri KL, Chowdhury AN, Kulshreshtha NP, & Punuru AR (1989) The sulfation of marble and the treatment of gypsum crusts. Studies in Conservation, 34(4):201–206. Taylor & Francis on behalf of The International Institute for Conservation of Historic and Artistic Works (IIC). https://doi.org/10.1179/sic.1989.34.4.201
Gerbersdorf SU, Wieprecht S (2015) Biostabilization of cohesive sediments: revisiting the role of abiotic conditions, physiology and diversity of microbes, polymeric secretion, and biofilm architecture. Geobiology 13(1):68–97
Geweely NS (2023) New frontiers review of some recent conservation techniques of organic and inorganic archaeological artefacts against microbial deterioration. Front Microbiol 14:1146582
Giacomini F (2024) Some considerations on recent approaches to cleaning, conservation, and display of ancient marble sculptures from historical collections. Stud Conserv 69(4):221–229
Gilbert JA, Stephens B (2018) Microbiology of the built environment. Nat Rev Microbiol 16(11):661–670
Gioventù E, Lorenzi P (2013) Bio-removal of black crust from marble surface: comparison with traditional methodologies and application on a sculpture from the Florence's English Cemetery. Proc Chem 8:123–129. https://doi.org/10.1016/j.proche.2013.03.017

Goins ES (1995) Alkoxysilane stone consolidants: the effect of the stone substrate on the polymerization process. Doctoral dissertation, University of London, University College London, UK
Hamilton DL (1996) Basic methods of conserving underwater archaeological material culture. US Department of Defense, Legacy Resource Management Program. https://doi.org/10.1016/j.ibiod.2019.104816
Kodarkar M, Prabhudesai S, Monteiro V, Joshi R, Bakilapadavu G, Mutnuri S (2023) Bioremoval of sulfates from black crust: a case study of St. Augustine tower, Goa-India. Environ Sustain. https://doi.org/10.1007/s42398-023-00293-6
Konstantinou C, Wang Y (2023) Unlocking the potential of microbially induced calcium carbonate precipitation (MICP) for hydrological applications: a review of opportunities, challenges, and environmental considerations. Hydrology 10(9):178
Lin W, Lin W, Cheng X, Chen G, Ersan YC (2021) Microbially induced desaturation and carbonate precipitation through denitrification: a review. Appl Sci 11(17):7842
Ludewig J (2024) The architectures of Australian resource imperialism portfolio
Martino M, Balloi A, Palla F (2022) Biocleaning. In: Palla F, Barresi G (eds) Biotechnology and conservation of cultural heritage. Springer, Cham. https://doi.org/10.1007/978-3-030-97585-2_4
Mazzoli R, Giuffrida MG, Pessione E (2018) Back to the past—forever young: cutting-edge biochemical and microbiological tools for cultural heritage conservation. Appl Microbiol Biotechnol 102:6815–6825
Meng Q, Li X, Geng J, Liu C, Ben S (2023) A biological cleaning agent for removing mold stains from paper artifacts. Herit Sci 11(1). https://doi.org/10.1186/s40494-023-01083-3
Moncmanová A (2007) Environmental factors that influence the deterioration of materials. WIT Trans State Art Sci Eng 28:1–25
Murugan R, Suraishkumar GK, Mukherjee A, Dhami NK (2021) Insights into the influence of cell concentration in design and development of microbially induced calcium carbonate precipitation (MICP) process. PLoS One 16(7):e0254536
Otero J (2022) Heritage conservation future: where we stand, challenges ahead, and a paradigm shift. Global Chall 6(1):2100084
Paolino B, Sorrentino MC, Pacifico S (2024) Greener solutions for biodeterioration of organic-media cultural heritage: where are we? Herit Sci 12(1):334
Park SS, Le TT, Nong Z, Moon HD, Lee DE (2020) Chemically induced calcium carbonate precipitation for improving strength of sand. J Mater Civ Eng 32(9):04020238
Parulekar-Berde C, Ghoble SS, Salvi SP, Berde VB (2020a) Microorganisms and their enzymes as biorestoration agents. In: Microbial biotechnology approaches to monuments of cultural heritage, pp 71–86. https://doi.org/10.1007/978-981-15-3401-0_5
Parulekar-Berde C, Surve RR, Salvi SP, Rawool PP, Chari PVB, Berde VB (2020b) Bioremediation of cultural heritage: removal of organic substances. In: Yadav A, Rastegari A, Gupta V, Yadav N (eds) Microbial biotechnology approaches to monuments of cultural heritage. Springer, Singapore. https://doi.org/10.1007/978-981-15-3401-0_6
Perito B, Marvasi M, Mastromei G (2020) Bacterial mineralization of calcium carbonate for conservation of stone artworks. IOP Conf Ser Mater Sci Eng 949(1):012090. https://doi.org/10.1088/1757-899X/949/1/012090
Pinzari F, Gutarowska B (2021) Extreme colonizers and rapid profiteers: the challenging world of microorganisms that attack paper and parchment. In: Microorganisms in the deterioration and preservation of cultural heritage. https://doi.org/10.1007/978-3-030-69411-1_4
Putra H, Yasuhara H, Kinoshita N, Neupane D, Lu CW (2016) Effect of magnesium as substitute material in enzyme-mediated calcite precipitation for soil-improvement technique. Front Bioeng Biotechnol 4:37
Ranalli G, Alfano G, Belli C, Lustrato G, Colombini MP, Bonaduce I, Zanardini E, Abbruscato P, Cappitelli F, Sorlini C (2005) Biotechnology applied to cultural heritage: biorestoration of frescoes using viable bacterial cells and enzymes. J Appl Microbiol 98(1):73–83. https://doi.org/10.1111/j.1365-2672.2004.02429.x

Ranalli G et al (2006) Biocleaning of black crusts on European cultural heritage by *Desulfovibrio vulgaris*. Appl Environ Microbiol 72(5):3733–3737. https://doi.org/10.1128/AEM.72.5.3733-3737.2006

Rivera LE, C, Ramos AP, Sánchez JIC, & Serrano ME. D (2018) Origin and control strategies of biofilms in the cultural heritage. In A. Méndez-Vilas (Ed.), Antimicrobials, Antibiotic Resistance, Antibiofilm Strategies and Activity Methods (pp. 51–74). IntechOpen, London, United Kingdom. https://doi.org/10.5772/intechopen.71269

Roig PB, Regidor Ros JL, Estellés RM (2013) Biocleaning of nitrate alterations on wall paintings by *Pseudomonas stutzeri*. Int Biodeterior Biodegrad 84:266–274. https://doi.org/10.1016/j.ibiod.2012.09.009

Roman C, Diaconescu R, Scripcariu L, Grigoriu A (2013) Biocides used in preservation, restoration and conservation of the paper. Eur J Sci Theol 9(4):263–271

Ruginescu R, Enache M, Popescu O, Gomoiu I, Cojoc R, Batrinescu-Moteau C, Maria G, Dumbravician M, Neagu S (2022) Characterization of some salt-tolerant bacterial hydrolases with potential utility in cultural heritage bio-cleaning. Microorganisms 10(3):1–15. https://doi.org/10.3390/microorganisms10030644

Samrot AV, Mun CY, Qi NX, Saigeetha S, Wilson S, Rajesh VV, Pachiyappan S, Rajalakshmi D, Chinni SV (2022) Plant latex: phytochemistry, medicinal properties and application – a review. J Pharm Negative Results 13(7):5351–5375. https://doi.org/10.47750/pnr.2022.13.S07.658

Silva M, Rosado T, Teixeira D, Candeias A, & Caldeira AT (2015) Production of green biocides for cultural heritage: Novel biotechnological solutions. International Journal of Conservation Science, 6, 519–530. Editura Universităţii "Alexandru Ioan Cuza" din Iaşi, Romania.

Soffritti I, D'Accolti M, Lanzoni L, Volta A, Bisi M, Mazzacane S, Caselli E (2019) The potential use of microorganisms as restorative agents: an update. Sustainability (Switzerland) 11(14):1–17. https://doi.org/10.3390/su11143853

Sridhar S, Bhatt N, Suraishkumar GK (2021) Mechanistic insights into ureolysis mediated calcite precipitation. Biochem Eng J 176:108214. https://doi.org/10.1016/j.bej.2021.108214

Strlič M, Thickett D, Taylor J, Cassar M (2013) Damage functions in heritage science. Stud Conserv 58(2):80–87

Strzelczyk AB (2004) Observations on aesthetic and structural changes induced in Polish historic objects by microorganisms. Int Biodeterior Biodegrad 53(3):151–156

Tiano P, Cantisani E, Sutherland I, Paget JM (2006) Biomediated reinforcement of weathered calcareous stones. J Cult Herit 7(1):49–55. https://doi.org/10.1016/j.culher.2005.10.003

Unković N, Grbić ML, Dimkić I (2023) Beneficial bacteria as natural biocontrol agents for preservation of fungal infested works of art – a review. Microbiology 44:35–41

Vokic D, Berovic M (2005) Use of lipase to remove oil-based overpaints. 14th Triennial Meeting, The Hague, 12–16 September 2005: Preprints (ICOM Committee for Conservation), pp 862–868

Wendelbo O (1976) The use of proteolytic enzymes in the restoration of paper and papyrus. University Library of Bergen, Bergen

Weththimuni ML, Licchelli M (2023) Heritage conservation and restoration: surface characterization, cleaning and treatments. Coatings 13(2):457

Wu F, Gu J-D, Li J, Feng H, Wang W (2022) Microbial colonization and protective management of wall paintings. In: Cultural heritage microbiology: recent developments. Archetype Publications, London, pp 57–84

Yadav AN, Rastegari AA, Gupta VK, Yadav N (2020) Microbial biotechnology approaches to monuments of cultural heritage. https://doi.org/10.1007/978-981-15-3401-0

Yu L, Zhang H (2020) Separation and purification of papain crude extract from papaya latex using quaternary ammonium ionic liquids as adjuvants in PEG-based aqueous two-phase systems. Food Anal Methods 13(7):1462–1474. https://doi.org/10.1007/s12161-020-01761-z

Part III
Industrial Applications for Food Safety and Human Health

Chapter 10
Cutting-Edge Antimicrobials in Food Packaging Applications

Neetu Singh, Alka Nanda, Ayushi Singh, Puja Kumari, and Tanya Singh

Abstract The application of antimicrobials in food packaging is a novel, multidisciplinary technique that mainly aims to enhance sustainability and ensure food safety. Antimicrobial (AM) food packaging is a significant alternative to safeguard food goods throughout their manufacturing process, including preparation, storage, and distribution. Its purpose is to increase the longevity of edible items by minimizing bacterial activity. Antimicrobial agents, such as natural extracts, nanoparticles, and synthetic compounds, are integrated into the packaging film using advanced techniques. This method enables the deliberate and regulated release of antimicrobial substances, thereby inhibiting the proliferation of detrimental microorganisms and prolonging the product's durability. The polymer coating inhibits the development of contaminating microbial activity by including antimicrobial compounds. This chapter examines the progress, uses, and challenges related to the use of antimicrobial compounds in the packaging of food products. Additionally, a comprehensive analysis of antimicrobial agents, delving into their mechanisms of action, integration into packaging systems, and their capacity to hinder the growth of harmful microorganisms is also targeted. This chapter also emphasizes current advancements in intelligent packaging technologies, such as the use of nanoparticles, which provide enhanced antibacterial effectiveness and sustainability. This chapter concludes by examining the future possibilities of antimicrobial packaging, highlighting the need for ongoing research and development to address current obstacles and improve food safety.

Keywords Antimicrobial packaging · Food safety · Natural antimicrobials · Synthetic antimicrobials · Sustainability · Food packaging

---

N. Singh (✉) · A. Nanda · A. Singh · P. Kumari · T. Singh
Department of Food and Nutrition, School of Home Science, Babasaheb Bhimrao Ambedkar University, Lucknow, Uttar Pradesh, India

N. K. Arora, J. Mishra (eds.), *Biotechnological Solutions for a Sustainable Future*, Microorganisms for Sustainability 55,
https://doi.org/10.1007/978-981-95-3989-5_10

10.1 Introduction

The Food and Agriculture Organization (FAO) estimates that nearly one-third of the worldwide food production is squandered each year due to microbial degradation, involving bacteria such as *Campylobacter jejuni*, *Escherichia coli*, *Listeria monocytogenes*, *Salmonella* spp., *Staphylococcus aureus*, and fungi like *Penicillium citrinum* and *Aspergillus niger* (Food and Nations 2019). This microbial activity results in the deterioration of several food categories, with significant waste rates in fresh produce (40%–50%), seafood (35%), grains (30%), and animal products (20%).

In response, antimicrobial packaging has arisen as a vital advancement in food preservation. This packaging integrates microbial suppression with conventional barrier qualities to prolong food preservation and mitigate the dangers associated with pathogenic microorganisms. Antimicrobial packaging seeks to inhibit microbial proliferation by using active agents or reactive polymers that address the critical growth factors of microorganisms (Han 2003). Research indicates that antimicrobial packaging improves food safety and security by efficiently managing pathogenic bacteria and spoilage microorganisms (Espitia et al. 2013; Floros et al. 1997).

Antimicrobial packaging has numerous benefits compared to conventional approaches, including the direct application of antimicrobial agents to food. This method ensures a continuous, steady release of active substances, preserving their potency over prolonged durations, in contrast to sprays or drips that may quickly diminish efficacy (Quintavalla and Vicini 2002). Furthermore, antimicrobial packaging can be utilized for various food items, such as vegetables, fruits, dairy products, and baked goods, to inhibit microbial degradation (Leistner 2007). Antimicrobial agents utilized in packaging may be classified as organic or chemical based. Consumers like organic antimicrobials, including essential oils and bacteriocins, for their safety profiles, despite the necessity for more intricate extraction methods. Conversely, chemical agents such as triclosan and metal nanoparticles provide enhanced efficacy and reduced costs; however, they may exhibit increased toxicity (Huang et al. 2019).

This chapter examines the progress, applications, and obstacles of antimicrobial packaging in improving food safety and prolonging shelf life. This study examines both natural and synthetic antimicrobial agents and their use in packaging materials, highlighting the potential for advancements in food preservation technologies (Appendini and Hotchkiss 2002).

10.2 Antimicrobial Food Packaging

Active packaging has numerous uses, including antimicrobial packaging (Floros et al. 1997). Active packaging refers to a packaging method that goes beyond basic barrier qualities by incorporating active substances or using polymers that have active functions (Han 2002). Foodstuffs can be protected against spoilage and harmful microbes by using antimicrobial packaging. Using antibacterial polymers that meet standard packaging standards or incorporating antibacterial substances into the packaging system can accomplish the new antimicrobial purpose. When a packaging system develops antimicrobial properties, it slows or stops the expansion of microbes by increasing the time it takes for the system to function, decreasing the growth rate, or lowering the number of active microbes (Han 2003).

The desired outcomes of traditional food wrapping comprise prolonging the period of preservation, maintaining quality, and ensuring safety, which can be accomplished in various ways. However, antimicrobial packing is particularly intended to regulate microbes that typically negatively impact these three goals. There may be some products that do not require the antimicrobial packaging method since they are not susceptible to deterioration or contamination by microbes. However, most food items deteriorate quickly, and the majority of hygienic equipment can become contaminated easily. Ensuring safety and security, maintaining quality, and extending shelf life are the major objectives of an antibacterial packaging system. Compared with traditional packaging—which prioritizes shelf-life and quality—antimicrobial packaging primarily targets microbial control to achieve these outcomes (Han 2003).

Every antibacterial substance has a unique action that uniquely impacts microbes. Unfortunately, there is no single antimicrobial agent effective against all spoilage and pathogenic microorganisms. The diverse physiologies of the microbes and the distinctive antibacterial mechanisms are to blame for this. When choosing which antimicrobial drugs to use, it may be advantageous to classify microbes simply. Bacteria can be classified based on various characteristics, including their oxygen requirements, cell wall composition, growth stage (spore-forming or vegetative cells), optimal growth temperature (thermophilic, mesophilic, or psychrotrophic), and resistance to acid or osmotic stress. Understanding the effectiveness and limitations of an antimicrobial agent requires knowledge of both its microbiological properties and its unique antimicrobial function (Han 2005). The structure of cell membranes and walls can be altered by some antimicrobial drugs, while others block vital metabolic or reproductive genetic pathways of microbes. The microbicidal and microbistatic effects are the two main purposes of microbial inhibition. The packing method must actively keep levels beyond the threshold of inhibition during the whole storage duration or shelf life to avoid the regrowth of the target microbes in case of microbial-static effects.

Release, absorption, and immobilization are the three main modes of action for antibacterial properties that are attained by including antimicrobial agents in the packaging system or by employing antibacterial polymeric materials. Inhibiting the growth of bacteria, the release type permits the movement of antimicrobial chemicals into meals or the headspace inside containers, as shown in Fig. 10.1. Both solutes and gases can serve as antibacterial agents. Gaseous antimicrobial agents, in contrast to their solute counterparts, can permeate any given space, even those between food products and their packaging. To inhibit the development of microbes, the absorption mechanism of antibacterial systems eliminates key components of food systems that microbes need for growth. For example, the oxygen-absorbing

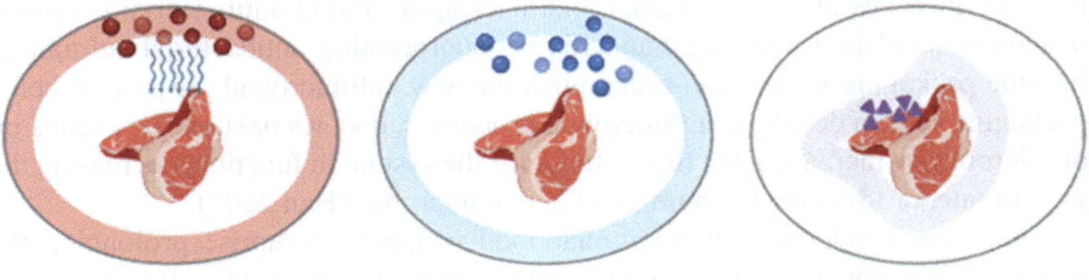

Fig. 10.1 Antimicrobial release mechanisms in food systems: (**a**) nonvolatile antimicrobial; (**b**) volatile antimicrobial; and (**c**) non-migration and absorption

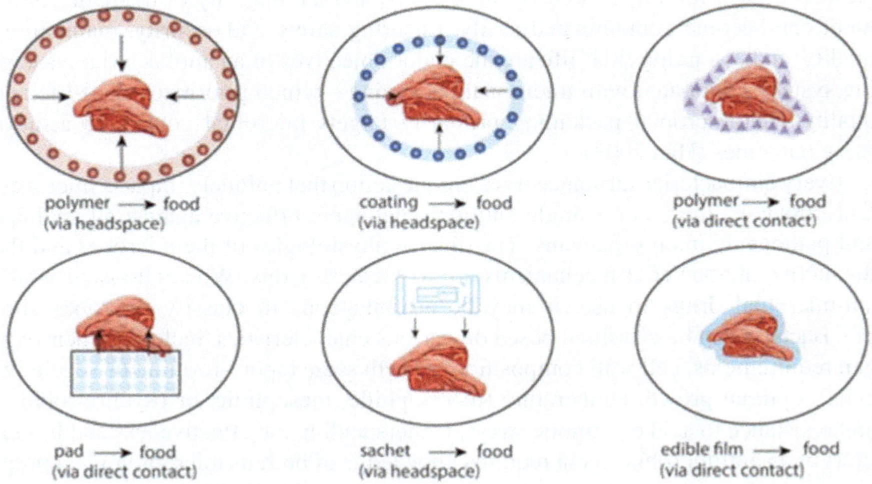

Fig. 10.2 Potential antimicrobial release systems for food packaging applications

device effectively prevents the growth of package mold. The immobilization technique inhibits microbial growth on the contact surface without releasing antimicrobial chemicals. Since the antimicrobial package is less likely to contact the full food products when dealing with solid foods, immobilization systems may not be as successful as they are with liquid foods (Han 2003).

Antimicrobial agents can be employed in many polymeric forms. Han (2005) has proposed a comprehensive definition of the many forms of antimicrobial packaging solutions, which is illustrated in Fig. 10.2. One potential approach is the integration of an antimicrobial agent into a polymer or coating, which can subsequently be dispersed in the headspace if the material exhibits volatility or on the surface of the food if it lacks instability. Another potential approach is the incorporation of the active chemical into a sachet that is composed of food-grade materials capable of facilitating the transportation of the substance and regulating its dispersion into the headspace.

10.3 Components of Antimicrobial Food Packaging Films and Coatings

Biodegradable packaging (which breaks down) and nonbiodegradable packaging (which does not break down) are the two main types of AM packaging (Nanda and Singh 2022). Several man-made polymers are nonbiodegradable as they cannot be decomposed by biological processes. Nevertheless, they are very appropriate for utilization as food packaging material owing to their economical price, lightweight nature, chemical inertness, exceptional ability to prevent the passage of substances, robust mechanical durability, excellent clarity, heat-sealability, and convenience for printing. The predominant polymers utilized in packaging comprise linear low-density polyethylene (LLDPE) $(C_2H_4)_n$, low-density polyethylene (LDPE), polypropylene (PP) $(C_3H_6)_n$, high-density polyethylene (HDPE), polystyrene (PS) $(C_8H_8)_n$, polyethylene terephthalate (PET) $(C_{10}H_8O_4)_n$, polyvinyl chloride (PVC) $(C_2H_3Cl)n$, and ethylene vinyl acetate (EVA) $(C_6H_{10}O_2)_n$. Nevertheless, these petroleum-derived commodities exert an adverse impact on the ecosystem. It leads to the degradation of landfills, adverse environmental effects, and significant energy consumption throughout the production procedure. The migration of chemicals from plastics into food items could potentially pose a risk to human wellness. Consequently, there has been a notable transition toward a higher utilization of biodegradable substances over the past few decades (Sung et al. 2013). Packaging films and coatings must be comprised of at least two elements: a matrix, often composed of a macromolecule capable of forming a cohesive framework, and a plasticizer, which is typically necessary to decrease the inherent rigidity and brittleness of most composites. In addition, supplementary constituents can be integrated to enhance the physical attributes of films and coatings, such as the barrier and mechanical attributes, as well as their moisture resistance. Figure 10.3 illustrates the fundamental and additional constituents of degrading antimicrobial wrappings (De Azeredo et al. 2014).

10.3.1 Antimicrobial Agents

An antibacterial agent is a substance that has the ability to kill bacteria or prevent their growth. Antimicrobial compounds can have different effects on different types of microorganisms. Antibacterial drugs specifically target bacteria, while antifungal agents are effective against fungi. Microbiocides are substances that kill many types of microorganisms, and microbiostatic agents prevent the growth of germs. It is commonly characterized as substances that can disinfect, eradicate, or impede the proliferation of germs, or serve as antiseptics or antibiotics (Han 2005). Antimicrobial substances can disrupt the formation of nucleic acids at various stages. They have the ability to impede the production or transformation of nucleotides; they can hinder DNA from acting as an accurate blueprint; and they can disrupt the polymerases responsible for both the transcription and replication of DNA (Neu and Gootz 1996).

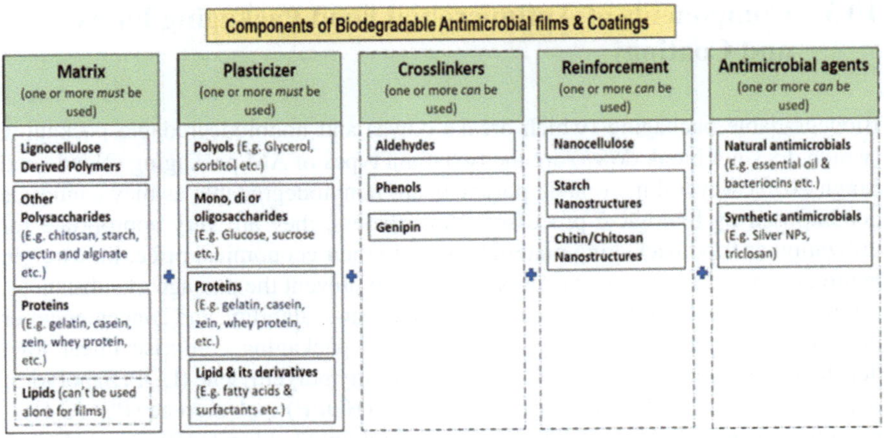

Fig. 10.3 Key components and functional additives for coatings and films in food packaging

Antimicrobials used in food packaging are designed to guarantee the safety of food and extend its shelf life. These packing chemicals inhibit the development of microbes to prevent food spoilage and illness. Antimicrobial packaging is crucial as the demand for minimally processed, preservative-free foods increases. Antimicrobials inhibit bacterial proliferation, hence guaranteeing the safety of food throughout its storage and transportation. This practice decreases food waste and enhances the shelf life of perishable products, resulting in advantages for both customers and the food industry. The use of antimicrobials produced from plants, animals, and microbes is growing in popularity since they are considered safe and environmentally friendly. Various natural antimicrobials, including bacteriocins, enzymes, and essential oils, exhibit a broad spectrum of antibacterial efficacy. Oils extracted from oregano, thyme, and rosemary include thymol and carvacrol, which can efficiently eradicate bacteria and fungi. Lactic acid bacteria produce bacteriocins, such as nisin, while lysozyme is found in egg whites. Both substances effectively attack infections without adversely affecting beneficial microbes. Various synthetic antimicrobials, including benzoic and sorbic acids and their salts, can impede the growth of mold and yeast. Silver nanoparticles, a synthetic substitute, are extensively used for their antimicrobial properties and ability to be incorporated into films and coatings (Vasile and Baican 2021).

10.3.1.1 Natural Antimicrobials

These agents are active organic compounds derived from microorganisms, animals, or plants and include flavonoids, polyphenols, alkaloids, tannins, isothiocyanates, terpenoids, polypeptides, and oxidized derivatives of these substances (Singh et al. 2023). Consumers prefer natural antimicrobial agents over synthetic antimicrobials due to safety concerns and the search for healthier alternatives.

Essential Oils

Essential oils (EOs) from plant parts like flowers, leaves, stems, bark, and seeds are known for their antibacterial properties. Composed mainly of terpenoids, esters, aldehydes, ketones, acids, and alcohols, the EOs composition varies based on plant cultivar, growth stage, and environment. Their antibacterial effect is due to their ability to incorporate into microbial cell membranes, increasing permeability and causing cell lysis by disrupting the protein–lipid interactions and membrane functions. EOs, such as thymol and carvacrol, are commonly used in antimicrobial food packaging for their volatility, which allows them to function without direct contact with food. Studies show that the mechanism of action involves targeting the phospholipid bilayer of bacterial membranes, interfering with enzyme systems, damaging genetic material, and oxidizing fatty acids. Carvacrol and thymol, for example, cause the outer membranes of *Escherichia coli* and *Salmonella typhimurium* to dissolve, while *Listeria monocytogenes* shows cell wall thickening and cytoplasm loss when exposed to thyme oil (Cid-Pérez et al. 2024). Oregano and cinnamon also exhibit similar effects. Terpenes destabilize bacterial cell wall lipids, leading to cell death, while non-phenolic isothiocyanates inactivate extracellular enzymes. Carvacrol specifically breaks down Gram-negative bacteria's outer membranes, releasing lipopolysaccharides and altering membrane permeability (Ribeiro-Santos et al. 2017; Yin et al. 2020).

Bacteriocins

Bacteriocins are antimicrobial peptides or proteins synthesized by diverse bacterial species, first identified in *Escherichia coli* in 1925 by Andre Gratia and his colleagues, who originally designated them as colicins (Thapar and Salooja 2022). These diminutive, cationic peptides are often synthesized by lactic acid bacteria (LAB) and exhibit heat resistance along with broad-spectrum efficacy against pathogenic bacterial strains. Bacteriocins are categorized according to their generating strains, resistance mechanisms, and the particular method by which they induce cell death (Wang et al. 2024). Their antibacterial mechanism operates by disrupting the proton motive force (PMF) across bacterial membranes, forming pores that upset the membrane potential and result in the discharge of cellular contents. Certain bacteriocins, including nisin, selectively target lipid II to impede cell wall synthesis and facilitate pore formation. Moreover, Gram-negative bacteria utilize proteins to translocate bacteriocins via their membranes. Bacteriocins can block nucleic acid and protein synthesis, demonstrating their flexibility. Bacteriocin-producing bacteria utilize immunity proteins to inhibit the function of their own-produced bacteriocins, hence preventing self-damage (Simons et al. 2020).

Nisin

Nisin, an antimicrobial peptide generated by *Lactococcus lactis* ribosomes, is a common food preservative. *L. lactis* fermentation in a modified milky environment yields 34 amino acids, including the uncommon lanthionine and dehydroalanine. Nisin primarily targets Gram-positive bacteria; activity against Gram-negative bacteriaincluding *E. coli*, *Pseudomonas aeruginosa*, and *Salmonella typhimurium*, typically requires permeabilizers such as EDTA (Santos et al. 2018). Nisin is active against *Listeria monocytogenes* in poultry, *Alicyclobacillus acidoterrestris* in orange juice, and *Staphylococcus aureus* in fermented foods (Eghbal et al. 2020). When coupled with ethylenediaminetetraacetic acid (EDTA), it effectively kills Gram-negative bacteria. It performs better in refrigerated milk and other settings when encapsulated in Arabic gum or alginate matrices (Kaur and Kaur 2021). Although it cannot destroy Gram-negative bacteria and fungi, nisin is an excellent preservative for meat, juices, and cheese, providing a natural alternative to synthetic antimicrobials (Kallinteri et al. 2013).

Pediocins

Pediocins, produced by *Pediococcus acidilactici* and *Pediococcus pentosaceus*, are safe bacteriocins used in food preservation (Porto et al. 2017). Pediocin PA-1 is an effective antimicrobial peptide, serving as an alternative biopreservative in the food industry (Rodríguez et al. 2002). Pediocin PA-1, when used in winemaking, has shown promise in effectively reducing bacterial growth and preventing spoilage. This antimicrobial compound may be used alone or in combination with sulfur dioxide and ethanol. Additionally, pediocin AcH from *Lactobacillus plantarum* WHE92 effectively reduces cheese contamination by *Listeria* spp. Thermostable and functional across various pH levels, medicines are effective against pathogens like *L. monocytogenes*, *Enterococcus faecalis*, *S. aureus*, and *Clostridium perfringens*. Generally recognized as safe (GRAS), pediocins also exhibit potential as probiotics, anticancer agents, and in reducing body odor, highlighting their diverse health benefits (Espitia et al. 2016a).

Lacticin

Lacticin is a powerful lantibiotic bacteriocin that effectively targets a range of harmful microorganisms. Lacticin 3147, synthesized by *Lactococcus lactis* subsp. lactic DPC3147, inhibits cell wall synthesis by binding to lipid II, leading to the formation of apertures and consequent cell death. Both lacticin 3147 and lacticin 481 have GRAS status, making them safe for food biopreservation. Studies by Scannell et al. (2000) demonstrated that lacticin and nisin outperform sodium metabisulfite in preserving fresh pork sausage, especially when combined with organic acids (Scannell

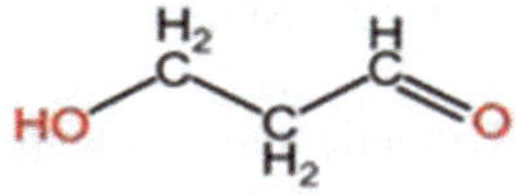

Fig. 10.4 Chemical structure of reuterin (Jeong et al. 2022)

et al. 2000). Lacticin 3147 also shows potential in reducing sulfite content in wine by controlling wine LAB growth in the presence of ethanol (Berbegal et al. 2019).

Reuterin

Lactobacillus reuteri produces reuterin, a strong antibacterial molecule (Fig. 10.4) that inhibits Gram-negative and Gram-positive bacteria, yeasts, molds, and protozoa during glycerol metabolism. It efficiently works against infections caused by *L. monocytogenes*, *S. aureus*, *E. coli*, and *Salmonella choleraesuis*, both when used alone and in combination with nisin. Studies indicate that reuterin has the potential to efficiently inhibit the growth of *L. monocytogenes* and *E. coli* in cooked pig and beef sausage, therefore successfully preventing contamination in meat and dairy products. Reuterin works better against different bacterial strains when combined with nisin, pediocin, and organic acids (Duda-Chodak et al. 2023).

Sakacins

Sakacins are a collection of lactic acid bacteriocins, namely sakacin A, G, K, P, and Q, which are generated by *Lactobacillus sakei*. Sakacin A and P are tiny polypeptides that are very resistant to heat. These bacteriocins are being investigated for their efficacy in combating *Listeria* in sausages, cheeses, cured meats, cold cuts, and other goods that undergo lactic acid fermentation. Recent research has shown that the use of *L. sakei* C2 and sakacin C2 enhances the safety and quality of vacuum-packed sliced cooked ham by effectively inhibiting lipid oxidation. In addition, *L. sakei* demonstrated substantial suppression of *L. monocytogenes* in a synbiotic cheese spread during storage and efficiently outcompeted *L. monocytogenes* in a simulated beef gravy environment (Barbiroli et al. 2017).

Enterocins

Enterocins A, B, and AS-48 are broad-spectrum antibacterial bacteriocins from *Enterococcus faecium*. These bacteriocins inhibit *Salmonella* and *E. coli* strains and are effective against salmonellosis. Enterocin AS-48-activated coatings on apple cubes reduce viable counts by 2.0–2.7 log cycles against *L. monocytogenes*. The study conducted by Baños et al. (2016) showed that the combination of Enterocin AS-48 and *Listeria* lytic bacteriophage P100 is very successful in controlling *L. monocytogenes* in fish preservation (Baños et al. 2016).

10.3.1.2 Synthetic Antimicrobials

Synthetic antimicrobials are engineered compounds designed to inhibit or destroy harmful microorganisms, including bacteria, viruses, and fungi. These agents are integral to various applications, from healthcare and personal hygiene products to industrial processes and food preservation. By mimicking or enhancing natural antimicrobial mechanisms, synthetic antimicrobials offer targeted, effective solutions to combat infections and contamination, especially in an era of increasing antimicrobial resistance.

Antibacterial Nanofillers

Metal and metal-oxide nanoparticles (NPs) have varying levels of antimicrobial activity against Gram-positive and Gram-negative bacteria and may even elicit distinct effects on bacterial strains belonging to the same Gram-positive or Gram-negative category (Auger et al. 2018; Cavassin et al. 2015; Zanet et al. 2019). This phenomenon is likely attributed to the existence of diverse membrane constituents inside various bacterial species, which serve as the initial barrier and primary target for nanoparticles (NPs). As an illustration, Gram-negative bacteria possess phospholipid and lipoprotein on their outer covering, whereas Gram-positive bacteria have a surface characterized by numerous holes and a significantly larger negative charge on the surface (Sánchez-López et al. 2020; Wang et al. 2017). One of the primary mechanisms linked to the antibacterial effects of metal- and metal oxide nanoparticles (NPs) is the induction of oxidative stress via reactive oxygen species (ROS). The primary forms of ROS that are predominantly generated include the superoxide radical (O_2•−), hydroxide radical (•OH), hydrogen peroxide (H_2O_2), and singlet oxygen. The overproduction of one or many reactive oxygen species (ROS) can be triggered by various nanoparticles (NPs). For example, zinc oxide nanoparticles (ZnO NPs) primarily produce H_2O_2, while titanium dioxide nanoparticles (TiO_2 NPs) generate O_2•− and •OH. The generation of H_2O_2 and •OH is observed in the presence of silver nanoparticles (Ag NPs) (Dakal et al. 2016; Durán et al. 2016). Elevated levels of reactive oxygen species (ROS) generation result in the occurrence of oxidative stress, compromised biomolecules and membrane integrity, inhibition of enzymes, and damage to nucleic acids. Reactive oxygen species (ROS) have the capability to induce perforations in the bacterial membrane, resulting in the release of cellular contents (Gold et al. 2018).

Silver Nanoparticles (Ag-NPs)

Silver, in all its many forms, has been used as a standalone or combined antibacterial agent. The antimicrobial properties of silver nitrate or silver sulfadiazine have been extensively researched for their ability to impede bacterial proliferation. These compounds are often used in creams and dressings for the treatment of burns and ulcers, as well as in food packaging to avoid contamination. Silver nanoparticles (Ag-NPs) are nanomaterials that have dimensions ranging from 1 to 100 nm. These possess superior capacity and surface area-to-volume ratio compared to bulk silver. The nanoscale electrical, optical, and catalytic capabilities of this material have led to the development of tailored medication delivery, diagnostic, detection, and imaging devices. Due to their unique features and widespread use in biomaterials, food, cosmetics, clothes, and pharmaceuticals, Ag-NPs have been extensively explored. Silver is also the least harmful to mammalian cells. Thus, current food packaging antibacterial nanocompound research uses Ag-NPs as shown in Table 10.1. Although Ag-NPs are designated as GRAS by the FDA, excessive ingestion levels induced by the relocation of these elements from packaging to foods may pose health risks (Simbine et al. 2019).

Kanmani and Lim (2013) found that silver nanoparticles have the potential to impede the proliferation of bacteria, viruses, and fungi. Initially, the bactericidal efficacy of silver nanoparticles (Ag-NPs) against various bacteria and algae was assessed. The bacteria tested included *Enterococcus faecalis*, *Escherichia coli*, *Staphylococcus aureus*, *Vibrio cholerae*, *S. epidermidis*, *Pseudomonas aeruginosa*, *Bacillus anthracis*, *Shigella flexneri*, *B. subtilis*, *Proteus mirabilis*, *B. cereus*, *Salmonella enterica*, *Micrococcus luteus*, and *Listeria monocytogenes*. *Von Naegeli's* algae were also included in the assessment. The bactericidal efficacy was evaluated by using silver ions. The mechanism by which Ag-NPs exhibit antibacterial activity is not yet fully understood. It is uncertain if Ag-NPs have a unique mode of action or whether their ability to eliminate bacteria is exclusively attributed to the

Table 10.1 Application of Ag-NPs in active food packaging

Composites nanoparticle	Composites biopolymer	Nanoparticle concentration	Target micro-organism	Application	References
Ag	Nanostructured starch	0.3, 0.5, 0.8, and 1.0 mM	*S. aureus*, *E. coli*, and *C. albicans*	AP[a]	Abreu et al. (2015)
Ag	Agar	0.5 mM	*E. coli* and *L. monocytogenes*	Ultraviolet screening and AP[a]	Shankar and Rhim (2015)
Ag–Cu	Fish skin gelatin	2% w/w	*L. monocytogenes* and *Salmonella typhimurium*	UV-screening and AP[a]	Arfat et al. (2017a)
Ag–Cu	Agar	0.5–4 wt%	*L. monocytogenes* and *S. typhimurium*	UV-screening and AP[a]	Arfat et al. (2017b)
Ag	PVA[a]	1%, 3%, 5%, and 7%	*S. aureus* and *E. coli*	AP[a]	Sarwar et al. (2018)
Ag	Chitosan	5% w/w	*E. coli*, *S. aureus*, *A. niger*, and *Penicillium citrinum*	AP[a] for litchi fruits	Wu et al. (2018)
Ag	Cellulose nanofibrils	100 mL	E. coli and *L. monocytogenes*	AP[a]	Yu et al. (2019a, b)

(continued)

Table 10.1 (continued)

Composites nanoparticle	Composites biopolymer	Nanoparticle concentration	Target micro-organism	Application	References
Ag	PVA[a]	40 μL	*S. typhimurium* and *S. aureus*	AP[a] for chicken meat	Mathew et al. (2019)
Ag	Carrageenan	0.075, 0.15, 0.22, and 0.3 wt%	*E. coli* and *L. monocytogenes*	AP[a]	Roy et al. (2019)
Ag	Chitosan	2 wt%	*E. coli, Salmonella, S. aureus*, and *L. monocytogenes*	Active and intelligent food packaging	Qin et al. (2019)
Ag-Cu	PLA[a]	96/4%	*S. typhimurium, L. monocytogenes*, and *Campylobacter jejuni*	UV-screening and AP[a] for chicken meat	Ahmed et al. (2018)

[a]Read AP as active food packaging; PVA as polyvinyl alcohol; PLA as polylactic acid

release of bioactive Ag^+ ions. Research has been conducted on the antibacterial properties of silver, silver ions, and silver nanoparticles against various microbes. A study conducted by Morones et al. (2005) identified three bactericidal mechanisms of silver nanoparticles (Ag-NPs), i.e., silver nanoparticles (Ag-NPs) with sizes ranging from 1 to 10 nm adhere to the cell membrane's surface and cause considerable disturbances to its functions, such as permeability and respiration, silver nanoparticles, Ag-NPs can penetrate bacterial cells and cause damage, perhaps by interacting with sulfur and phosphorus molecules such as DNA and Ag-NPs produce silver ions that can interact with the negatively charged cell membrane (Morones et al. 2005).

Polymer-Coated Nanoparticles

Polymer coatings, such as polyethylene glycol (PEG) or chitosan, enhance the stability of silver nanoparticles and decrease the possibility of aggregation. These coatings further aid in regulating the release of silver ions, thereby extending their antibacterial efficacy. Polymer-coated silver nanoparticles (Ag-NPs) provide advantages in preserving the functionality of the nanoparticles over extended periods and guaranteeing their secure use in food packaging (Mohamad Kasim et al. 2020).

Enhancing the biocompatibility of silver nanoparticles may be achieved by applying a protein coating, such as casein or whey protein. This coating increases their safety for use in food packaging. Protein coatings further improve the attachment of nanoparticles to packaging surfaces, thereby ensuring continuous antimicrobial protection. Protein-coated silver nanoparticles provide a more organic method for integrating antibacterial substances into packaging materials (Valenzuela-Salas et al. 2021).

Plant-derived extracts could be used as efficient coatings for silver nanoparticles (Ag-NPs), providing improved antibacterial properties and decreased ecological footprint in comparison to conventional synthetic coatings. The incorporation of plant-based coatings greatly enhances the biocompatibility and sustainability of Ag-NPs. The presence of phytochemicals in these natural coatings not only adds an extra layer of antimicrobial protection but also enhances their eco-friendly characteristics, which aligns with the growing demand for sustainable food packaging solutions. This method utilizes the natural characteristics of plant extracts to reduce the possible negative impacts of artificial substances, providing a more ecologically conscious option while yet retaining the functional effectiveness of Ag-NPs (Jain et al. 2021).

Zinc Oxide Nanoparticles (ZnO-NPs)

ZnO-NPs have antibacterial characteristics and UV-blocking capabilities. They are used in packaging films and coatings. ZnO is widely used for its antibacterial and photocatalytic characteristics. ZnO-NPs are effective semiconductors with high biocompatibility, stability, and biocidal properties. They are cost-effective. They also have antibacterial effects against foodborne diseases and microorganisms (Table 10.2). ZnO-NPs are used in food packaging coatings to maintain color, prevent degradation, and improve material qualities such as mechanical quality, barrier attributes, and stability (Al-Tayyar et al. 2020; Espitia et al. 2016a, b).

Titanium Dioxide Nanoparticles (TiO_2 NPs)

TiO_2-NPs are a kind of white metallic oxide with multiple applications in the food industry. These applications include their usage as food additives and as components of nanocomposites used in food packaging. TiO_2 nanoparticles are used in food packaging polymers to selectively block ultraviolet (UV) radiation while simultaneously enhancing the chemical, mechanical, and barrier attributes of the films. TiO_2-NPs are often used as a coloring agent in processed foods because of their high refractive index, brightness, and color stability. Titanium dioxide has antibacterial properties by generating free radicals and reactive oxygen species (ROS) that effectively interact with and eradicate bacterial cells. TiO_2 is a good photocatalyst because it is chemically stable, less expensive, nontoxic, and ecologically benign (Jaiswal et al. 2019). The photocatalytic reaction of TiO_2 has been used to inactivate a wide spectrum of microorganisms (Altın and Sökmen 2014; Gupta et al. 2013; Long et al. 2014). Recent years have witnessed a significant emphasis on the advancement of food packaging that is coated or combined with TiO_2-NPs, as outlined in Table 10.3 (Zhou et al. 2009).

Table 10.2 Application of ZnO-NPs in active food packaging

Composites nanoparticle	Composites biopolymer	Nanoparticle concentration	Target microorganism	Application	References
ZnO	Ethylcellulose/gelatin nanofibers	1.5 wt%	*E. coli* and *S. aureus*	AP[a]	Abreu et al. (2015)
ZnO	Corn starch, guanidine-based starch	1%, 3%, 5%, 7%, and 9%	*E. coli*	Green-based AP[a]	Ni et al. (2018)
ZnO	Soy protein isolate	2 mg/mL	*A. niger*	AP[a]	Wu et al. (2019)
ZnO	PBS[a]	2–10 wt%	*E. coli* and *S. aureus*	AP[a]	Petchwattana et al. (2016)
ZnO	Polylactide/polyethylene glycol/polycaprolactone	3% w/w	*S. aureus* and *E. coli*	AP[a] for eggs	Ahmed et al. (2019)
ZnO	Chitosan/carboxymethyl cellulose	0.5, 1, and 2%	Yeasts and molds	AP[a] for bread slice	(Noshirvani et al. 2017)
ZnO	Methylcellulose/chitosan nanofibers	1% and 2% w/v	*E. coli*, *S. aureus*, *Aspergillus* sp., and *Penicillium* sp.	AP[a]	Sani et al. (2023)

[a]Read AP as active food packaging; PBS as poly (butylene succinate)

10.4 Recent Advances in Antimicrobial Packaging Technologies

The integration of antimicrobial compounds into packaging materials' composition is a critical step toward producing effective antimicrobial packaging (Table 10.4). These agents are incorporated in a variety of ways; each of these options possesses its distinct pros and downsides.

10.4.1 Incorporation Methods

10.4.1.1 Direct Incorporation Method

Direct incorporation is a packaging technology that includes embedding or integrating certain materials, chemicals, or technologies directly into the packaging material during the production process. This method may be used to modify the usefulness, shelf life, and safety of the packaging. To develop an antimicrobial packaging system, antimicrobial compounds can be directly incorporated into wrapping films, coated using numerous methods as shown in Fig. 10.5, or developed from polymers. It is more effective in achieving antimicrobial activities (Irkin and Esmer 2015).

Table 10.3 Application of TiO_2-NPs in active food packaging

Composites nanoparticle	Composites biopolymer	Nanoparticle concentration	Target micro-organism	Application	References
TiO_2	Wheat gluten and cellulose nanocrystals	0.6%	*S. cerevisiae*, *E. coli*, and *S. aureus*	AP[a]	El-Wakil et al. (2015)
TiO_2	Whey protein isolate / cellulose nanofiber	1.0% (w/w)	*Pseudomonas* spp., *Enterobacteriaceae*, *lactic bacteria*, *S. aureus*, *L. monocytogenes*, and *E. coli*	AP[a] for meat preservation	Sani et al. (2017)
TiO_2	Zein/sodium alginate	0.5 wt.%	*S. aureus* and *E. coli*	AP[a] for meat, cheese, nuts, cereal	Amjadi et al. (2020)
TiO_2	Chitosan/ PVA	2, 4, 8 wt.%	*S. aureus*, *Pseudomonas aeruginosa*, *E. coli (O157:H7)*, and *C. albicans*	AP[a] of soft white cheese	Youssef et al. (2015)
TiO_2	Chitosan/ alginate	0.1-0.3%	*E. coli*, *S. aureus*, *S. typhi*, and *L. monocytogene*	AP[a] for fresh produce	Perera et al. (2022)

[a]Read AP as Active food packaging

10.4.1.2 Casting Method

The casting process, commonly referred to as solvent casting, is the predominant technique used for producing films in laboratory and pilot settings. In order to fabricate biopolymer films, the biopolymer must first be dissolved in the proper solvent, then the solution must be poured into a mould, and finally the solution must be dried (Suhag et al. 2020).

10.4.1.3 Blending

Antimicrobial compounds are blended directly into the wrapping items during the manufacturing procedure. This technology is widely utilized for polymer-based packaging materials. The antimicrobial ingredient is spread consistently throughout the material, resulting in constant antibacterial action. The polymer structure is formed by intermolecular electrostatic and hydrogen bond development (Cuervo-Rodríguez et al. 2017).

Table 10.4 Overview of advanced antimicrobials for food packaging innovations

Type of antimicrobial	Compounds	Source	Target microorganism	Mechanism of action	Advantages	Limitations
Natural Antimicrobials	Essential oils	Thyme, Clove, Cinnamon, etc.	Gram Positive: *S. aureus*, *L. Shigella son nei*, *S. flexneri*. Gram-negative: *E. coli O157:H7 E. coli S. enteritidis S. typhimurium* Fungi: *Candida albicans Penicillium digitatum*	Increased permeability and cell lysis due to integration into cell membrane lipid layers	Approved for US and EU food contact. Volatile acts without food contact.	Problems with stability. Responses to allergens. Limited availability and higher cost
	Bacteriocins	Lactic Acid Bacteria	Gram +ve[a]: *L. monocytogenes* Gram −ve[a]: *E. coli* pathogens: *S. aureus*, *Salmonella* spp.	Cell membrane disruption, cell wall synthesis suppression, pore formation	Safe for food storage. Effective bio preservative	Little impact on Gram-ve bacteria and fungus
	Lysozyme	Chicken eggs, Vegetables	Gram-positive: *Bacillus*, *Clostridium*, *Listeria* spp. Weak effect on Gram-negative bacteria	Hydrolyses bacterial cell wall peptidoglycan	Safe to eat. Effective against some germs	Spectrum limited Allergen potential
	Chitosan	Shellfish	Gram +vea: *S. aureus*, *L. monocytogenes* Gram −ve: *Y. enterocolitica* Fungi: *A. flavus*	Cell membrane disruption, function disruption	Biodegradable	Possible allergens Environment-dependent activity

Synthetic Antimicrobials	Silver Nanoparticles	Engineered	Gram +ve[a] and Gram −ve[a] bacteria, fungi, viruses	Breaking cell membranes, entering cells, releasing bioactive silver ions	Widely effective against many bacteria. High surface-to-volume ratio	Excessive use may pose health risks and environmental issues
	Silver Nanoparticles (PEG, Chitosan coatings)	Engineered	Gram +ve[a] and Gram −ve[a] bacteria, fungi, viruses	Enhanced stability, reduced aggregation, regulated ion release	Extended antibacterial efficacy. Secure use in food packaging	Potential for increased complexity and cost
	Silver Nanoparticles (Casein, Whey coatings)	Engineered	Gram +ve[a] and Gram −ve[a] bacteria, fungi, viruses	Improved biocompatibility and attachment to surfaces	Organic integration into packaging materials	High cost
	Silver Nanoparticles (Plant extracts)	Plant-derived	Gram +ve[a] and Gram −ve[a] bacteria, fungi, viruses	Enhanced biocompatibility, sustainability, and antimicrobial properties	Eco-friendly and sustainable	Silver Nanoparticles (Plant extracts)
	Zinc oxide Nanoparticles	Engineered	Gram +ve[a]: *B. subtilis* and *S. aureus* Gram −ve[a]: *P. aeruginosa*, *C. jejuni* and *E. coli*	Cell wall disruption due to ZnO-localized interaction, enhanced membrane permeability, mitochondria weakness, intracellular outflow, and release in gene expression of oxidative stress	Nontoxic, Photo-oxidizing, photocatalysis impacts & large surface area-to-volume ratio	Sensitive to intense processing conditions
	Titanium dioxide Nanoparticles	Engineered	Gram +ve[a]: *S. aureus* and Gram −ve[a]: *E. coli* & *P. putida* bacteria & fungi	Kills bacteria by producing free radicals like superoxide & singlet oxygen	Photocatalytic activity, Non-toxic & approved by American Food & Drug Administration (FDA) for use in human food	High intake is harmful although considered safe by the FDA (Food and Drug Administration)

[a]Read Gram +ve as Gram-positive bacteria and Gram −ve as Gram-negative bacteria

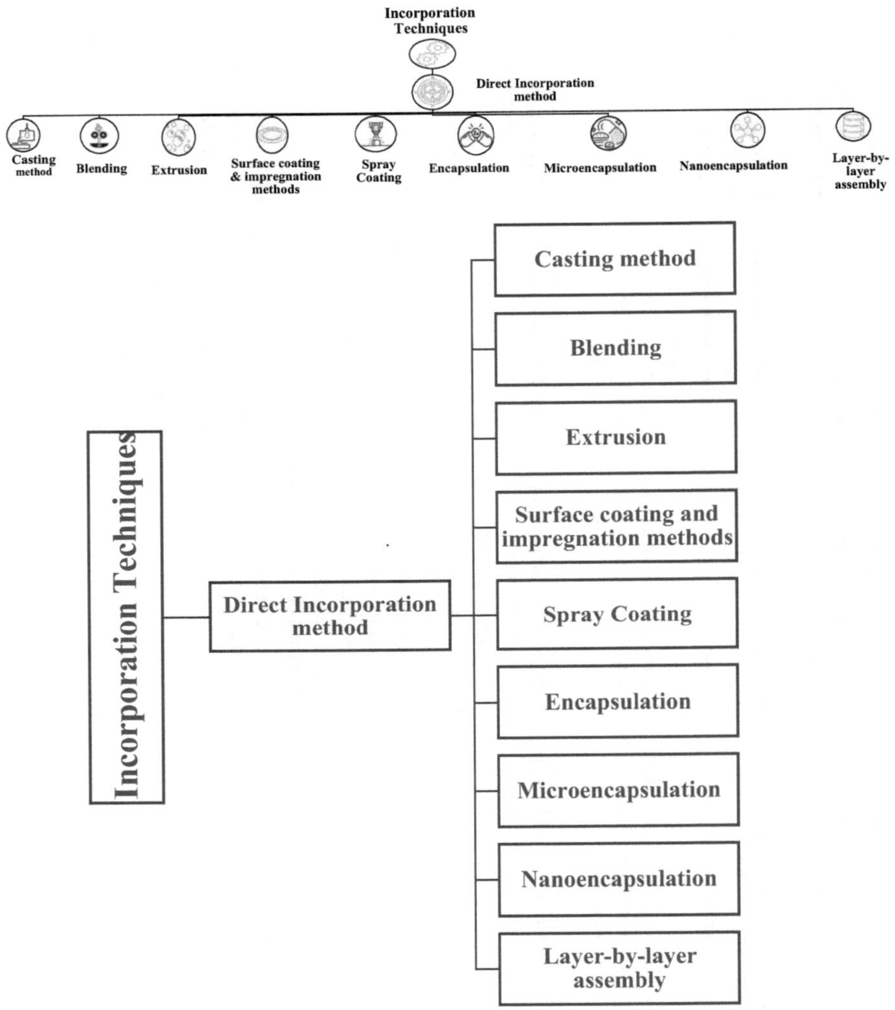

Fig. 10.5 Methods of direct antimicrobial incorporation in food packaging

10.4.1.4 Extrusion

Antimicrobial compounds are combined with the polymer matrix and then extruded to make films or other packaging shapes. This process is commonly used to create antimicrobial films and coatings. The extrusion process involves a series of consecutive phases. The process involves preparing formulations by blending various raw materials, combining the ingredients in an extruder to create pellets, cutting the extrudates into pellets using a pelletizer, drying the pellets in a hot-air oven, extruding the pellets into sheets using a second extruder, and ultimately converting the blended resins into a film using a blown film extruder (Kumar et al. 2020).

10.4.1.5 Surface Coating and Impregnation Methods

Coating is often used to cover the surface of fresh goods, including fruits, vegetables, fish, and meat. Chitosan-based composite solutions may be used to increase the shelf life of food goods by applying them by spreading, spraying, dipping, or immersing. The coating is an excellent approach to limit the growth of germs, resulting in a longer shelf life and helping to retain the quality of food.

10.4.1.6 Spray Coating

Spray coating involves spraying an antimicrobial solution or suspension over the packaging material's surface. This approach is ideal for applying thin, homogeneous coatings of antimicrobial chemicals. Spraying methods minimize layer contamination, provide temperature control for coating solutions, and enable continuous manufacturing. It provides consistent coatings ideal for complex forms and surfaces.

10.4.1.7 Encapsulation

Encapsulation is the act of confining a material (active agent) inside another substance, leading to tiny particles that gradually release their contents at controlled rates over extended durations under certain circumstances. Encapsulating antimicrobial compounds in food packaging enhances the efficiency of the wrapping substances by safeguarding the antimicrobial compounds from degradation, volatilization, or undesirable interactions with the packaging materials, enhancing the compatibility between the packaging polymer and the antimicrobial substance, improving the accessibility of the antimicrobial agent, and enabling controlled dispersion or responsive release to prolong the effectiveness of the active material (Becerril et al. 2020).

10.4.1.8 Microencapsulation

Microencapsulation is the process of encapsulating antimicrobial chemicals in microscopic capsules, which are subsequently put into packaging material. These capsules can be modified to eventually release antibacterial chemicals over time. Microcapsules can be liquid, solid, or gas, and range in size from 1 to 1000 μm, depending on the technology utilized. Physical approaches generate microparticles by introducing physical changes to the shell material, such as drying, gelling, or freezing it. Physical methods used to create microspheres include atomization techniques such as spray-drying and spray-chilling, lyophilization, spray coating using fluid bed, as well as extrusion- and emulsion-based processes (Arenas-Jal et al. 2020).

10.4.1.9 Nanoencapsulation

This method produces nano-sized particles or carriers (1–100 nm). Nanoencapsulation involves two steps: loading antimicrobial compounds into carriers and reducing their size to the nanoscale. Nanoencapsulation systems have a higher surface area than big particles, resulting in improved solubility, adsorption, controlled release, and bioavailability. Nanoencapsulation improves antimicrobial solubility, prolongs release, and preserves volatile antimicrobials in food packaging. Nanoencapsulation aims to create bioactive nanocarriers in various forms such as nanocapsules, nanospheres, nanoshells, branching nanocubes, nanostars, nanorods, and nanoclusters. Encapsulated antimicrobials consist of core and wall components. Antimicrobial agents are inserted in the core and covered with wall materials such as lipids, proteins, and polysaccharides (Bahrami et al. 2020).

10.4.1.10 Layer-by-Layer Assembly

Layer-by-layer assembly includes successively applying antimicrobial agents and other components onto the package surface. Each layer can be tailored to give certain characteristics, such as controlled release or increased stability. The LBL assembly is a versatile method that has been extensively studied in the production of nanocomposite films. It permits the integration of functional attributes from different polymers, facilitating efficient regulation of material properties and functioning. This method enables the production of multicomponent films without requiring sophisticated equipment. The surface modification in LBL assembly is mainly determined by the deposition and mutual attraction of alternating polyelectrolytes with opposite net charges onto a solid substrate (Kumar et al. 2020).

10.4.2 Smart Packaging

10.4.2.1 Intelligent Packaging Systems That Respond to Environmental Changes

IP, or intelligent packaging, is a nascent technology in the food packaging sector. This technology has the capacity to improve the traceability, safety, and quality of food (Nanda et al. 2024) (Fig. 10.6). The term "IP" refers to the scientific and technological methods employed to monitor and transmit the condition of a food packaging system to key stakeholders, such as manufacturers, retailers, and consumers.

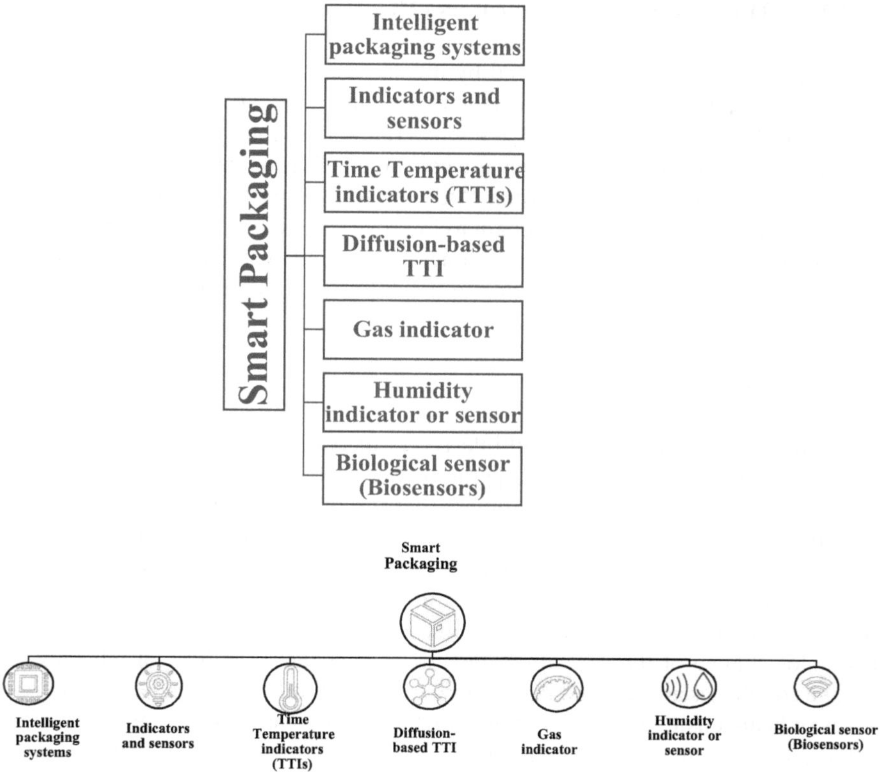

Fig. 10.6 Smart packaging methods: innovations for food preservation and monitoring

IP improves decision-making, safety, and quality by delivering valuable information and timely alerts regarding potential risks. Intellectual property relies on sensors, indicators, and identification systems to accurately assess and communicate the attributes of food products to customers (Firouz et al. 2021).

10.4.2.2 Indicators and Sensors for Real-Time Monitoring

Indicators provide consumers with details regarding the existence or nonexistence of a chemical, the interaction between two or more compounds, or the concentration of a certain substance or class of substances. Visual signs, such as color shifts or dye dispersion, are commonly used to convey information. Indicators typically use qualitative or semiquantitative information. There are three main sorts of indicators: time, temperature, freshness, and gas (Ghaani et al. 2016).

10.4.2.3 Time Temperature Indicators (TTIs)

Food quality is significantly affected by its temperature history throughout manufacturing and eating. Cold chain management solutions rely heavily on temperature control. TTIs may be used to track the temperature chronology of food items from the time they are packaged until they are consumed. TTIs quantify the overall effect of temperature on the quality of food. Attached to food packaging, they track temperature exposure along the cold chain to improve food safety. These indicators are operated by reacting two or more components (e.g., enzymatic, mechanical, chemical) to cause irreversible discoloration of the indicator (Firouz et al. 2021).

10.4.2.4 Diffusion-Based TTI

This indication is made up of multiple layers of paper, film, and glue. When the conditions for storing food, like temperature, exceed the permitted limits, a notification will be triggered to alert customers that the product is no longer in optimal condition. This indicator features a permeable wick that holds pigmented fatty acid esters. When the temperature surpasses a specific threshold, the ink migrates to different layers, resulting in a shift in the color of the indicator. This indicator assesses the impact of unfavorable temperatures during long storage periods on the product. The color change of the indication remains permanent, regardless of whether the storage conditions are restored to appropriate levels. Indicators on packing or export boxes of heat-sensitive products offer valuable insights into storage and transit conditions. These signs are designed to track and display the storage conditions rather than offering details about the product itself (Firouz et al. 2021).

10.4.2.5 Gas Indicator

The adhesive label affixed to the surface of the package reveals changes in the composition of gases. This indicator can ascertain the existence or nonexistence of carbon dioxide (CO_2), oxygen (O_2), or ethylene (C_2H_4). These indicators need to establish and sustain touch with both the edible items and the surrounding environment within the packaging. The film's color varies with the level of CO_2 gas in the package's headspace, providing a reliable indication of the quality of the packed food. These indicators can detect subtle changes in food quality, leading to noticeable results (Firouz et al. 2021).

10.4.2.6 Humidity Indicator or Sensor

Humidity sensors often use electronic hygrometers to measure variations in capacitive or resistance qualities that are influenced by humidity levels. Most colorimetric humidity indicators primarily rely on inorganic salts. The production of these markers often involves the use of cobalt (II) chloride, which is frequently used for dehydrated food items that are vulnerable to moisture (Firouz et al. 2021).

10.4.3 Biological Sensor (Biosensors)

Sensors provide great potential for enhancing future IP systems. The sensors are composed of three main components: the sensor, signal conditioning (processing), and interface, which is responsible for displaying the measured values. The first segment detects physical or chemical attributes and converts them into an electrical signal. The signal conditioning device manipulates the generated signal. This device carries out signal amplification, linearization, and scaling. The ultimate element might manifest as an adapter, mechanical connection, wireless sensor, or similar device (Firouz et al. 2021; Ghaani et al. 2016).

10.5 Consumer Acceptance and Market Trends

Consumer acceptance is crucial for the adoption of antimicrobial food packaging technologies that improve food safety and extend shelf life. Consumers are becoming increasingly wary of synthetic chemicals in their food, resulting in a demand for antimicrobial remedies that mitigate common diseases, such as food poisoning. Antarctic krill oil is promoted as devoid of artificial ingredients, whereas fish oil is more popular compared to other types containing additives. Consumers exhibit loyalty toward suppliers who provide organic items aligning with their interests. The need for natural feed ingredients has compelled food firms to offer sustainable, healthy alternatives designed for these consumer categories. Moreover, customers are advocating for restrictions on specific packaging materials (Siddiqui et al. 2022). Market dynamics are propelling the adoption of antimicrobial food packaging technologies, especially those using natural antimicrobials promoted as "clean label" products. These attract people desiring products devoid of synthetic additives. Simultaneously, research on smart packaging, which incorporates antimicrobial films alongside functionalities such as freshness indicators and active environmental management, is accelerating. These developments are revolutionizing food packaging and are anticipated to impact future laws in the industry (Ghoshal 2018).

10.6 Future Prospects and Challenges

The future of antimicrobial agents in food packaging looks auspicious, yet has numerous hurdles. Conventional techniques for administering antimicrobial drugs via spraying or dripping are constrained by their swift dispersion and contact with food, which diminishes their efficacy (Quintavalla and Vicini 2002). Moreover, these substances might adversely affect the flavor of food. Polymers, however, provide a more regulated release of antimicrobial agents to the surface, ensuring a consistent presence that inhibits the proliferation of foodborne bacteria (Gharsallaoui et al. 2016). Progress in biotechnology and nanotechnology is facilitating the creation of more efficient and safer natural and synthetic antimicrobial agents, which can be included in intelligent packaging systems for real-time food safety surveillance. A significant movement is underway toward biodegradable and sustainable packaging materials, advancing global environmental objectives. Nonetheless, obstacles persist, such as the elevated expense of natural antimicrobials, rigorous regulatory standards, and apprehensions regarding the toxicity and environmental consequences of synthetic agents. Moreover, consumer adoption of antimicrobial packaging, particularly synthetic alternatives, is inconsistent, necessitating educational initiatives and clear information on their safety and advantages. The industry must also tackle challenges concerning the scalability of these technologies and their efficacy against new infections. Notwithstanding these hurdles, continued research and collaboration can transform food safety and sustainability in packaging.

10.7 Conclusion

Innovative antimicrobial solutions in food packaging are revolutionizing food preservation and protection against microbial infection. These sophisticated materials provide substantial advantages for food safety, prolongation of shelf life, and the mitigation of foodborne illnesses. As the food industry transitions to sustainable and eco-friendly packaging solutions, the advancement of antimicrobial packaging that is both efficacious and environmentally responsible will be pivotal in fulfilling customer needs for safer and more sustainable food items. The prospects for antimicrobial food packaging are favorable, supported by ongoing research into innovative antimicrobial agents, nanotechnology, and intelligent packaging technologies. It is essential to tackle regulatory, safety, and environmental challenges to guarantee that these technologies are secure for customers and the environment.

References

Abreu AS, Oliveira M, de Sá A, Rodrigues RM, Cerqueira MA, Vicente AA, Machado A (2015) Antimicrobial nanostructured starch based films for packaging. Carbohydr Polym 129:127–134

Ahmed J, Arfat YA, Bher A, Mulla M, Jacob H, Auras R (2018) Active chicken meat packaging based on polylactide films and bimetallic Ag–Cu nanoparticles and essential oil. J Food Sci 83(5):1299–1310

Ahmed J, Mulla M, Jacob H, Luciano G, Bini T, Almusallam A (2019) Polylactide/poly (ε-caprolactone)/zinc oxide/clove essential oil composite antimicrobial films for scrambled egg packaging. Food Packag Shelf Life 21:100355

Al-Tayyar NA, Youssef AM, Al-Hindi RR (2020) Antimicrobial packaging efficiency of ZnO-SiO2 nanocomposites infused into PVA/CS film for enhancing the shelf life of food products. Food Packag Shelf Life 25:100523

Altın İ, Sökmen M (2014) Preparation of TiO2-polystyrene photocatalyst from waste material and its usability for removal of various pollutants. Appl Catal B Environ 144:694–701

Amjadi S, Almasi H, Ghorbani M, Ramazani S (2020) Preparation and characterization of TiO2NPs and betanin loaded zein/sodium alginate nanofibers. Food Packag Shelf Life 24:100504

Appendini P, Hotchkiss JH (2002) Review of antimicrobial food packaging. Innov Food Sci Emerg Technol 3(2):113–126

Arenas-Jal M, Suñé-Negre J, García-Montoya E (2020) An overview of microencapsulation in the food industry: opportunities, challenges, and innovations. Eur Food Res Technol 246:1371–1382

Arfat YA, Ahmed J, Hiremath N, Auras R, Joseph A (2017a) Thermo-mechanical, rheological, structural and antimicrobial properties of bionanocomposite films based on fish skin gelatin and silver-copper nanoparticles. Food Hydrocoll 62:191–202

Arfat YA, Ahmed J, Jacob H (2017b) Preparation and characterization of agar-based nanocomposite films reinforced with bimetallic (Ag-Cu) alloy nanoparticles. Carbohydr Polym 155:382–390

Auger S, Henry C, Péchoux C, Suman S, Lejal N, Bertho N et al (2018) Exploring multiple effects of Zn0. 15Mg0. 85O nanoparticles on *Bacillus subtilis* and macrophages. Sci Rep 8(1):12276

Bahrami A, Delshadi R, Assadpour E, Jafari SM, Williams L (2020) Antimicrobial-loaded nanocarriers for food packaging applications. Adv Colloid Interface Sci 278:102140

Baños A, García-López JD, Núñez C, Martínez-Bueno M, Maqueda M, Valdivia E (2016) Biocontrol of *Listeria monocytogenes* in fish by enterocin AS-48 and *Listeria* lytic bacteriophage P100. LWT-Food Sci Technol 66:672–677

Barbiroli A, Musatti A, Capretti G, Iametti S, Rollini M (2017) Sakacin-A antimicrobial packaging for decreasing *Listeria* contamination in thin-cut meat: preliminary assessment. J Sci Food Agric 97(3):1042–1047

Becerril R, Nerín C, Silva F (2020) Encapsulation systems for antimicrobial food packaging components: an update. Molecules 25(5):1134

Berbegal C, Fragasso M, Russo P, Bimbo F, Grieco F, Spano G, Capozzi V (2019) Climate changes and food quality: the potential of microbial activities as mitigating strategies in the wine sector. Fermentation 5(4):85

Cavassin ED, de Figueiredo LFP, Otoch JP, Seckler MM, de Oliveira RA, Franco FF et al (2015) Comparison of methods to detect the in vitro activity of silver nanoparticles (AgNP) against multidrug resistant bacteria. J Nanobiotechnol 13:1–16

Cid-Pérez TS, Munguía-Pérez R, Nevárez-Moorillón GV, Ochoa-Velasco CE, Navarro-Cruz AR, Avila-Sosa R (2024) Carvacrol and thymol effect in vapor phase on *Escherichia coli* and *Salmonella* serovar Typhimurium growth inoculated in a fresh salad. Heliyon 10(9):e29638

Cuervo-Rodríguez R, López-Fabal F, Gómez-Garcés JL, Muñoz-Bonilla A, Fernández-García M (2017) Contact active antimicrobial coatings prepared by polymer blending. Macromol Biosci 17(11):1700258

Dakal TC, Kumar A, Majumdar RS, Yadav V (2016) Mechanistic basis of antimicrobial actions of silver nanoparticles. Front Microbiol 7:1831

De Azeredo H, Rosa M, De Sá M, Souza Filho M, Waldron K (2014) The use of biomass for packaging films and coatings. In: Advances in biorefineries. Cambridge, United Kingdom: Woodhead Publishing Limited, pp 819–874

Duda-Chodak A, Tarko T, Petka-Poniatowska K (2023) Antimicrobial compounds in food packaging. Int J Mol Sci 24(3):2457

Durán N, Durán M, De Jesus MB, Seabra AB, Fávaro WJ, Nakazato G (2016) Silver nanoparticles: A new view on mechanistic aspects on antimicrobial activity. Nanomed Nanotechnol Biol Med 12(3):789–799

Eghbal N, Chihib NE, Gharsallaoui A (2020) Nisin. In: Antimicrobials in food. CRC Press, pp 309–338

El-Wakil NA, Hassan EA, Abou-Zeid RE, Dufresne A (2015) Development of wheat gluten/nanocellulose/titanium dioxide nanocomposites for active food packaging. Carbohydr Polym 124:337–346

Espitia PJP, Soares NDFF, Teófilo RF, dos Reis Coimbra JS, Vitor DM, Batista RA et al (2013) Physical–mechanical and antimicrobial properties of nanocomposite films with pediocin and ZnO nanoparticles. Carbohydr Polym 94(1):199–208

Espitia P, Batista R, Otoni C, Soares N (2016a) Antimicrobial food packaging incorporated with triclosan: potential uses and restrictions. In: Antimicrobial food packaging (Second Edition). Amsterdam, Netherlands: Academic Press, pp 417–423

Espitia P, Otoni C, Soares N (2016b) Zinc oxide nanoparticles for food packaging applications. In: Antimicrobial food packaging (Second Edition). Amsterdam, Netherlands: Elsevier, pp 425–431

Firouz MS, Mohi-Alden K, Omid M (2021) A critical review on intelligent and active packaging in the food industry: research and development. Food Red Int 141:110113

Floros JD, Dock LL, Han JH (1997) Active packaging technologies and applications. Food Cosmet Drug Packag 20(1):10–17

Food and Nations AO (2019) Technical platform on the measurement and reduction of food loss and waste. FAO, Rome

Ghaani M, Cozzolino CA, Castelli G, Farris S (2016) An overview of the intelligent packaging technologies in the food sector. Trends Food Sci Technol 51:1–11

Gharsallaoui A, Joly C, Oulahal N, Degraeve P (2016) Nisin as a food preservative: part 2: antimicrobial polymer materials containing nisin. Crit Rev Food Sci Nutr 56(8):1275–1289

Ghoshal G (2018) Recent trends in active, smart, and intelligent packaging for food products. In: Food packaging and preservation. Amsterdam, Netherlands: Elsevier, pp 343–374

Gold K, Slay B, Knackstedt M, Gaharwar AK (2018) Antimicrobial activity of metal and metal-oxide based nanoparticles. Adv Ther 1(3):1700033

Gupta K, Singh R, Pandey A, Pandey A (2013) Photocatalytic antibacterial performance of TiO_2 and Ag-doped TiO_2 against *S. aureus*, *P aeruginosa* and *E coli*. Beilstein J Nanotechnol 4(1):345–351

Han J (2002) Protein-based edible films and coatings carrying antimicrobial agents. In: Protein-based films and coatings. CRC Press, Boca Raton, pp 485–499

Han JH (2003) Antimicrobial food packaging, vol 8. Woodhead Publishing, Cambridge, pp 50–70

Han JH (2005) Antimicrobial packaging systems. In: Innovations in food packaging. Amsterdam, Netherlands:Elsevier, pp 80–107

Huang T, Qian Y, Wei J, Zhou C (2019) Polymeric antimicrobial food packaging and its applications. Polymers 11(3):560

Irkin R, Esmer OK (2015) Novel food packaging systems with natural antimicrobial agents. J Food Sci Technol 52:6095–6111

Jain N, Jain P, Rajput D, Patil UK (2021) Green synthesized plant-based silver nanoparticles: therapeutic prospective for anticancer and antiviral activity. Micro Nano Syst Lett 9(1):5

Jaiswal L, Shankar S, Rhim J-W (2019) Applications of nanotechnology in food microbiology. In: Methods in microbiology, vol 46. Amsterdam, Netherland: Elsevier, pp 43–60

Jeong YR, Kim YE, Lee SH (2022) Review on the structural features and biological activities of reuterin. Yakhak Hoeji 66(4):169–174

Kallinteri L, Kostoula O, Savvaidis I (2013) Efficacy of nisin and/or natamycin to improve the shelf-life of Galotyri cheese. Food Microbiol 36(2):176–181

Kanmani P, Lim ST (2013) Synthesis and structural characterization of silver nanoparticles using bacterial exopolysaccharide and its antimicrobial activity against food and multidrug resistant pathogens. Process Biochem 48(7):1099–1106

Kaur R, Kaur L (2021) Encapsulated natural antimicrobials: a promising way to reduce microbial growth in different food systems. Food Control 123:107678

Kumar S, Mukherjee A, Dutta J (2020) Chitosan based nanocomposite films and coatings: emerging antimicrobial food packaging alternatives. Trends Food Sci Technol 97:196–209

Leistner L (2007) Update on hurdle technology for mild and effective preservation of foods. In: Handbook of food preservation. Boca Raton, CRC press, pp 913–922

Long M, Wang J, Zhuang H, Zhang Y, Wu H, Zhang J (2014) Performance and mechanism of standard nano-TiO2 (P-25) in photocatalytic disinfection of foodborne microorganisms—*Salmonella typhimurium* and *Listeria monocytogenes*. Food Control 39:68–74

Mathew S, Snigdha S, Mathew J, Radhakrishnan E (2019) Biodegradable and active nanocomposite pouches reinforced with silver nanoparticles for improved packaging of chicken sausages. Food Packag Shelf Life 19:155–166

Mohamad Kasim AS, Ariff AB, Mohamad R, Wong FWF (2020) Interrelations of synthesis method, polyethylene glycol coating, physico-chemical characteristics, and antimicrobial activity of silver nanoparticles. Nanomaterials 10(12):2475

Morones JR, Elechiguerra JL, Camacho A, Holt K, Kouri JB, Ramírez JT, Yacaman M (2005) The bactericidal effect of silver nanoparticles. Nanotechnology 16(10):2346

Nanda A, Singh N (2022) Preparation and antimicrobial activity of corn cob and coir reinforced biodegradable starch biocomposite films for food packaging application. Asian Food Sci J 21(7):53–66

Nanda A, Pandey P, Rajinikanth P, Singh N (2024) Revolution of nanotechnology in food packaging: harnessing electrospun zein nanofibers for improved preservation—a review. Int J Biol Macromol 260:129416

Neu HC, Gootz TD (1996) Antimicrobial chemotherapy. Medical Microbiology. 4th edition. Galveston, TX: University of Texas Medical Branch at Galveston.

Ni S, Zhang H, Dai H, Xiao H (2018) Starch-based flexible coating for food packaging paper with exceptional hydrophobicity and antimicrobial activity. Polymers 10(11):1260

Noshirvani N, Ghanbarzadeh B, Mokarram RR, Hashemi M (2017) Novel active packaging based on carboxymethyl cellulose-chitosan-ZnO NPs nanocomposite for increasing the shelf life of bread. Food Packag Shelf Life 11:106–114

Perera KY, Sharma S, Duffy B, Pathania S, Jaiswal AK, Jaiswal S (2022) An active biodegradable layer-by-layer film based on chitosan-alginate-TiO2 for the enhanced shelf life of tomatoes. Food Packag Shelf Life 34:100971

Petchwattana N, Covavisaruch S, Wibooranawong S, Naknaen P (2016) Antimicrobial food packaging prepared from poly (butylene succinate) and zinc oxide. Measurement 93:442–448

Porto MCW, Kuniyoshi TM, Azevedo POS, Vitolo M, Oliveira RS (2017) *Pediococcus* spp.: an important genus of lactic acid bacteria and pediocin producers. Biotechnol Adv 35(3):361–374

Qin Y, Liu Y, Yuan L, Yong H, Liu J (2019) Preparation and characterization of antioxidant, antimicrobial and pH-sensitive films based on chitosan, silver nanoparticles and purple corn extract. Food Hydrocoll 96:102–111

Quintavalla S, Vicini L (2002) Antimicrobial food packaging in meat industry. Meat Sci 62(3):373–380

Ribeiro-Santos R, Andrade M, de Melo NR, Sanches-Silva A (2017) Use of essential oils in active food packaging: recent advances and future trends. Trends Food Sci Technol 61:132–140

Rodríguez JM, Martínez MI, Kok J (2002) Pediocin PA-1, a wide-spectrum bacteriocin from lactic acid bacteria. Crit Rev Food Sci Nutr 42(2):91–121

Roy S, Shankar S, Rhim J-W (2019) Melanin-mediated synthesis of silver nanoparticle and its use for the preparation of carrageenan-based antibacterial films. Food Hydrocoll 88:237–246

Sánchez-López E, Gomes D, Esteruelas G, Bonilla L, Lopez-Machado AL, Galindo R et al (2020) Metal-based nanoparticles as antimicrobial agents: an overview. Nanomaterials 10(2):292

Sani MA, Ehsani A, Hashemi M (2017) Whey protein isolate/cellulose nanofibre/TiO_2 nanoparticle/rosemary essential oil nanocomposite film: Its effect on microbial and sensory quality of lamb meat and growth of common foodborne pathogenic bacteria during refrigeration. Int J Food Microbiol 251:8–14

Sani MA, Dabbagh-Moghaddam A, Jahed-Khaniki G, Ehsani A, Sharifan A, Khezerlou A (2023) Biopolymers-based multifunctional nanocomposite active packaging material loaded with zinc oxide nanoparticles, quercetin and natamycin; development and characterization. J Food Meas Charact 17(3):2488–2504

Santos JC, Sousa RC, Otoni CG, Moraes AR, Souza VG, Medeiros EA, Soares NF (2018) Nisin and other antimicrobial peptides: production, mechanisms of action, and application in active food packaging. Innov Food Sci Emerg Technol 48:179–194

Sarwar MS, Niazi MBK, Jahan Z, Ahmad T, Hussain A (2018) Preparation and characterization of PVA/nanocellulose/Ag nanocomposite films for antimicrobial food packaging. Carbohydr Polym 184:453–464

Scannell AG, Ross RP, Hill C, Arendt EK (2000) An effective lacticin biopreservative in fresh pork sausage. J Food Prot 63(3):370–375

Shankar S, Rhim J-W (2015) Amino acid mediated synthesis of silver nanoparticles and preparation of antimicrobial agar/silver nanoparticles composite films. Carbohydr Polym 130:353–363

Siddiqui SA, Zannou O, Bahmid NA, Fidan H, Alamou A-F, Nagdalian AA et al (2022) Consumer behavior towards nanopackaging—a new trend in the food industry. Future Foods 6:100191

Simbine EO, Rodrigues LDC, Lapa-Guimaraes J, Kamimura ES, Corassin CH, Oliveira CAFD (2019) Application of silver nanoparticles in food packages: a review. Food Sci Technol 39:793–802

Simons A, Alhanout K, Duval RE (2020) Bacteriocins, antimicrobial peptides from bacterial origin: overview of their biology and their impact against multidrug-resistant bacteria. Microorganisms 8(5):639

Singh A, Dikshit R, Singh N (2023) Valorization of agro-industrial wastes for biorefinery products. In: Green technologies for industrial waste remediation. Springer Nature Switzerland, Cham, pp 101–123

Suhag R, Kumar N, Petkoska AT, Upadhyay A (2020) Film formation and deposition methods of edible coating on food products: a review. Food Res Int 136:109582

Sung S-Y, Sin LT, Tee T-T, Bee S-T, Rahmat A, Rahman W (2013) Antimicrobial agents for food packaging applications. Trends Food Sci Technol 33(2):110–123

Thapar P, Salooja MK (2022) Bacteriocins: applications in food preservation and therapeutics. In: Lactobacillus—a multifunctional genus. IntechOpen

Valenzuela-Salas LM, Blanco-Salazar A, Perrusquía-Hernández JD, Nequiz-Avendaño M, Mier-Maldonado PA, Ruiz-Ruiz B et al (2021) New protein-coated silver nanoparticles: characterization, antitumor and amoebicidal activity, antiproliferative selectivity, genotoxicity, and biocompatibility evaluation. Pharmaceutics 13(1):65

Vasile C, Baican M (2021) Progresses in food packaging, food quality, and safety—controlled-release antioxidant and/or antimicrobial packaging. Molecules 26(5):1263

Wang L, Hu C, Shao L (2017) The antimicrobial activity of nanoparticles: present situation and prospects for the future. Int J Nanomedicine 12:1227–1249

Wang Y, Wang Y, Sun T, Xu J (2024) Bacteriocins in cancer treatment: mechanisms and clinical potentials. Biomolecules 14(7):831

Wu Z, Huang X, Li Y-C, Xiao H, Wang X (2018) Novel chitosan films with laponite immobilized Ag nanoparticles for active food packaging. Carbohydr Polym 199:210–218

Wu J, Sun Q, Huang H, Duan Y, Xiao G, Le T (2019) Enhanced physico-mechanical, barrier and antifungal properties of soy protein isolate film by incorporating both plant-sourced cinnamaldehyde and facile synthesized zinc oxide nanosheets. Colloids Surf B Biointerfaces 180:31–38

Yin J, Meng Q, Cheng D, Fu J, Luo Q, Liu Y (2020) Mechanisms of bactericidal action and resistance of polymyxins for Gram-positive bacteria. Appl Microbiol Biotechnol 104:3771–3780

Youssef AM, El-Sayed SM, Salama HH, El-Sayed HS, Dufresne A (2015) Evaluation of bionanocomposites as packaging material on properties of soft white cheese during storage period. Carbohydr Polym 132:274–285

Yu H, Liu L, Li X, Zhou R, Yan S, Li C et al (2019a) Fabrication of polylysine based antibacterial coating for catheters by facile electrostatic interaction. Chem Eng J 360:1030–1041

Yu Z, Wang W, Kong F, Lin M, Mustapha A (2019b) Cellulose nanofibril/silver nanoparticle composite as an active food packaging system and its toxicity to human colon cells. Int J Biol Macromol 129:887–894

Zanet V, Vidic J, Auger S, Vizzini P, Lippe G, Iacumin L et al (2019) Activity evaluation of pure and doped zinc oxide nanoparticles against bacterial pathogens and *Saccharomyces cerevisiae*. J Appl Microbiol 127(5):1391–1402

Zhou J, Wang S, Gunasekaran S (2009) Preparation and characterization of whey protein film incorporated with TiO_2 nanoparticles. J Food Sci 74(7):N50–N56

Chapter 11
The Silent Menace of Fumonisins: Cancer Risks, Outbreak Lessons, and Mitigation Pathways

Livpreet Kaur, Gurleen Kaur, Shiv Kumar Dwivedi, and Jitendra Mishra

Abstract Fumonisins, primarily produced by *Fusarium verticillioides*, are a group of mycotoxins commonly found in maize and maize-based products, posing a significant global food safety concern due to their carcinogenic and systemic toxic potential. Fumonisin B_1 (FB_1), the most prevalent and toxic variant, has been classified as a Group 2B carcinogen by the IARC and is strongly associated with esophageal and liver cancers, as well as nephrotoxicity. Mechanistically, FB_1 disrupts sphingolipid metabolism by inhibiting ceramide synthase, resulting in apoptotic imbalance, oxidative stress, and epigenetic dysregulation—hallmarks of carcinogenesis. Epidemiological studies in high-risk regions, such as northeastern Iran and southern Africa, support a strong correlation between dietary FB_1 exposure and elevated cancer incidence. Moreover, FB_1 has been linked to renal carcinogenesis in rodent models, where the kidney's proximal tubules display heightened sensitivity to sphingolipid perturbation. Given its environmental resilience and food chain persistence, effective control of fumonisins demands an integrated farm-to-fork approach. Strategies include pre- and postharvest interventions, biological controls using *Trichoderma* and *Bacillus* species, and emerging chemical detoxification techniques like ozonation and enzymatic hydrolysis. Despite promising advances, challenges remain—particularly in regulatory approval, affordability, and field scalability. This chapter also highlights rapid detection technologies and potential therapeutic responses for fumonisin-induced liver cancer. India's experience with mycotoxin outbreaks underscores the urgency of strengthening food safety surveillance and public health awareness. Collectively, this work reinforces fumonisin's significant public health implications and the necessity for multidisciplinary mitigation efforts that span agriculture, toxicology, biotechnology, and public policy.

L. Kaur · G. Kaur · J. Mishra (✉)
Amity School of Biological Sciences, Amity University Punjab, Mohali, Punjab, India
e-mail: jmishra@pb.amity.edu

S. K. Dwivedi
Department of Environmental Science, School of Earth and Environmental Sciences, Babasaheb Bhimrao Ambedkar University, Lucknow, India

N. K. Arora, J. Mishra (eds.), *Biotechnological Solutions for a Sustainable Future*, Microorganisms for Sustainability 55,
https://doi.org/10.1007/978-981-95-3989-5_11

Keywords Mycotoxin · Fungi · Carcinogens · Biocontrol · Maize

11.1 Introduction

Mycotoxins, toxic secondary metabolites produced by certain species of filamentous fungi (Moss 1992), represent a significant global public health and food safety challenge. These compounds readily contaminate a wide range of agricultural commodities—including cereals, nuts, dried fruits, and milk—both pre- and postharvest, particularly under warm, damp, and humid conditions (Adeyeye 2016). Critically, their inherent chemical stability allows them to persist through many food processing stages, ensuring their entry into the human food chain (Cano et al. 2018). Exposure to mycotoxins can induce mycotoxicosis in both animals and humans, with the severity of health effects dependent on multiple factors. These include the specific mycotoxin(s) involved, the dose and duration of exposure, the age and nutritional status of the individual, and potential synergistic interactions with other chemical agents (Hussein and Brasel 2001; Peraica et al. 1999). Mycotoxins exert diverse toxicities, classified clinically by target organ (e.g., nephrotoxins like ochratoxin A, hepatotoxins like aflatoxins, neurotoxins like fumonisins, immunotoxins like zearalenone) (Bennett and Klich 2003; El-Sayed et al. 2022) and by cellular biologists based on mechanistic effects (e.g., carcinogens, mutagens, teratogens, allergens) (Zain 2011). Among the most concerning mycotoxins are aflatoxins, ochratoxins, trichothecenes, zearalenone, ergot alkaloids, and fumonisins.

The global disease burden attributed to mycotoxin exposure is substantial, with chronic low-level intake posing a major concern due to the carcinogenic potential of several key types (Khan et al. 2024). Due to their established risks, the International Agency for Research on Cancer (IARC) categorizes aflatoxins as Group 1 carcinogens (with conclusive evidence linking them to liver cancer in humans), while fumonisin B_1 is classified as a Group 2B agent (indicating possible carcinogenicity, primarily associated with esophageal cancer (IARC 2012). Mycotoxins, especially aflatoxins, are recognized as critical environmental carcinogens, contributing significantly to liver cancer incidence in high-exposure regions such as sub-Saharan Africa and parts of Asia (Liu and Wu 2010). Countries like India face a rising challenge, with studies detecting unsafe levels of fumonisins, particularly in maize from states like Bihar, Uttar Pradesh, and Maharashtra (Navale and Vamkudoth 2025). Despite these risks, routine mycotoxin monitoring in food is often lacking, public awareness is low, and inadequate storage infrastructure coupled with limited regulatory enforcement heightens exposure risks, especially in local markets and among subsistence farmers (Navale and Vamkudoth 2025).

Cancer, a leading cause of global morbidity and mortality, arises from complex interactions between genetic predisposition, lifestyle factors (e.g., smoking, diet), infections, and environmental exposures (Mahmood and Srivastava 2022). Mycotoxins constitute a significant class of pervasive environmental carcinogens within the global food supply, contributing measurably to the cancer burden, particularly in regions with high staple food contamination and weak regulatory frameworks.

This chapter delves into fumonisins, primarily produced by *Fusarium* species. It examines their chemical structure, toxicological mechanisms, and synthesizes epidemiological, in vivo, and in vitro evidence supporting their role in carcinogenesis. Additionally, this chapter highlights current and emerging strategies to reduce fumonisin contamination across the food chain, as well as advances in detection technologies for improved monitoring.

11.2 Fumonisins: Structure and Type

Fumonisin B_1 is a mycotoxin primarily produced by *Fusarium verticillioides* and *Fusarium proliferatum*, with other *Fusarium* species (e.g., *F. nygamai*, *F. oxysporum*) also capable of its synthesis (Frisvad et al. 2007). While *Fusarium* is the dominant producer, rare cases of fumonisin B_1 production by *Aspergillus* species (e.g., *A. niger*) have also been reported (Frisvad et al. 2007). Fumonisin was first identified in the late 1980s following reports of a fatal neurological disease characterized by brain tissue liquefaction known as equine leukoencephalomalacia (ELEM) in horses and pulmonary edema in pigs consuming moldy corn. The compound was discovered in *F. verticillioides* (formerly *F. moniliforme*) cultures in 1988 (Gelderblom et al. 1988). *Fusarium* belongs to the Nectriaceae family and thrives as a saprophyte in plants and soil (Kamle et al. 2019). Since then, more than 15 fumonisin homologs have been identified, grouped into four main series based on structural variations: fumonisin A (FA), B (FB), C (FC), and P (FP). Among them, fumonisin B-series (particularly FB_1, FB_2, and FB_3) are the most studied and widely occurring, with FB_1 being the most toxic and prevalent form (Kamle et al. 2019). Fumonisin P-series (FP), isolated from *Aspergillus niger*, although structurally distinct, are included under the fumonisin umbrella due to their similar backbones and toxicological relevance (Frisvad et al. 2007).

Fumonisins (Fig. 11.1) consist of several key functional groups: two methyl groups ($-CH_3$), one amine group ($-NH_3$), one to four hydroxyl groups (–OH), and two tricarboxylic ester groups attached to a polyketide-derived backbone (Kamle et al. 2019). Fumonisin B_1 (FB_1) is a diester formed by propane-1,2,3-tricarboxylic acid (TCA) and 2-amino-12,16-dimethyl-3,4,10,14,15-pentahydroxy-eicosane. Hydroxyl groups at the C-14 and C-15 positions are involved with the carboxyl groups (–COOH) of TCA to form ester linkages. Fumonisin B_2 and B_3 are the C-5 and C-10 dehydroxy analogs of FB_1 (Shephard 1998).

Fumonisin B_1 comprises 70% of the total fumonisins found in food and feed samples. These mycotoxins primarily target the liver and kidneys, disrupting sphingolipid metabolism (Voss et al. 2007). Chronic exposure to fumonisins has been linked to several serious health conditions in humans and animals. In humans, these mycotoxins are associated with neural tube defects (NTDs) in developing fetuses,

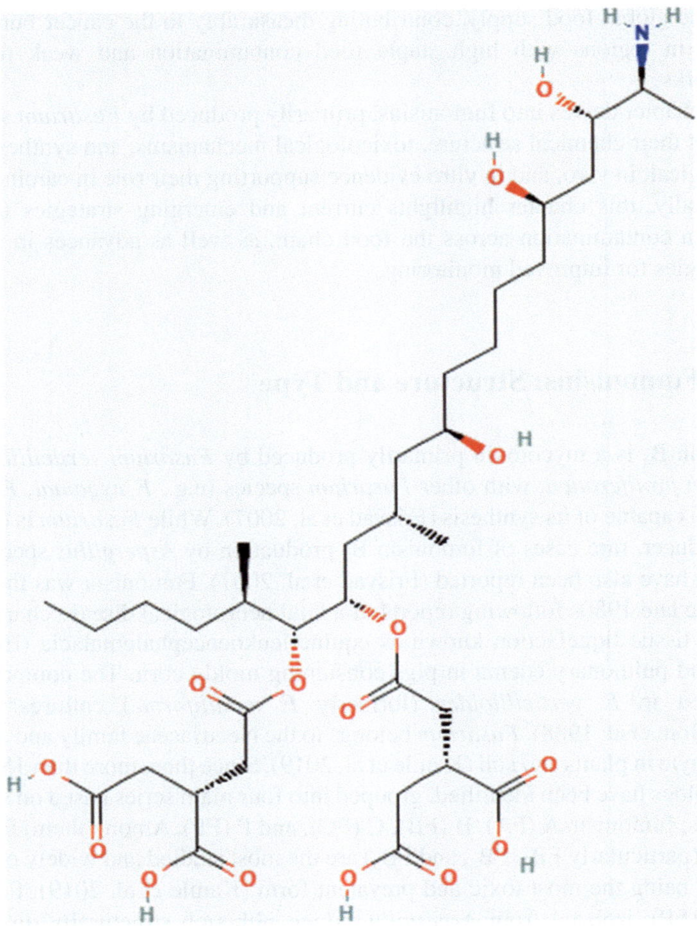

Fig. 11.1 Chemical structure of Fumonisin B_1 (C34H59NO15). IUPAC Name: (2R)-2-[2-[(5R,6R,7S,9S,11R,16R,18S,19S)-19-amino-6-[(3R)-3,4-dicarboxybutanoyl] oxy-11,16,18-trihydroxy-5,9-dimethylicosan-7-yl]oxy-2-oxoethyl] butanedioic acid. This molecule consists of a long carbon backbone with multiple hydroxyl (–OH), amino (–NH_2), and carboxyl (–COOH) groups that contribute to its biological activity. It includes a tetrahydrofuran ring, a central carbon chain, and a terminal tricarboxylic acid group. Fumonisin B_1 is characterized by its stereochemistry, with functional groups positioned at specific chiral centers, influencing its toxicity. The compound primarily targets sphingolipid metabolism, disrupting normal cellular functions, and is classified as a possible human carcinogen (Group 2B) by the IARC. (Source: PubChem Compound Summary for CID 2733487, Fumonisin B_1. Retrieved June 9, 2025, from https://pubchem.ncbi.nlm.nih.gov/compound/Fumonisin-B1)

primarily through their disruption of sphingolipid metabolism and interference with folate absorption—a critical nutrient for proper neural tube formation (Ross et al. 1992; Marasas et al. 2004). Studies from high-exposure regions like South Africa and Guatemala have demonstrated this connection. Fumonisins also compromise

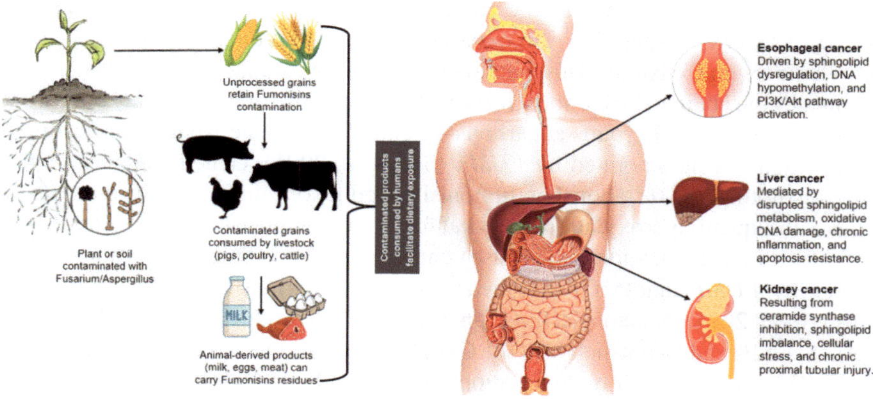

Fig. 11.2 How fumonisins enter the food chain. Fumonisins contamination, starting from fungal-infected crops (*Fusarium*/*Aspergillus*), poses risks to human health. Contaminated grains consumed by livestock lead to toxin residues in animal products (milk, eggs, meat), which, when ingested by humans, cause dietary exposure. The graphic highlights three cancer types linked to Fumonisins: esophageal cancer (via sphingolipid dysregulation, DNA hypomethylation, and PI3K/Akt activation), liver cancer (through oxidative DNA damage, chronic inflammation, and apoptosis resistance), and kidney cancer (due to ceramide synthase inhibition and sphingolipid imbalance)

immune function by altering cytokine production and impairing macrophage activity, increasing susceptibility to infections and porcine pulmonary edema (Harrison et al. 1990). Furthermore, regular consumption of fumonisin-contaminated maize increases the risk of esophageal and liver cancers, as well as growth retardation in children (Sun et al. 2007; Shirima et al. 2015) (Fig. 11.2).

11.3 Fumonisin B_1 Biosynthesis

The production of fumonisins is a highly regulated biochemical process involving a specific cluster of genes and enzymes encoded by the *Fusarium* genome (Proctor et al. 1999). Research has established that the genes responsible for fumonisin biosynthesis are located in a single locus: 17 gene FUM cluster (Alexander et al. 2009), a well-defined genomic region containing 15–16 core biosynthetic, transporter, and regulatory genes in *Fusarium* species like *F. verticillioides* and *F. proliferatum*; some strains of *F. oxysporum* and other species also possess functional clusters (Proctor et al. 2004, 2013; Brown and Proctor 2016). The first step is catalyzed by *FUM1*, which encodes a polyketide synthase (PKS). This enzyme assembles an 18-carbon linear polyketide chain using acetyl-CoA as a starter and malonyl-CoA (and methylmalonyl-CoA) as extenders. Critically, this initial polyketide product lacks the amine group characteristic of fumonisins (Proctor et al. 1999; Bojja et al. 2004).

The incorporation of the amine group occurs in the next key step, catalyzed by *FUM8*. *FUM8*, resembling an oxoamine synthase, condenses the 18-carbon polyketide with L-alanine. This reaction introduces the amine group at carbon-2 and elongates the chain to 20 carbons, forming the first stable fumonisin backbone intermediate. The methyl groups at carbons-12 and -16 originate from methylmalonyl-CoA extenders incorporated by *FUM1* (Seo et al. 2001). Genetic studies confirm that disrupting *FUM8* abolishes fumonisin production, underscoring its essential role. Subsequent hydroxylations are vital for structural integrity and toxicity. *FUM7* (a cytochrome P450 monooxygenase) catalyzes hydroxylation at carbon-15, followed by *FUM6* (another P450) hydroxylating at carbon-14 (Uhlig et al. 2012; Susca et al. 2014). Mutants lacking functional *FUM6* produce FB_5 and FB_6 (deoxyfumonisin B_1 and B_2), which lack the C-14 hydroxyl group and exhibit significantly reduced toxicity. Further hydroxylations determine the specific B-series analog produced: FUM2 encodes a P450 responsible for hydroxylation at carbon-10, while *FUM3* encodes a P450 hydroxylating at carbon-5 (Proctor et al. 2003). The presence or absence of these hydroxyl groups yields fumonisins B_1 (OH at C5 and C10), B_2 (no OH at C10), B_3 (no OH at C5), or B_4 (no OH at C5 or C10), with FB_1 being the most prevalent and toxic.

The final critical modifications involve the addition of tricarballylic acid (TCA) ester groups. This complex process requires multiple enzymes, including *FUM9* (a predicted dehydrogenase), *FUM10* (a cytochrome P450 monooxygenase), *FUM13* (a transferase), and other cluster genes (like *FUM18*/*FUM19*) implicated in TCA synthesis or attachment (Proctor et al. 2008). These ester groups are essential for the compound's biological activity. Once synthesized, *FUM12*, which encodes a major facilitator superfamily (MFS) transporter protein, plays a crucial role in exporting FB_1 from the fungal cell, preventing intracellular accumulation and releasing the toxin into the environment to contaminate plants or food (Bojja et al. 2004; Sweeney and Dobson 1999).

The expression of the FUM cluster is tightly regulated. The pathway-specific transcription factor FUM21, a Zn(II)2Cys6 protein encoded within the cluster, is essential for activating the biosynthetic genes (Brown et al. 2007). Additionally, global regulators responsive to environmental cues integrate external signals: these include phosphate acquisition Control 1 (PacC/Pac1 pH response), nitrogen response area (AreA, nitrogen regulation), carbon repression adaptor (CREA, carbon catabolite repression), and Stunted A (STUA, developmental regulation (Flaherty and Woloshuk 2004; Shim and Woloshuk 2001). Environmental factors significantly influence production. Carbon and nitrogen source availability, water activity (humidity), temperature, and pH profoundly affect FUM gene expression and fumonisin output (Marín et al. 2013). Plant-derived signals during infection also play a major regulatory role. This complex regulation allows *Fusarium* to produce toxins in response to favorable or stressful conditions, explaining contamination variability in agriculture. Studies confirm that disruptions in specific *FUM* genes (e.g., *FUM1*, *FUM6*, *FUM8* and *FUM21*) abolish or drastically alter fumonisin production (Proctor et al. 2004).

11.4 Fumonisins and Cancer

11.4.1 Esophageal Cancer

Esophageal cancer (EC) poses a substantial global health challenge, with an estimated 604,000 new cases reported worldwide in 2020 (Sung et al. 2021). Projections suggest an alarming rise of over 50% in incidence by 2040, potentially exceeding 900,000 annual cases, driven by aging populations, persistent risk factors (e.g., tobacco/alcohol use, dietary carcinogens), and regional disparities in early detection (International Agency for Research on Cancer (IARC) 2002). The disease exhibits striking geographical variation, with high incidence "corridors" in Eastern Asia (notably China) and Eastern Africa, where specific subtypes (e.g., esophageal squamous cell carcinoma) are linked to mycotoxin exposure, including Fumonisins (Chen et al. 2022; Morgan et al. 2022). While established risk factors include tobacco use, alcohol consumption, GERD, obesity, and dietary habits, exposure to environmental carcinogens like fumonisins is increasingly recognized as a contributory factor, particularly in specific high-incidence regions (Kamangar et al. 2009). Epidemiological studies strongly associate fumonisin B_1 (FB_1) exposure with elevated rates of esophageal squamous cell carcinoma (ESCC). Research from high-ESCC areas in Iran and China demonstrates this link: staple foods like corn and rice show significantly higher FB_1 contamination levels in these regions. Crucially, biomarker studies in China revealed urinary FB_1 levels approximately threefold higher in ESCC cases compared to controls (Alizadeh et al. 2012; Xue et al. 2019). These consistent geographical and biomarker patterns suggest that chronic dietary FB_1 exposure contributes to esophageal carcinogenesis, especially within vulnerable populations.

Unlike many other carcinogens, fumonisin B_1 (FB_1) exerts toxicity primarily through its structural similarity to sphingoid bases (sphinganine [Sa] and sphingosine [So]) (Gardner et al. 2016). This mimicry allows FB_1 to competitively inhibit key enzymes in the de novo ceramide biosynthesis pathway, specifically ceramide synthases (e.g., CerS2 and CerS5) (Riley et al. 1996). This inhibition disrupts sphingolipid homeostasis, leading to the pathological accumulation of free sphinganine, sphingosine, and their bioactive phosphorylated derivatives (sphinganine-1-phosphate [Sa1P] and sphingosine-1-phosphate [So1P]). As ceramide and its metabolites are critical regulators of apoptosis, proliferation, and differentiation, their depletion shifts the cellular balance toward pro-survival signaling and uncontrolled proliferation (Riley et al. 1996). Furthermore, FB_1-induced accumulation of sphingoid bases (e.g., sphingosine [So] and sphingosine-1-phosphate [S1P]) promotes oxidative stress and membrane dysfunction. In vivo studies confirm that this buildup triggers excessive reactive oxygen species (ROS) production, leading to DNA strand breaks, lipid peroxidation, and chromatin instability (Yu et al. 2021; Zitomer et al. 2009). The mechanism involves S1P-mediated activation of NADPH oxidases and mitochondrial dysfunction, exacerbating cellular damage (Arumugam et al. 2021). Both oxidative DNA damage and chromatin disruption are well-established initiating events and contributors to carcinogenesis (Klaunig and Wang 2018).

Beyond its metabolic disruption, FB_1 induces epigenetic alterations conducive to oncogenesis. A key mechanism involves the dysregulation of histone acetylation (Sugiyama et al. 2021). Studies in primary human esophageal epithelial cells (HEECs) demonstrate that FB_1 exposure upregulates class IIa histone deacetylases (HDACs), particularly HDAC4 and HDAC6 (Yu et al. 2021). These HDACs catalyze the removal of acetyl groups from histones, promoting chromatin compaction and transcriptional repression of tumor suppressor genes (e.g., p21, PTEN) (Li and Seto 2016). This FB_1-induced hyper-deacetylation disrupts cellular checkpoints that regulate proliferation and apoptosis, thereby promoting malignant transformation (Yu et al. 2021). Mechanistically, FB_1's disruption of sphingolipid metabolism synergizes with HDAC overexpression to amplify oncogenic signaling, such as PI3K/Akt activation (Yu et al. 2021).

Fumonisin B_1 (FB_1) exerts its carcinogenic effects through multiple interconnected mechanisms, including sphingolipid disruption, epigenetic modifications, and direct activation of the PI3K/Akt/mTOR pathway, a central oncogenic signaling cascade that regulates cell survival, proliferation, and metabolic reprogramming in cancers (Fresno Vara et al. 2004; Yu et al. 2021). By inhibiting ceramide synthase, FB_1 creates a sphingolipid imbalance characterized by reduced proapoptotic ceramide and elevated pro-survival sphingosine-1-phosphate (S1P), which subsequently dysregulates Akt phosphorylation (Arumugam, et al. 2021). This hyperactivation of PI3K/Akt signaling suppresses apoptosis through Bcl-2 upregulation and Bad inactivation while promoting neoplastic transformation, particularly in high-exposure tissues such as the esophagus and liver (IARC 2002). Emerging evidence suggests that FB_1's activation of PI3K/Akt may be further amplified by its induction of HDAC expression and activity (Arumugam et al. 2020). Once activated, Akt drives oncogenesis by promoting cell cycle progression through the inactivation of tumor suppressors like FoxO transcription factors and by impairing the DNA damage response through CHK1 dysfunction, which collectively reduce DNA repair fidelity and increase genomic instability (Arumugam et al. 2020). These coordinated mechanisms create a permissive environment for mutation accumulation, ultimately contributing to FB_1-associated carcinogenesis through sustained proliferative signaling, evasion of apoptosis, and genomic instability.

11.4.2 Liver Cancer

High dietary exposure to fumonisin B_1 (FB_1) has been associated—rather than definitively proven to be causative—with increased hepatocellular carcinoma (HCC) incidence in several maize-consuming populations, including the Transkei region of South Africa, certain provinces of China, and parts of Central America (WHO 2002). Comparable ecological and case-control studies in Qidong (China) and Guatemala corroborate this association while also highlighting confounding factors such as aflatoxin B_1 exposure and chronic hepatitis B infection (Alvarez et al. 2020; Chu et al. 2018).

FB_1 is a highly polar molecule; less than 4% of an oral dose is absorbed, and ≥80% is excreted unchanged in feces within 72 h (Norred et al. 1996). Rodent studies show that ~0.3–1% of unmetabolized FB_1 persists in tissues, with kidneys accumulating more than the liver (Harrer et al. 2014). There is no convincing evidence of classic phase I/II hepatic biotransformation in humans. Gut microbiota can hydrolyze FB_1 to partially de-esterified forms (HFB_1), which still retain some toxicity (Harrer et al. 2014). Ceramide synthase–mediated N-acyl-FB_1 metabolites are detectable in liver and kidneys, but their in vivo toxicological significance remains under investigation (Zhang et al. 2024).

The primary molecular target of FB_1 is ceramide synthase, leading to the accumulation of free sphinganine and sphinganine-1-phosphate, perturbing the sphingolipid rheostat and triggering mitogenic and stress-response signaling (Faure and Molino 2015). An elevated sphinganine:sphingosine ratio in serum or urine is therefore a validated biomarker of FB_1 exposure (Régnier et al. 2017). Prolonged FB_1 exposure induces oxidative stress in hepatocytes; FB_1-treated cells generate excessive reactive oxygen species (ROS) and undergo lipid peroxidation. ROS elevate the DNA adduct 8-hydroxy-2′-deoxyguanosine (8-OH-dG), a marker of oxidative DNA damage (Chen et al. 2021). Activation of the Nrf2–ARE pathway has been observed as an adaptive response but may become overwhelmed under chronic exposure (Arumugam et al. 2019).

FB_1 upregulates pro-inflammatory cytokines (TNF-α, IL-1β, IL-8) and activates NF-κB signaling in hepatocytes (Chen et al. 2021; Régnier et al. 2017; Mahmoodi et al. 2012). Chronic NF-κB activation fosters a microenvironment conducive to compensatory cell proliferation and survival, thereby promoting tumorigenesis (Aggarwal and Sung 2011). FB_1 modulates apoptosis in a dose-dependent manner. High doses precipitate apoptosis, whereas low, chronic doses inhibit it. In HepG2 cells, FB_1 upregulates inhibitor-of-apoptosis proteins (IAPs) such as BIRC family members and downregulates the proapoptotic mitochondrial protein Smac/DIABLO, reducing caspase activity (Chuturgoon et al. 2015). Inhibition of ceramide synthase concurrently lowers proapoptotic ceramide levels, reinforcing apoptosis evasion (Mullen et al. 2011). FB_1 decreases hepatic miR-27b and thereby increases CYP1B1 expression in HepG2 cells (Chuturgoon et al. 2014). Upregulated CYP1B1 can bio-activate environmental pro-carcinogens such as aflatoxin B_1, suggesting synergistic interactions in regions where multiple mycotoxins co-occur. Emerging evidence also points to Wnt/β-catenin pathway activation and epigenetic alterations—including global DNA hypomethylation—contributing to hepatocarcinogenesis (Song et al. 2024).

The International Agency for Research on Cancer (IARC) classifies FB_1 as Group 2B (possibly carcinogenic to humans), and the JECFA provisional maximum tolerable daily intake (PMTDI) is 2 μg kg^{-1} bw day^{-1} for the sum of FB_1, FB_2, and

FB_3.[1] Regulatory limits for raw maize in the European Union are set at 2000 μg kg^{-1}, while Codex Alimentarius provides similar guidance (Shephard et al. 2019). Effective risk-mitigation strategies include nixtamalization, density segregation, biological control of *Fusarium verticillioides*, and breeding of resistant maize varieties (Ortega-Beltran et al. 2023). Chronic-diet studies show that 50 ppm FB_1 in male BD IX rats leads to HCC incidences exceeding 60%, and National Toxicology Program studies (TR-496) confirm dose-dependent liver tumor formation in $B6C3F_1$ mice (Shephard et al. 2019).

11.4.3 Kidney Cancer

The kidney is a primary target organ for fumonisin B_1 (FB_1) toxicity, exhibiting high sensitivity to both acute and chronic exposure. Studies indicate that male rodents develop severe renal injury at lower doses and earlier timepoints compared to females, demonstrating a significantly greater sensitivity (Riley et al. 1994). Beyond impairing nephron function, FB_1 promotes renal carcinogenesis through sustained cytotoxicity, regenerative hyperplasia, and progression to neoplasia; it can stimulate cell proliferation, damage DNA, and ultimately lead to renal tubular adenomas and carcinomas following long-term exposure in rodent models (Voss et al. 2002). Chronic FB_1 exposure is carcinogenic in rodents, consistently inducing kidney tumors (Voss et al. 2002).

FB_1 is poorly absorbed systemically after ingestion, resulting in relatively low distribution to organs like the liver and kidneys. Within the renal tissue, FB_1 undergoes metabolism via N-acylation with long-chain fatty acids, forming N-acyl-FB_1 derivatives. These ceramide-like metabolites, found at elevated levels in rat kidneys, exhibit cytotoxic effects in vitro (Harrer et al. 2014). The nephron's structure, particularly the proximal tubule, heightens the renal vulnerability. Approximately 60–70% of the glomerular filtrate is reabsorbed in the proximal tubule. While water and soluble nutrients follow this bulk reabsorption, toxins like fumonisin B_1 are typically taken up via specific organic anion/cation transporters, leading to their accumulation in the proximal tubular cell (Otani et al. 2017). This reabsorption concentrates FB_1 intracellularly, prolonging exposure and increasing the potential for disrupting lipid signaling pathways (e.g., sphingolipid metabolism) and cellular redox balance (Domijan et al. 2007).

Histopathological progression in FB_1-exposed kidneys begins with cytoplasmic swelling and degeneration of the proximal tubular epithelium. Electron microscopy reveals characteristic ultrastructural damage, including mitochondrial swelling, endoplasmic reticulum disruption, and loss of brush border microvilli (illustrative findings supported by Müller et al. 2012). Advanced lesions include tubular dilation, interstitial inflammation, glomerular basement membrane thickening, and,

[1] https://inchem.org/documents/jecfa/jecmono/v47je01.htm

notably, karyomegaly (enlargement of tubular cell nuclei), a recognized preneoplastic alteration (Müller et al. 2012). Chronic exposure drives the sequence from initial cellular injury and adaptive hyperplasia toward the development of renal tubular adenomas and carcinomas (Voss et al. 2002). FB_1-induced renal neoplasms in rats are frequently aggressive, exhibiting poor differentiation and local tissue invasion (Bucci et al. 1998).

The pronounced renal sensitivity to FB_1 stems largely from its potent inhibition of ceramide synthase (key enzyme: sphinganine N-acyltransferase), the committed step in de novo ceramide and complex sphingolipid biosynthesis (Riley et al. 2001). This inhibition causes significant accumulation of sphingoid bases (sphinganine, sphingosine) and depletion of complex sphingolipids, dramatically elevating the sphinganine-to-sphingosine (Sa/So) ratio. The kidney consistently exhibits the greatest magnitude of increase in the Sa/So ratio following FB_1 exposure, establishing it as a highly sensitive and specific biomarker of effect in preclinical studies (Riley et al. 1994). This disruption of sphingolipid homeostasis alters membrane properties (e.g., fluidity, function), impairs calcium signaling and ion transport, and contributes directly to nephron dysfunction (Riley et al. 1996). Furthermore, elevated sphingoid bases induce organelle stress, promoting lysosomal membrane permeabilization, endoplasmic reticulum stress, and mitochondrial dysfunction within renal cells (Riley et al. 1996).

Direct epidemiological evidence linking FB_1 exposure to human kidney cancer remains limited (Rubagotti et al. 2006). Studies in FB_1-exposed animals identify elevated urinary biomarkers of proximal tubular damage, such as β2-microglobulin and N-acetyl-β-D-glucosaminidase (NAG) (Bucci et al. 1998). Therefore, monitoring such biomarkers, potentially combined with the Sa/So ratio (though technically challenging in humans), could provide indicators of FB_1-induced renal tubular injury in exposed populations; however, their specificity and utility for routine human risk assessment require further validation (Van Den Brand et al. 2022). Given the high consumption of maize, a common vehicle for fumonisins, in certain regions, chronic low-level exposure poses a plausible risk for FB_1-associated nephrotoxicity in humans; however, extrapolating the observed rodent carcinogenicity to human kidney cancer risk necessitates considerable caution and is not yet supported by direct evidence (Torres et al. 2015).

11.5 Mycotoxins Consumption Outbreak in India: Learning from the Past

India's history with mycotoxin-related illnesses offers stark, life-saving lessons on food safety failures, particularly concerning aflatoxins. One of the most severe documented outbreaks globally occurred in 1974 across villages in Rajasthan and neighboring Gujarat. This tragedy stemmed from the consumption of maize (corn) heavily contaminated with *Aspergillus flavus* and exceptionally high levels of aflatoxin B_1 (reaching up to 15,000 μg/kg, far exceeding safe limits) (Krishnamachari et al. 1975a; Bhat and Krishnamachari 1977). The maize had been harvested after unseasonal rains and stored under damp, unventilated conditions, creating an ideal environment for toxigenic mold proliferation (Krishnamachari et al. 1975a; Bhat and Krishnamachari 1977). Affected individuals, primarily from impoverished tribal communities with limited alternative food sources, rapidly developed acute aflatoxicosis characterized by severe toxic hepatitis leading to liver failure. Symptoms included intense jaundice (yellowing of skin and eyes), painful abdominal swelling due to ascites (fluid accumulation), portal hypertension, gastrointestinal hemorrhage, and lower limb edema (Krishnamachari et al. 1975a). The outbreak resulted in at least 397 documented cases and 108 fatalities within a few months, with a disturbingly high mortality rate among young adults (Krishnamachari et al. 1975b). Decades later, in 1995, a similar but smaller outbreak struck Adivasi (tribal) villages in the Panchmahals district of Gujarat (Reddy and Raghavender 2007). Here, the culprit was rain-damaged sorghum found to contain high levels of aflatoxin B_1 (Reddy and Raghavender 2007). Victims presented with identical symptoms of acute liver failure—jaundice, ascites, portal hypertension, and hemorrhagic necrosis of the liver—leading to 50 illnesses and 8 deaths (Reddy and Raghavender 2007).

These acute outbreaks shared critical, preventable factors: reliance on mold-susceptible staple grains (maize, sorghum) as dietary mainstays, crop damage due to erratic weather (unseasonal rains), inadequate postharvest drying and storage infrastructure (damp, warm conditions), and profound poverty that left vulnerable populations with no choice but to consume visibly spoiled grain despite awareness of the risks (Sharma and Parisi 2017). While these dramatic events highlight the potential for acute poisoning, the far greater, insidious public health burden in India stems from chronic, low-level dietary exposure to aflatoxins and other mycotoxins (e.g., fumonisins, ochratoxin A) (Sharma and Parisi 2017). Contamination is pervasive in staples like groundnuts (peanuts), maize, sorghum, millets, spices (especially chilies and turmeric), and dried coconut (Reddy et al. 2010). Regular, long-term intake of even moderate levels of aflatoxin B_1 is a major risk factor for hepatocellular carcinoma (HCC—liver cancer). Wenndt et al. (2020) investigated aflatoxin B_1 contamination in 160 Indian households and found 30–80% had contaminated food, especially groundnuts and maize. The key risk factors included postharvest washing, sack storage, and groundnut presence. Surprisingly, larger landholdings showed higher contamination. A household-level risk index was developed with moderate predictive accuracy. Furthermore, substantial epidemiological evidence links chronic childhood exposure to aflatoxins with impaired growth, stunting, and weakened immune function, contributing to long-term developmental setbacks (Visser et al. 2020). The chronic effects are rarely diagnosed as "mycotoxicosis" but represent a significant, silent contributor to India's disease burden, particularly affecting resource-poor communities with limited dietary diversity and suboptimal storage conditions.

11.6 Prevention Strategies

Fumonisins—carcinogenic mycotoxins linked to hepatotoxicity and nephrotoxicity—pose a critical global food safety challenge. As contamination can occur at any stage, a multi-barrier approach integrating agricultural, technological, and regulatory strategies is essential for effective prevention.

11.6.1 Pre- and Postharvest Approaches

Effective control of fumonisin begins with robust preharvest strategies, as environmental conditions such as high humidity (>70%), warm temperatures (22–28 °C), and insect damage create ideal circumstances for *Fusarium* infection and fumonisin synthesis (Duvick 2001). Genetic interventions like deploying Bt maize hybrids can significantly reduce fumonisin levels (by 60–90%) by limiting insect damage that facilitates fungal entry (Duvick 2001). This can be further enhanced by marker-assisted breeding to improve host resistance (Li et al. 2024). Agronomic practices play a vital role—rotating crops (especially with non-cereal crops), optimizing planting and harvesting windows to minimize field exposure, maintaining balanced fertilization (particularly nitrogen), and eliminating infected residues are all critical steps in breaking the pathogen's life cycle (Li et al. 2024). Integrated pest management (IPM) is indispensable in this context, combining insecticide application, biological control agents, and monitoring tools like pheromone traps to prevent kernel damage from insect vectors (Alberts et al. 2016).

However, even the best preharvest practices can be undermined by poor postharvest handling. To prevent fungal growth and toxin formation, maize should be dried immediately after harvest to below 14% moisture content—especially important in humid regions (Magan 2006). While sun-drying remains a practical option for smallholder farmers, controlled drying and proper storage conditions are ideal. Poor ventilation and residual moisture can create microclimates conducive to *Fusarium* persistence and fumonisin accumulation. Hermetic storage systems, such as PICS Bag™ and airtight silos, help maintain grain quality by limiting oxygen availability (Kabak et al. 2006). Additionally, simple techniques like manual sorting and washing of kernels can further reduce toxin levels, particularly in rural or resource-limited settings (Alberts et al. 2016). Reinforcing these technical interventions with strict enforcement of mycotoxin regulatory thresholds (e.g., EU limits of 800–4000 μg/kg) and improved supply chain traceability ensures a more comprehensive and sustainable fumonisin control strategy (European Commission 2007; Liu et al. 2025).

11.6.2 Biological Control Strategies

Biological control strategies offer a sustainable and eco-friendly alternative to chemical methods by incorporating beneficial microorganisms to suppress *F. verticillioides* and inhibit fumonisin production. These microbes act through mechanisms such as competitive exclusion, antibiosis, mycoparasitism, and induced systemic resistance (ISR), effectively disrupting fungal colonization and toxin biosynthesis (Lee et al. 2023). Notably, *Trichoderma* species (e.g., *T. harzianum*, *T. asperellum*) secrete hydrolytic enzymes like chitinases and β-1,3-glucanases that degrade fungal cell walls while producing antifungal metabolites such as peptaibols (Markovich and Kononova 2003). Field applications of *Trichoderma*-based biofungicides have led to significant reductions in fumonisin B_1 levels (Altinok 2009; Ferrigo et al. 2014; Alberts et al. 2016). Likewise, endophytic bacteria such as *Bacillus subtilis* and *Pediococcus pentosaceus* colonize maize tissues and produce bioactive compounds—lipopeptides (e.g., surfactin, iturin) and bacteriocins—that inhibit fungal growth and reduce toxin accumulation in planta (Alberts et al. 2016). Recent insights also highlight the role of bacterial volatile organic compounds (VOCs), like 2,3-butanediol, in priming maize defense pathways, thereby further suppressing fumonisin biosynthesis (Yasmin et al. 2021).

For successful implementation, formulation technologies such as *polymer-based encapsulation* are crucial to protect microbial viability under variable field conditions (John et al. 2011). *Seed coating* with microbial consortia—e.g., *Trichoderma* combined with *Bacillus*—can boost root colonization by 3–5 times compared to single-strain applications, providing synergistic and prolonged protection (Raza et al. 2024). When integrated with resistant maize hybrids, these biological agents have demonstrated higher reductions in fumonisin contamination (Chiuraise et al. 2016).

However, several challenges hinder large-scale adoption. Biological efficacy can decline under environmental stressors such as drought or extreme temperatures. Regulatory approvals for microbial biocontrol agents often involve complex and time-consuming processes. Additionally, *shelf-life limitations*, particularly for non-spore-forming strains like *Pediococcus*, present formulation and storage challenges that need to be addressed through technological innovation.

11.6.3 Chemical Approaches for Fumonisin Detoxification

When contamination occurs, chemical interventions can degrade or sequester fumonisins, though their application requires stringent safety validation. Alkaline processing (nixtamalization) remains a cornerstone method, where lime treatment (calcium hydroxide) hydrolyzes fumonisin B_1's tricarballylic acid groups, reducing toxicity by 50–80% in maize products (Kabak et al. 2006). However, this process demands precise control to prevent nutritional degradation (e.g., loss of niacin) or formation of potentially hazardous byproducts like *Fusarium* reaction metabolites. In animal feed, mycotoxin-binding adsorbents—particularly bentonite clays and hydrated sodium calcium aluminosilicate (HSCAS)—applied at inclusion rates of 0.5–2% can effectively sequester fumonisins in the gastrointestinal tract, reducing their bioavailability by 40–60% (Piva et al. 2005; Jouany 2007). HSCAS, a phyllosilicate clay, exhibits high affinity for fumonisin B_1 due to its porous structure and cation-exchange capacity, forming stable complexes that prevent intestinal absorption (Siri-anusornsak et al. 2023). Bentonite achieves similar effects through electrostatic interactions, though its efficacy varies with clay mineralogy (Diaz et al. 2004). These reductions in bioavailability correlate with mitigated toxicity in livestock, including decreased sphinganine-to-sphingosine ratios in pigs and poultry (Boudergue et al. 2009). However, binding efficiency depends on pH, toxin concentration, and feed composition, with optimal performance in diets containing <10 ppm fumonisins (Kabak et al. 2006; Authority et al. 2018). Despite their efficacy, these clays are prohibited in human food due to risks of heavy metal contamination (e.g., lead, dioxins) and nutrient interference.

Emerging chemical strategies present promising advancements in the precise and safe detoxification of fumonisins in food and feed systems. Among these, *ozonation*—applied either in gaseous or aqueous form—has demonstrated high efficacy, achieving over 90% degradation of fumonisins through oxidative cleavage. When applied at optimized doses (20–50 ppm) for short durations (30 min), it effectively reduces toxin levels without significantly compromising the nutritional quality of grains (Ribeiro et al. 2021). *Enzymatic detoxification*, particularly using recombinant carboxylesterases like FumD, offers a targeted approach by hydrolyzing the ester bonds of fumonisin B_1 (FB_1), converting it into the much less toxic hydrolyzed form (HFB_1). This method has been successfully validated in maize-based food products (Alberts et al. 2021). Additionally, *advanced oxidation processes (AOPs)*—which synergistically combine UV light with hydrogen peroxide or ozone—accelerate fumonisin degradation while minimizing damage to the grain matrix, enhancing both safety and quality (Sujayasree et al. 2022).

Despite their potential, these technologies face critical implementation barriers. Regulatory approval remains a key hurdle, especially due to the need for thorough toxicological evaluation of detoxification byproducts to ensure consumer safety. Moreover, the cost and scalability of these interventions pose significant challenges for widespread industrial adoption, particularly in low-resource settings. Lastly, consumer perception and acceptance of chemically treated food demand transparent labeling and effective public communication to build trust and ensure market viability.

11.6.4 Future in Rapid Detection and Innovative Treatment Strategies

Rapid detection of fumonisins is advancing through biosensors and predictive modeling, though significant validation gaps persist. Nucleic acid aptamers (e.g., gold-nanoparticle electrodes with 0.14 ng/mL LOD; Naghshbandi et al. 2023) and antibody-based microfluidics (e.g., PDMS chips at 97 pg/mL; Lu and Gunasekaran 2019) enable ultrasensitive FB_1 quantification, while Ag-CeO_2 nanocomposites extend detection limits to 1.5×10^{-3} ng/mL (Dhiman et al. 2021). However, these platforms lack field validation in complex grain matrices and face scalability barriers due to cost and shelf-life constraints. Machine learning models show promise in predicting contamination using climate/crop data (Gbashi et al. 2024), yet their real-world utility hinges on integration with low-cost IoT sensors and edge computing—currently unrealized.

While fumonisins are classified as *possibly carcinogenic to humans* (IARC Group 2B), their strongest scientific link is to liver cancer (hepatocellular carcinoma, or HCC), where they disrupt the body's natural fat metabolism pathways. This connection is crucial because it guides which modern cancer treatments are most relevant for fumonisin-exposed populations. For advanced HCC, breakthrough combination therapies have emerged: pairing immune-boosting drugs like atezolizumab (which blocks cancer's "invisibility cloak") with bevacizumab (which starves tumors by cutting off blood supply) significantly extends life compared to older drugs like sorafenib (Hack et al. 2020). Another promising drug, lenvatinib, works by blocking signals that fuel tumor growth—signals that may become overactive *because* of fumonisin's damage to liver cells.

New technologies aim to make treatments safer and more targeted. For example, encapsulating natural compounds like curcumin or berberine in microscopic "fat bubbles" (liposomes) allows higher doses to reach liver tumors while sparing healthy tissue, dramatically shrinking tumors in animal studies (Wu et al. 2022). Though not yet proven in humans, this approach holds significant potential. Similarly, epigenetic therapies—drugs designed to reverse harmful chemical changes to DNA caused by toxins like fumonisin—are being explored. While early lab results are intriguing, we lack evidence that these drugs specifically counteract fumonisin's effects in patients.

11.7 Conclusion

Fumonisins represent a persistent and insidious threat to global food security and public health, especially in regions heavily reliant on maize as a dietary staple. Their ability to induce carcinogenic, nephrotoxic, and hepatotoxic effects through multifaceted molecular pathways—including disruption of sphingolipid metabolism, oxidative damage, inflammation, and epigenetic remodeling—warrants serious concern. Epidemiological and experimental data highlight the profound cancer risks linked to chronic FB_1 exposure, particularly esophageal and liver cancers in vulnerable populations. While promising control technologies exist, including biological agents, ozonation, enzymatic detoxification, and AI-based detection models, translating these innovations into practice requires substantial policy, infrastructure, and educational support. Regulatory enforcement, especially in low- and middle-income countries like India, remains inadequate, often exacerbated by poor postharvest practices, inadequate storage, and lack of awareness. Integrated approaches that couple scientific innovation with policy reform, stakeholder training, and community-level engagement are imperative. Moreover, continued research is essential to bridge current knowledge gaps—especially regarding long-term human risk, safe detoxification thresholds, and synergistic mycotoxin interactions. Ultimately, managing fumonisin risks is not solely a scientific challenge but a societal imperative demanding coordinated action across the food chain to ensure health equity and sustainable food systems.

References

Adeyeye SA (2016) Fungal mycotoxins in foods: a review. Cogent Food Agric 2(1):1213127. https://doi.org/10.1080/23311932.2016.1213127

Aggarwal BB, Sung B (2011) NF-κB in cancer: a matter of life and death. Cancer Discov 1(6):469. https://doi.org/10.1158/2159-8290.CD-11-0260

Alberts JF, van Zyl WH, Gelderblom WCA (2016) Biologically based methods for control of fumonisin-producing Fusarium species and reduction of the fumonisins. Front Microbiol 7(APR):187686. https://doi.org/10.3389/FMICB.2016.00548/BIBTEX

Alberts JF, Davids I, Moll W-D, Schatzmayr G, Burger H-M, Shephard GS, Gelderblom WCA (2021) Enzymatic detoxification of the fumonisin mycotoxins during dry milling of maize. Food Control 123:107726. https://doi.org/10.1016/j.foodcont.2020.107726

Alexander NJ, Proctor RH, McCormick SP (2009) Genes, gene clusters, and biosynthesis of trichothecenes and fumonisins in *Fusarium*. Toxin Rev 28(2–3):198–215. https://doi.org/10.1080/15569540903092142

Alizadeh AM, Roshandel G, Roudbarmohammadi S, Roudbary M, Sohanaki H, Ghiasian SA, Taherkhani A, Semnani S, Aghasi M (2012) Fumonisin B1 contamination of cereals and risk of esophageal cancer in a high risk area in Northeastern Iran. Asian Pac J Cancer Prev 13(6):2625–2628. https://doi.org/10.7314/APJCP.2012.13.6.2625

Altinok HH (2009) In vitro production of fumonisin B 1 and B 2 by *Fusarium moniliforme* and the biocontrol activity of *Trichoderma harzianum*. Ann Microbiol 59:509–516

Alvarez CS, Hernández E, Escobar K, Villagrán CI, Kroker-Lobos MF, Rivera-Andrade A, Smith JW, Egner PA, Lazo M, Freedman ND, Guallar E, Dean M, Graubard BI, Groopman JD, Ramírez-Zea M, McGlynn KA (2020) Aflatoxin B1 exposure and liver cirrhosis in Guatemala: a case–control study. BMJ Open Gastroenterol 7(1):e000380. https://doi.org/10.1136/BMJGAST-2020-000380

Arumugam T, Pillay Y, Ghazi T, Nagiah S, Abdul NS, Chuturgoon AA (2019) Fumonisin B1-induced oxidative stress triggers Nrf2-mediated antioxidant response in human hepatocellular carcinoma (HepG2) cells. Mycotoxin Res 35(1):99–109. https://doi.org/10.1007/S12550-018-0335-0

Arumugam T, Ghazi T, Chuturgoon AA (2021) Molecular and epigenetic modes of Fumonisin B1 mediated toxicity and carcinogenesis and detoxification strategies. Crit Rev Toxicol 51(1):76–94

Arumugam T, Ghazi T, Chuturgoon A (2020) Fumonisin B1 epigenetically regulates PTEN expression and modulates DNA damage checkpoint regulation in HepG2 liver cells. Toxins 12(10). https://doi.org/10.3390/TOXINS12100625

Authority EFS, Brancato A, Brocca D, De Lentdecker C, Erdos Z, Ferreira L, Greco L, Jarrah S, Kardassi D, Leuschner R, Lythgo C, Medina P, Miron I, Molnar T, Nougadere A, Pedersen R, Reich H, Sacchi A, Santos M et al (2018) Review of the existing maximum residue levels for bispyribac according to Article 12 of Regulation (EC) No 396/2005. EFSA J 16(1):e05142. https://doi.org/10.2903/j.efsa.2018.5142

Bennett JW, Klich M (2003) Mycotoxins. Clin Microbiol Rev 16(3):497. https://doi.org/10.1128/CMR.16.3.497-516.2003

Bhat RV, Krishnamachari KA (1977) Follow-up study of aflatoxic hepatitis in parts of western India. Indian J Med Res 66(1):55–58

Bojja RS, Cerny RL, Proctor RH, Du L (2004) Determining the biosynthetic sequence in the early steps of the fumonisin pathway by use of three gene-disruption mutants of *Fusarium verticillioides*. J Agric Food Chem 52(10):2855–2860

Boudergue C, Burel C, Dragacci S, Favrot MC, Fremy JM, Massimi C et al (2009) Review of mycotoxin-detoxifying agents used as feed additives: mode of action, efficacy and feed/food safety. EFSA Supporting Publ 6(9):22E. https://doi.org/10.2903/sp.efsa.2009.EN-22

Brown DW, Proctor RH (2016) Insights into natural products biosynthesis from analysis of 470 polyketide synthases from *Fusarium*. Fungal Genet Biol 89:37–51

Brown DW, Butchko RAE, Busman M, Proctor RH (2007) The *Fusarium verticillioides FUM* gene cluster encodes a Zn (II)2Cys6 protein that affects *FUM* gene expression and fumonisin production. Eukaryot Cell 6. https://doi.org/10.1128/ec.00400-06

Bucci TJ, Howard PC, Tolleson WH, Laborde JB, Hansen DK (1998) Renal effects of fumonisin mycotoxins in animals. Toxicol Pathol 26(1):160–164. https://doi.org/10.1177/019262339802600119

Cano PM, Puel O, Oswald IP (2018) Mycotoxins: fungal secondary metabolites with toxic properties. In: Fungi. CRC Press, pp 318–371

Chen J, Wen J, Tang Y, Shi J, Mu G, Yan R, Cai J, Long M (2021) Research progress on fumonisin B1 contamination and toxicity: a review. Molecules 26(17):5238. https://doi.org/10.3390/MOLECULES26175238

Chen W, Zheng R, Baade PD, Zhang S, Zeng H, Bray F et al (2022) Cancer statistics in China, 2022. J Natl Cancer Center 2(1):1–9. https://doi.org/10.1016/j.jncc.2022.02.002

Chiuraise N, Derera J, Yobo KS, Magorokosho C, Nunkumar A, Qwabe NFP (2016) Progress in stacking aflatoxin and fumonisin contamination resistance genes in maize hybrids. Euphytica 207:49–67

Chu YJ, Yang HI, Wu HC, Lee MH, Liu J, Wang LY, Lu SN, Jen CL, You SL, Santella RM, Chen CJ (2018) Aflatoxin B1 exposure increases the risk of hepatocellular carcinoma associated with hepatitis C virus infection or alcohol consumption. Eur J Cancer (Oxford, England: 1990) 94:37. https://doi.org/10.1016/J.EJCA.2018.02.010

Chuturgoon AA, Phulukdaree A, Moodley D (2014) Fumonisin B1 modulates expression of human cytochrome P450 1b1 in human hepatoma (Hepg2) cells by repressing Mir-27b. Toxicol Lett 227(1):50–55. https://doi.org/10.1016/j.toxlet.2014.02.026

Chuturgoon AA, Phulukdaree A, Moodley D (2015) Fumonisin B1 inhibits apoptosis in HepG2 cells by inducing Birc-8/ILP-2. Toxicol Lett 235(2):67–74. https://doi.org/10.1016/j.toxlet.2015.03.006

Dhiman TK, Lakshmi GBVS, Dave K, Roychoudhury A, Dalal N, Jha SK, Kumar A, Han K-H, Solanki PR (2021) Rapid and label-free electrochemical detection of fumonisin-B1 using microfluidic biosensing platform based on Ag-CeO2 nanocomposite. J Electrochem Soc 168(7):077510. https://doi.org/10.1149/1945-7111/AC13CF

Diaz DE, Hagler WM, Blackwelder JT et al (2004) Aflatoxin binders II: reduction of aflatoxin M1 in milk by sequestering agents of cows consuming aflatoxin in feed. Mycopathologia 157:233–241. https://doi.org/10.1023/B:MYCO.0000020587.93872.59

Domijan AM, Želježić D, Milić M, Peraica M (2007) Fumonisin B1: oxidative status and DNA damage in rats. Toxicology 232(3):163–169. https://doi.org/10.1016/j.tox.2007.01.007

Duvick J (2001) Prospects for Reducing Fumonisin Contamination of Maize through Genetic Modification. Environ Health Perspect, 109 Suppl 2, 337–342. https://doi.org/10.1289/ehp.01109s2337

El-Sayed RA, Jebur AB, Kang W, El-Demerdash FM (2022) An overview on the major mycotoxins in food products: characteristics, toxicity, and analysis. J Future Foods 2(2):91–102. https://doi.org/10.1016/J.JFUTFO.2022.03.002

European Commission (2007) Commission Regulation (EC) No 1126/2007 of 28 September 2007 amending Regulation (EC) No 1881/2006 setting maximum levels for certain contaminants in foodstuffs as regards Fusarium toxins in maize and maize products. Off J Eur Union L 255:14–17

Faure JD, Molino D (2015) Sphingolipid variety, biosynthesis and regulation. In: Sphingolipids: biology, synthesis and functions. Nova Science Publishers, pp 39–66

Ferrigo D, Raiola A, Rasera R, Causin R (2014) *Trichoderma harzianum* seed treatment controls *Fusarium verticillioides* colonization and fumonisin contamination in maize under field conditions. Crop Prot 65:51–56

Flaherty JE, Woloshuk CP (2004) Regulation of fumonisin biosynthesis in *Fusarium verticillioides* by a zinc binuclear cluster-type gene, ZFR1. Appl Environ Microbiol 70(5):2653–2659

Fresno Vara JÁ, Casado E, de Castro J, Cejas P, Belda-Iniesta C, González-Barón M (2004) PI3K/Akt signalling pathway and cancer. Cancer Treat Rev 30(2):193–204. https://doi.org/10.1016/J.CTRV.2003.07.007

Frisvad JC, Smedsgaard J, Samson RA, Larsen TO, Thrane U (2007) Fumonisin B2 production by *Aspergillus niger*. J Agric Food Chem 55(23):9727–9732. https://doi.org/10.1021/JF0718906

Gardner NM, Riley RT, Showker JL, Voss KA, Sachs AJ, Maddox JR, Gelineau-van Waes JB (2016) Elevated nuclear sphingoid base-1-phosphates and decreased histone deacetylase activity after fumonisin B1 treatment in mouse embryonic fibroblasts. Toxicol Appl Pharmacol 298:56–65. https://doi.org/10.1016/j.taap.2016.02.018

Gbashi S, Adelusi OA, Njobeh PB (2024) Insights from modelling sixteen years of climatic and fumonisin patterns in maize in South Africa. Sci Rep 14(1):1–12. https://doi.org/10.1038/s41598-024-60904-y

Gelderblom WC, Jaskiewicz K, Marasas WF, Thiel PG, Horak RM, Vleggaar R, Kriek NP (1988) Fumonisins--novel mycotoxins with cancer-promoting activity produced by *Fusarium moniliforme*. Appl Environ Microbiol 54(7):1806–1811. https://doi.org/10.1128/aem.54.7.1806-1811.1988

Hack SP, Spahn J, Chen M, Cheng AL, Kaseb A, Kudo M, Lee HC, Yopp A, Chow P, Qin S (2020) IMbrave 050: a phase III trial of atezolizumab plus bevacizumab in high-risk hepatocellular carcinoma after curative resection or ablation. Future Oncol 16(15):975–989. https://doi.org/10.2217/FON-2020-0162

Harrer H, Humpf HU, Voss KA (2014) In vivo formation of N-acyl-fumonisin B1. Mycotoxin Res 31(1):33. https://doi.org/10.1007/S12550-014-0211-5

Harrison LR, Colvin BM, Greene JT, Newman LE, Cole JR (1990) Pulmonary edema and hydrothorax in swine produced by fumonisin B1, a toxic metabolite of Fusarium moniliforme. J Vet Diagn Invest 2(3):217–221. https://doi.org/10.1177/104063879000200312

Hussein HS, Brasel JM (2001) Toxicity, metabolism, and impact of mycotoxins on humans and animals. Toxicology 167(2):101–134

IARC Working Group on the Evaluation of Carcinogenic Risks to Humans. (2002). Some traditional herbal medicines, some mycotoxins, naphthalene and styrene. IARC Monographs on the Evaluation of Carcinogenic Risks to Humans/World Health Organization, International Agency for Research on Cancer, 82, 1–556. https://www.ncbi.nlm.nih.gov/books/NBK326619/

John RP, Tyagi RD, Brar SK, Surampalli RY, Prévost D (2011) Bio-encapsulation of microbial cells for targeted agricultural delivery. Crit Rev Biotechnol 31(3):211–226

Jouany JP (2007) Methods for preventing, decontaminating and minimizing the toxicity of mycotoxins in feeds. Anim Feed Sci Technol 137(3–4):342–362. https://doi.org/10.1016/j.anifeedsci.2007.06.009

Kabak B, Dobson ADW, Var I (2006) Strategies to prevent mycotoxin contamination of food and animal feed: a review. Crit Rev Food Sci Nutr 46(8):593–619. https://doi.org/10.1080/10408390500436185

Kamangar F, Chow WH, Abnet CC, Dawsey SM (2009) Environmental causes of esophageal cancer. Gastroenterol Clin N Am 38(1):27. https://doi.org/10.1016/J.GTC.2009.01.004

Kamle M, Mahato DK, Devi S, Lee KE, Kang SG, Kumar P (2019) Fumonisins: impact on agriculture, food, and human health and their management strategies. Toxins 11(6):328. https://doi.org/10.3390/TOXINS11060328

Khan R, Anwar F, Ghazali FM (2024) A comprehensive review of mycotoxins: toxicology, detection, and effective mitigation approaches. Heliyon 10(8):e28361. https://doi.org/10.1016/J.HELIYON.2024.E28361

Klaunig JE, Wang Z (2018) Oxidative stress in carcinogenesis. Curr Opin Toxicol 7:116–121. https://doi.org/10.1016/J.COTOX.2017.11.014

Krishnamachari KA, Bhat RV, Nagarajan V, Tilak TB (1975a) Hepatitis due to aflatoxicosis. An outbreak in Western India. Lancet (London, England), 1(7915):1061–1063. https://doi.org/10.1016/s0140-6736(75)91829-2

Krishnamachari KA, Bhat RV, Nagarajan V, Tilak TB (1975b) Investigations into an outbreak of hepatitis in part of Western India, Ind J Med Res 63: 1036. https://doi.org/10.1016/s0140-6736(75)91829-2

Lee J, Kim S, Jung H et al (2023) Exploiting bacterial genera as biocontrol agents: mechanisms, interactions and applications in sustainable agriculture. J Plant Biol 66:485–498. https://doi.org/10.1007/s12374-023-09404-6

Li Y, Seto E (2016) HDACs and HDAC inhibitors in cancer development and therapy. Cold Spring Harb Perspect Med 6(10):a026831. https://doi.org/10.1101/CSHPERSPECT.A026831

Li T, Li J, Wang J, Xue KS, Su X, Qu H, Duan X, Jiang Y (2024) The occurrence and management of fumonisin contamination across the food production and supply chains. J Adv Res 60:13–26. https://doi.org/10.1016/J.JARE.2023.08.001

Liu Y, Wu F (2010) Global burden of aflatoxin-induced hepatocellular carcinoma: a risk assessment. Environ Health Perspect 118(6):818. https://doi.org/10.1289/EHP.0901388

Liu Z, Yu X, Liu N, Liu C, Jiang A, Chen L (2025) Integrating AI with detection methods, IoT, and blockchain to achieve food authenticity and traceability from farm-to-table. Trends Food Sci Technol 158:104925. https://doi.org/10.1016/j.tifs.2025.104925

Lu L, Gunasekaran S (2019) Dual-channel ITO-microfluidic electrochemical immunosensor for simultaneous detection of two mycotoxins. Talanta 194:709–716. https://doi.org/10.1016/J.TALANTA.2018.10.091

Magan N (2006) Mycotoxin contamination of food in Europe: early detection and prevention strategies. Mycopathologia 162(3):245–253. https://doi.org/10.1007/S11046-006-0057-2/METRICS

Mahmood A, Srivastava R (2022) Etiology of cancer. In: Understanding cancer. Academic Press, pp 37–62

Mahmoodi M, Alizadeh AM, Sohanaki H, Rezaei N, Amini-Najafi F, Khosravi AR, Hosseini S-K, Safari Z, Hydarnasab D, Khori V, Mahmoodi M (2012) Impact of fumonisin B1 on the production of inflammatory cytokines by gastric and colon cell lines. Iran J Allergy Asthma Immunol 11(2):165–173. https://ijaai.tums.ac.ir/index.php/ijaai/article/view/341

Marasas WF, Riley RT, Hendricks KA, Stevens VL, Sadler TW, Gelineau-van Waes J, Missmer SA, Cabrera J, Torres O, Gelderblom WC, Allegood J, Martínez C, Maddox J, Miller JD, Starr L, Sullards MC, Roman AV, Voss KA, Wang E, Merrill AH Jr (2004) Fumonisins disrupt sphingolipid metabolism, folate transport, and neural tube development in embryo culture and in vivo: a potential risk factor for human neural tube defects among populations consuming fumonisin-contaminated maize. J Nutr 134(4):711–716. https://doi.org/10.1093/jn/134.4.711

Marín S, Ramos AJ, Cano-Sancho G, Sanchis V (2013) Mycotoxins: occurrence, toxicology, and exposure assessment. Food Chem Toxicol 60:218–237. https://doi.org/10.1016/j.fct.2013.07.047

Markovich NA, Kononova GL (2003) Lytic enzymes of *Trichoderma* and their role in plant defense from fungal diseases: a review. Appl Biochem Microbiol 39:341–351

Morgan E, Soerjomataram I, Rumgay H, Coleman HG, Thrift AP, Vignat J, Laversanne M, Ferlay J, Arnold M (2022) The global landscape of esophageal squamous cell carcinoma and esophageal adenocarcinoma incidence and mortality in 2020 and projections to 2040: new estimates from GLOBOCAN 2020. Gastroenterology 163(3):649–658.e2. https://doi.org/10.1053/J.GASTRO.2022.05.054

Moss MO (1992) Secondary metabolism and food intoxication—moulds. J Appl Bacteriol 73:80s–88s

Mullen TD, Jenkins RW, Clarke CJ, Bielawski J, Hannun YA, Obeid LM (2011) Ceramide synthase-dependent ceramide generation and programmed cell death: involvement of salvage pathway in regulating postmitochondrial events. J Biol Chem 286(18):15929. https://doi.org/10.1074/JBC.M111.230870

Müller S, Dekant W, Mally A (2012) Fumonisin B1 and the kidney: modes of action for renal tumor formation by fumonisin B1 in rodents. Food Chem Toxicol 50(10):3833–3846. https://doi.org/10.1016/J.FCT.2012.06.053

Naghshbandi B, Adabi M, Pooshang Bagheri K, Tavakolipour H (2023) Design of a new electrochemical aptasensor based on screen printed carbon electrode modified with gold nanoparticles for the detection of fumonisin B1 in maize flour. J Nanobiotechnol 20(1):534. https://doi.org/10.1186/S12951-022-01745-7

Navale VD, Vamkudoth KR (2025) Development of multiplex PCR assay for detection of mycotoxigenic Fusarium species contamination in stored maize grains, India. The Microbe 6:100240. https://doi.org/10.1016/J.MICROB.2025.100240

Norred WP, Voss KA, Riley RT, Plattner RD (1996) Fumonisin toxicity and metabolism studies at the USDA: fumonisin toxicity and metabolism. Adv Exp Med Biol 392:225–236. https://doi.org/10.1007/978-1-4899-1379-1_20

Ortega-Beltran A, Bandyopadhyay R (2023) Aflatoxin biocontrol in practice requires a multidisciplinary, long-term approach. Frontiers in Sustain Food Syst 7. https://doi.org/10.3389/fsufs.2023.1110964

Otani N, Ouchi M, Hayashi K, Jutabha P, Anzai N (2017) Roles of organic anion transporters (OATs) in renal proximal tubules and their localization. Anat Sci Int 92(2):200–206. https://doi.org/10.1007/s12565-016-0369-3

Peraica M, Radić B, Lucić A, Pavlović M (1999) Toxic effects of mycotoxins in humans. Bull World Health Organ 77(9):754. https://pmc.ncbi.nlm.nih.gov/articles/PMC2557730/

Piva G, Casadei G, Pagliuca G, Cabassi E, Galvano F, Solfrizzo M et al (2005) Activated carbon does not prevent the toxicity of culture material containing fumonisin B1 when fed to weanling piglets. J Anim Sci 83(8):1939–1947. https://doi.org/10.2527/2005.8381939x

Proctor RH, Desjardins AE, Plattner RD, Hohn TM (1999) A polyketide synthase gene required for biosynthesis of fumonisin mycotoxins in *Gibberella fujikuroi* mating population A. Fungal Genet Biol 27(1):100–112

Proctor RH, Brown DW, Plattner RD, Desjardins AE (2003) Co-expression of 15 contiguous genes delineates a fumonisin biosynthetic gene cluster in *Gibberella moniliformis*. Fungal Genet Biol 38(2):237–249

Proctor RH, Plattner RD, Brown DW, Seo J-A, Lee Y-W (2004) Discontinuous distribution of fumonisin biosynthetic genes in the *Gibberella fujikuroi* species complex. Mycol Res 108:815

Proctor RH, Busman M, Seo J-A, Lee Y-W, Plattner RD (2008) A fumonisin biosynthetic gene cluster in *Fusarium oxysporum* strain O-1890 and the genetic basis for B versus C fumonisin production. Fungal Genet Biol 45:1016–1026

Proctor RH, Van Hove F, Susca A, Stea G, Busman M, van der Lee T et al (2013) Birth, death, and horizontal transfer of the fumonisin biosynthetic gene cluster during the evolutionary diversification of *Fusarium*. Mol Microbiol 90(2):290–306

Raza A, Hassan A, Akram W, Anjum T, Aftab ZEH, Ali B (2024) Seed coating with the synthetic consortium of beneficial *Bacillus* microbes improves seedling growth and manages Fusarium wilt disease. Sci Hortic 325:112645

Reddy KRN, Salleh B, Saad B, Abbas HK, Abel CA, Shier WT (2010) An overview of mycotoxin contamination in foods and its implications for human health. Toxin Rev 29(1):3–26. https://doi.org/10.3109/15569541003598553

Reddy BN, Raghavender CR (2007) Outbreaks ofaflatoxicoses in India. Afr J Food, Agric, Nutr Dev 7(5)

Régnier M, Gourbeyre P, Pinton P, Napper S, Laffite J, Cossalter AM, Bailly JD, Lippi Y, Bertrand-Michel J, Bracarense APFRL, Guillou H, Loiseau N, Oswald IP (2017) Identification of signaling pathways targeted by the food contaminant FB1: transcriptome and kinome analysis of samples from pig liver and intestine. Mol Nutr Food Res 61(12):1700433. https://doi.org/10.1002/MNFR.201700433

Ribeiro DF, Faroni LRDA, Pimentel MAG, Prates LHF, Heleno FF, De Alencar ER (2021) Ozone as a fungicidal and detoxifying agent to maize contaminated with fumonisins. Ozone Sci Eng 44(1):38–49. https://doi.org/10.1080/01919512.2021.1924616

Riley RT, Hinton DM, Chamberlain WJ, Bacon CW, Wang E, Merrill AH, Voss KA (1994) Dietary fumonisin B1 induces disruption of sphingolipid metabolism in Sprague-Dawley rats: A new mechanism of nephrotoxicity. J Nutr 124(4):594–603. https://doi.org/10.1093/jn/124.4.594

Riley RT, Wang E, Schroeder JJ, Smith ER, Plattner RD, Abbas N, Yoo HS, Merrill AH (1996) Evidence for disruption of sphingolipid metabolism as a contributing factor in the toxicity and carcinogenicity of fumonisins. Nat Toxins 4(1):3–15. https://doi.org/10.1002/19960401NT2

Riley RT, Enongene E, Voss KA, Norred WP, Meredith FI, Sharma RP, Spitsbergen J, Williams DE, Carlson DB, Merrill AH (2001) Sphingolipid perturbations as mechanisms for fumonisin carcinogenesis. Environ Health Perspect 109(Suppl 2):301. https://doi.org/10.1289/EHP.01109S2301

Ross PF, Rice LG, Osweiler GD, Nelson PE, Richard JL, Wilson TM (1992) A review and update of animal toxicoses associated with fumonisin-contaminated feeds and production of fumonisins by Fusarium isolates. Mycopathologia 117(1–2):109–114. https://doi.org/10.1007/BF00497286/METRICS

Rubagotti A, Martorana G, Boccardo FM (2006) Epidemiology of kidney cancer. Eur Urol Suppl 5(8):558–565

Seo JA, Proctor RH, Plattner RD (2001) Characterization of four clustered and coregulated genes associated with fumonisin biosynthesis in Fusarium verticillioides. Fungal Genet Biol 34(3):155–165

Sharma RK, Parisi S (2017) Aflatoxins in Indian food products. In: Toxins and contaminants in Indian food products. Springer International Publishing, pp 13–24. https://doi.org/10.1007/978-3-319-48049-7_2

Shephard GS (1998) Chromatographic determination of the fumonisin mycotoxins. J Chromatogr A, 815(1):31–39. https://doi.org/10.1016/S0021-9673(98)00187-3

Shephard GS, Burger HM, Rheeder JP, Alberts JF, Gelderblom WCA (2019) The effectiveness of regulatory maximum levels for fumonisin mycotoxins in commercial and subsistence maize crops in South Africa. Food Control 97:77–80. https://doi.org/10.1016/J.FOODCONT.2018.10.004

Shim WB, Woloshuk CP (2001) Regulation of fumonisin B(1) biosynthesis and conidiation in Fusarium verticillioides by a cyclin-like (C-type) gene, FCC1. Appl Environ Microbiol, 67(4), 1607–1612. https://doi.org/10.1128/AEM.67.4.1607-1612.2001

Shirima CP, Kimanya ME, Routledge MN, Srey C, Kinabo JL, Humpf HU, Wild CP, Tu YK, Gong YY (2015) A prospective study of growth and biomarkers of exposure to aflatoxin and fumonisin during early childhood in Tanzania. Environ Health Perspect 123(2):173–178. https://doi.org/10.1289/EHP.1408097

Siri-anusornsak W, Meneely J, Greer B, Vangnai K, Mahakarnchanakul W, Elliott C, Petchkongkaew A, Kolawole O (2023) In vitro assessment of commercial multi-mycotoxin binders to reduce the bioavailability of emerging mycotoxins in livestock. Emerging Contam 9(4):100256. https://doi.org/10.1016/j.emcon.2023.100256

Song P, Gao Z, Bao Y, Chen L, Huang Y, Liu Y, Dong Q, Wei X (2024) Wnt/β-catenin signaling pathway in carcinogenesis and cancer therapy. J Hematol Oncol 17(1):1–30. https://doi.org/10.1186/S13045-024-01563-4

Sugiyama KI, Kinoshita M, Furusawa H, Sato K, Honma M (2021) Epigenetic effect of the mycotoxin fumonisin B1 on DNA methylation. Mutagenesis 36(4):295–301. https://doi.org/10.1093/MUTAGE/GEAB019

Sujayasree OJ, Chaitanya AK, Bhoite R, Pandiselvam R, Kothakota A, Gavahian M, Mousavi Khaneghah A (2022) Ozone: an advanced oxidation technology to enhance sustainable food consumption through mycotoxin degradation. Ozone Sci Eng 44(1):17–37

Sun G, Wang S, Hu X, Su J, Huang T, Yu J, Tang L, Gao W, Wang JS (2007) Fumonisin B1 contamination of home-grown corn in high-risk areas for esophageal and liver cancer in China. Food Addit Contam 24(2):181–185. https://doi.org/10.1080/02652030601013471

Sung H, Ferlay J, Siegel RL, Laversanne M, Soerjomataram I, Jemal A, Bray F (2021) Global cancer statistics 2020: GLOBOCAN estimates of incidence and mortality worldwide for 36 cancers in 185 countries. CA Cancer J Clin 71(3):209–249. https://doi.org/10.3322/caac.21660

Susca A, Proctor RH, Butchko RA, Haidukowski M, Stea G, Logrieco A, Moretti A (2014) Variation in the fumonisin biosynthetic gene cluster in fumonisin-producing and nonproducing black aspergilli. Fungal Genet Biol 73:39–52

Sweeney MJ, Dobson ADW (1999) Mycotoxin production by Aspergillus, *Fusarium* and *Penicillium* species. Int J Food Microbiol 43(3):141–158

Torres O, Matute J, Gelineau-van Waes J, Maddox JR, Gregory SG, Ashley-Koch AE et al (2015) Human health implications from co-exposure to aflatoxins and fumonisins in maize-based foods in Latin America: Guatemala as a case study. World Mycotoxin J 8(2):143–160

Uhlig S, Busman M, Shane DS, Rønning H, Rise F, Proctor R (2012) Identification of early fumonisin biosynthetic intermediates by inactivation of the FUM6 gene in *Fusarium verticillioides*. J Agric Food Chem 60:10293–10301

Van Den Brand AD, Bajard L, Steffensen IL, Brantsæter AL, Dirven HA, Louisse J et al (2022) Providing biological plausibility for exposure–health relationships for the mycotoxins deoxynivalenol (DON) and fumonisin B1 (FB1) in humans using the AOP framework. Toxins 14(4):279

Visser ME, Schoonees A, Ezekiel CN, Randall NP, Naude CE (2020) Agricultural and nutritional education interventions for reducing aflatoxin exposure to improve infant and child growth in low- and middle-income countries. Cochrane Database Syst Rev *4*(4):Cd013376. https://doi.org/10.1002/14651858.CD013376.pub2

Voss KA, Howard PC, Riley RT, Sharma RP, Bucci TJ, Lorentzen RJ (2002) Carcinogenicity and mechanism of action of fumonisin B1: A mycotoxin produced by *Fusarium moniliforme* (= *F. verticillioides*). Cancer Detect Prev 26(1):1–9. https://doi.org/10.1016/S0361-090X(02)00011-9

Voss KA, Smith GW, Haschek WM (2007) Fumonisins: toxicokinetics, mechanism of action and toxicity. Anim Feed Sci Technol 137(3–4):299–325. https://doi.org/10.1016/J.ANIFEEDSCI.2007.06.007

Wenndt A, Sudini HK, Pingali P, Nelson R (2020) Exploring aflatoxin contamination and household-level exposure riskin diverse Indian food systems. PloS one, 15(10):e0240565

WHO (2002) Evaluation of Certain Mycotoxins in Food: Fifty-sixth Report of the Joint FAO/WHO Expert Committee on Food Additives. Geneva: World Health Organization (WHO Technical Report Series, No. 906)

Wu J, Qi C, Wang H, Wang Q, Sun J, Dong J, Yu G, Gao Z, Zhang B, Tian G (2022) Curcumin and berberine co-loaded liposomes for anti-hepatocellular carcinoma therapy by blocking the cross-talk between hepatic stellate cells and tumor cells. Front Pharmacol 13:961788. https://doi.org/10.3389/FPHAR.2022.961788/BIBTEX

Xue KS, Tang L, Sun G, Wang S, Hu X, Wang JS (2019) Mycotoxin exposure is associated with increased risk of esophageal squamous cell carcinoma in Huaian area, China. BMC Cancer 19(1):1218. https://doi.org/10.1186/S12885-019-6439-X

Yasmin H, Rashid U, Hassan MN, Nosheen A, Naz R, Ilyas N et al (2021) Volatile organic compounds produced by *Pseudomonas pseudoalcaligenes* alleviated drought stress by modulating defense system in maize (Zea mays L.). Physiol Plant 172(2):896–911

Yu S, Jia B, Liu N, Yu D, Zhang S, Wu A (2021) Fumonisin B1 triggers carcinogenesis via HDAC/PI3K/Akt signalling pathway in human esophageal epithelial cells. Sci Total Environ 787. https://doi.org/10.1016/j.scitotenv.2021.147405

Zain ME (2011) Impact of mycotoxins on humans and animals. J Saudi Chem Soc 15(2):129–144. https://doi.org/10.1016/J.JSCS.2010.06.006

Zhang Z, Fang Q, Xie T, Gong X (2024) Mechanism of ceramide synthase inhibition by fumonisin B1. Structure 32(9):1419–1428

Zitomer NC, Mitchell T, Voss KA, Bondy GS, Pruett ST, Garnier-Amblard EC et al (2009) Ceramide synthase inhibition by fumonisin B1 causes accumulation of 1-deoxysphinganine: a novel category of bioactive 1-deoxysphingoid bases and 1-deoxydihydroceramides biosynthesized by mammalian cell lines and animals. J Biol Chem 284(8):4786–4795. https://doi.org/10.1074/jbc.M808798200

Chapter 12
Microbial Carotenoids: Biosynthesis Pathways, Production Strategies, and Their Industrial Applications

Rashmi Shahi, N. Lohith Kumar, and Ankita Mishra

Abstract Carotenoids are yellow, orange, or red pigments, also known as accessory pigments, found in bacteria, fungi, cyanobacteria, algae, and higher plant parts such as flowers, fruits, and storage roots. They are classified into two groups: carotenes and xanthophylls. Carotenoids are synthesized through two distinct pathways—the mevalonate (MVA) pathway and the 2-C-methyl-D-erythritol 4-phosphate (MEP) pathway. Both pathways converge to produce isopentenyl diphosphate (IPP) and dimethylallyl diphosphate (DMAPP), which are essential precursors for various isoprenoids, including carotenoids. Carotenoids can be extracted from plant materials or produced through bioprocessing technologies using microorganisms. The fermentation process for microbial carotenoid biosynthesis requires meticulous control of nutritional and processing parameters and involves both upstream and downstream steps to ensure efficient extraction and recovery of microbial carotenoids. Several strategies have been employed to reduce costs and enhance yields, such as metabolic engineering and recombinant DNA technology. Among the commercially significant carotenoids for human health and medicine are astaxanthin, β-carotene, canthaxanthin, lutein, lycopene, and zeaxanthin. Carotenoids possess various biological properties, including providing coloration to plants and animals, acting as photoprotective agents, alleviating stress, functioning as hormones in plants and animals, serving as precursors to vitamin A, and contributing to the prevention of human diseases such as cancer and neurodegenerative disorders. The applications of carotenoids extend to their use as food

R. Shahi (✉)
Department of Microbiology, Mahayogi Gorakhnath University, Gorakhpur, India
e-mail: rashmishahi.ahs@mgug.ac.in

N. Lohith Kumar
Department of Agricultural Microbiology, University of Agricultural Sciences, Bengaluru, India

A. Mishra
Department of Biotechnology, Mahayogi Gorakhnath University, Gorakhpur, India

N. K. Arora, J. Mishra (eds.), *Biotechnological Solutions for a Sustainable Future*, Microorganisms for Sustainability 55,
https://doi.org/10.1007/978-981-95-3989-5_12

colorants, animal feed supplements for livestock and poultry, and health promotion in humans in the form of supplements in the pharmaceutical and nutraceutical industries.

Keywords Carotenoids · Lycopene · Zeaxanthin · DNA technology · Fermentation

12.1 Introduction

Plants and phototrophic microorganisms synthesize several pigments in addition to chlorophyll, such as carotenoids, phycocyanin, and phycoerythrin. These pigments, referred to as accessory pigments, play an important role in photosynthesis by absorbing light in the blue-green and yellow range (470–630 nm) and transferring it to chlorophyll. This enhances the efficiency of photosynthesis across a broader spectrum of wavelengths (Barreiro and Barredo 2018). Moreover, accessory pigments serve as a protective mechanism, shielding microorganisms or plants from intense sunlight that could potentially damage the photosynthetic apparatus through oxidation (Henríquez et al. 2016). In non-phototrophic microorganisms, they function as photo-protective and stress-protecting agents (Saini et al. 2019).

Carotenoids are yellow, orange, or red-colored, fat-soluble pigments that are widely distributed in nature. They are found in bacteria, fungi, cyanobacteria, algae, and various parts of higher plants such as flowers, fruits, and storage roots. The presence of carotenoids in such plant parts imbues them with characteristic colors and properties. Examples include pumpkins, tomatoes, carrots, corn, and daffodils (Heba et al. 2015; Saini and Keum 2017). Carotenoids also serve as natural colorants in animals, though this occurs solely due to dietary intake rather than endogenous carotenoid synthesis. Examples of such animals include flamingo plumage, lobster shells, salmon fish, canary birds, and pigments present in egg yolk (Yabuzaki 2017). Certain land-dwelling arthropods like aphids and spider mites have acquired carotenogenic genes from fungi, becoming capable of carotenoid production. Similarly, endosymbiotic bacteria within whiteflies contribute to carotenoid production (Landrum 2010).

In 1831, Heinrich W. F. Wackenroder was the first to isolate and crystallize a pigment from carrot roots and coined the term ‘carotene’ to describe it (Wackenroder 1831). In 1837, Berzelius extracted a yellow pigment from aging leaves and named it ‘xanthophyll’ to introduce this pigment. Later, in 1911, Tswett examined the chemical nature and similarities among these compounds and introduced the common term ‘carotenoids,’ encompassing both classes of pigments. More than 1100 natural carotenoids from 683 sources have been isolated and identified. These natural carotenoids are further categorized into two groups—carotenes and xanthophylls (Yabuzaki 2017). Carotenes are pure hydrocarbons, composed solely of hydrogen and carbon, while xanthophylls are oxygenated carotenoids containing at least one oxygen functional group, such as hydroxyl, keto, epoxy, methoxy, or carboxylic acid groups (Mata-Gómez et al. 2014). Based on nutritional properties, carotenoids are also classified as provitamin A, non-provitamin A, and ketocarotenoids (Olson 1999; Maldonade et al. 2007; Jayaraj et al. 2008).

These compounds exhibit significant importance in plants and microorganisms and are diverse in nature. In addition to their crucial role in photosynthesis, they serve as antioxidants, neutralizing free radicals and preventing oxidative damage to cells (Young 1991; Walter and Strack 2011). While primarily recognized for their natural coloring properties, carotenoids also offer health benefits and other important functions. Due to their extensive applications in the food industry, feed supplements, cosmetics, nutraceuticals, pharmaceuticals, and aquaculture, carotenoids have gained considerable industrial relevance (Barreiro and Barredo 2018).

Numerous medical studies have been conducted to explore the mechanisms of ROS-induced oxidative stress and strategies to reduce it (Bhosale and Berstein 2005). Various abnormalities associated with chronic diseases, including immune disorders, neurological conditions, cancer, cardiovascular diseases, diabetes, and photosensitive disorders, result from an imbalance between the detoxification process and ROS generation (Britton 1995; Berman et al. 2015). Carotenoids are recognized for their role in mitigating degenerative pathogenesis linked to oxidative stress, making them potential treatments for alleviating symptoms of degenerative diseases such as Parkinson's and Alzheimer's (Fiedor and Burda 2014; Vílchez et al. 2011).

In contemporary lifestyles, there is an increasing demand for nutritional supplements such as immunostimulants and vitamins, driven by a growing emphasis on disease prevention through diets with significant health benefits. Due to their antioxidant and anti-stress effects, carotenoids have gained importance as nutritional and health supplements (Sathasivam and Ki 2018; Granado et al. 2003).

In addition to these properties, carotenoids help reduce cell damage, making them valuable for enhancing skin health and appearance (Masaki 2010). Because of their nontoxic nature even at high concentrations, natural carotenoids are considered a suitable alternative to synthetic colorants in foods (Zaccarim et al. 2018). Additionally, in aquaculture, carotenoids are added to feed as supplements to enhance vibrant colors in aquatic animals (Vílchez et al. 2011).

Carotenoids are obtained either through extraction from plant sources or through bioprocessing technologies using microorganisms (Valduga et al. 2009a). The industrial interest in microbial carotenoids has surged in recent decades. Biotechnological methods for carotenoid production are considered sustainable, primarily due to their smaller production area requirements compared to plant extraction. Moreover, these methods are independent of climatic changes, seasonality, and varying soil compositions (Valduga et al. 2009b). Additionally, biotechnological processes offer complete control, ensuring higher production yields of desired compounds with lower processing costs through the utilization of inexpensive substrates and efficient downstream recovery methods (Cardoso et al. 2017a).

Given that most microbial carotenoids are intracellular, optimizing the extraction process is crucial for enhancing recovery yields. Among the commercially important carotenoids in human health and medicine are astaxanthin, β-carotene, canthaxanthin, lutein, lycopene, and zeaxanthin (Saini and Keum 2018). Considering the increasing interest in microbial carotenoids, this chapter explores key aspects of carotenoid production, including microbial sources, biosynthesis, extraction, commercialization, and market applications.

12.2 Global Market

The carotenoid market has emerged as a cornerstone of the nutraceutical and functional foods industry, experiencing strong growth in recent years. According to the Global Market Carotenoids Report from BCC Research (BCC Research, 2018), commercial carotenoid production was valued at 1.5 billion US dollars in 2017. It has grown from 1.81 billion dollars in 2023 to 1.9 billion dollars in 2024 at a compound annual growth rate (CAGR) of 5.8%. According to the Carotenoids Global Market Report 2024 from The Business Research Company, the carotenoid market is expected to achieve steady growth in the next few years, reaching 2.3 billion dollars in 2028 at a CAGR of 4.8%. This growth is driven by multiple factors: increasing awareness of carotenoids' health benefits, rising demand in the food and beverage sector, expansion of the nutraceutical industry, growing animal feed applications, and government regulations favoring natural ingredients. Particularly in health supplements, carotenoids are valued for their antioxidant properties, demonstrating efficacy in preventing conditions like age-related macular degeneration (AMD) while supporting overall wellness (Foong et al. 2021; Research and Markets 2024).

12.3 Chemical, Physical, and Biological Characteristics of Carotenoids

Carotenoids are polyisoprenoid compounds, specifically tetraterpenoids, composed of 40 carbon atoms per molecule and a variable number of hydrogen atoms (Britton 1995). The central region features a long-conjugated system of alternating double and single bonds. The intense colors exhibited by carotenoids—ranging from yellow to orange to red—result from their absorption spectrum (400–500 nm, violet to green light) (Britton 1995; Cardoso et al. 2017a).

This absorption spectrum depends on the number of conjugated double bonds in the polyene chain. When exceeding nine conjugated double bonds, the molecule exhibits pigment properties. Typically, carotenoids contain 9–13 conjugated double bonds, each capable of forming numerous geometric isomers (Olson and Krinsky 1995).

According to Britton (1995), the molecular color is determined by the following:

- The extent of conjugation
- Functional group presence/absence
- Saturation/unsaturation degree
- Structural modifications (e.g., cyclization or hydrogenation level changes)

Carotenoids are inherently hydrophobic molecules that readily aggregate and crystallize in aqueous media. In higher animals, they accumulate as microcrystalline aggregates within chromoplasts (Britton 1995). Their localization within membrane structures significantly influences membrane properties, including fluidity, rigidity, and mechanical strength.

Due to their pronounced hydrophobicity, carotenoids are predominantly confined to hydrophobic membrane domains, occurring in aqueous environments only when bound to proteins (Foong et al. 2021). The presence of polar functional groups modulates both their molecular interactions and overall polarity. Key physicochemical properties—including solubility and absorption characteristics—are further influenced by isomerization state and molecular geometry (Frank and Cogdell 1993). Trans-isomers demonstrate greater structural rigidity and an increased propensity for aggregation/crystallization compared to their cis counterparts (Kushwaha et al. 2014).

The molecular dimensions of carotenoids are principally determined by their terminal groups: (i) acyclic carotenoids adopt extended linear conformations with flexible termini and (ii) cyclization reduces both molecular length and end-group bulk (Britton 1995). Carotenoids perform diverse biological functions in both plant and animal systems. Their most visible role involves providing pigmentation to organisms, but their physiological importance extends far beyond coloration. As photo-protective compounds, carotenoids mitigate oxidative stress and prevent cellular damage. In invertebrates, they serve a protective function by shielding egg proteins from protease enzyme activity (Vershinin 1999). Carotenoid derivatives fulfill essential roles as visual pigments (retinal in animals), photosynthetic chromophores (bacteriorhodopsin), and hormonal regulators (retinoic acid in animals, abscisic acid in plants). These multifaceted functions highlight carotenoids' evolutionary significance across biological kingdoms (Vershinin 1999).

Beyond their biological roles, carotenoids possess significant nutritional value attributable to their unique photochemical properties. These compounds serve as essential vitamin A precursors and contribute to the prevention of various human diseases, particularly cancer and neurodegenerative disorders (Fiedor and Burda 2014). The photoprotective capacity of carotenoids stems from their distinctive electronic energy level configuration and their polyene structure, enabling energy dissipation (Frank and Cogdell 1993). Notably, carotenoids are the sole natural compounds that shield the reaction center from photodamage (Goss and Lepetit 2015). This protective capability arises from their unique arrangement of electronic levels due to the polyene structure. Furthermore, carotenoids exhibit the ability to transfer energy both within light-harvesting complexes—moving from the carotenoid excited state to chlorophyll—and within photosynthetic reaction centers—transferring energy from triplet chlorophyll or singlet oxygen to carotenoid (Vershinin 1999).

12.4 Classification of Carotenoids

Currently, more than 700 structurally distinct carotenoids have been identified, all derived from a C-40 hydrocarbon backbone (Armstrong 1994). These compounds exist in both cyclic and acyclic forms, often modified with various oxygen-containing functional groups (Goodwin 1980). Carotenoids can be divided into two major structural classes: (1) hydrocarbon carotenoids (carotenes) such as lycopene, α-carotene, and β-carotene and (2) oxygenated derivatives (xanthophylls) including lutein, astaxanthin, and canthaxanthin. Functionally, carotenoids are categorized as either primary or secondary. Primary carotenoids, including violaxanthin, antheraxanthin, and zeaxanthin (collectively called VAZ), serve as essential light-harvesting pigments in photosynthetic systems (Saini et al. 2015). In contrast, secondary carotenoids like astaxanthin and canthaxanthin accumulate during abiotic stress conditions, where they protect cellular components from damage (Berman et al. 2015).

Carotenoids are further classified by their nutritional properties into provitamin A carotenoids (e.g., α-carotene and β-carotene) and non-provitamin A carotenoids (e.g., lycopene, lutein, and astaxanthin) (Jaswir et al. 2011). This distinction arises from their structural differences: provitamin A carotenoids contain an unsubstituted β-ionone ring that enables conversion to retinal, while non-provitamin A carotenoids lack this critical structural feature. Of the approximately 700 known carotenoids, only about 50 demonstrate provitamin A activity (Kushwaha et al. 2014). The most nutritionally significant vitamin A precursors in humans include α-carotene, β-carotene, and β-cryptoxanthin (Park et al. 2009; Carrillo-Lopez et al. 2010). Notably, β-carotene—the most abundant provitamin A carotenoid—plays a particularly important role in preventing ocular disorders such as night blindness (Imamura et al. 2006).

Carotenoids also include two additional important categories: apocarotenoids and C50 carotenoids (Kushwaha et al. 2014). Apocarotenoids are cleavage products formed through the oxidative action of carotenoid cleavage dioxygenases (CCDs) on parent carotenoid molecules. These derivatives hold substantial biological and commercial value, including vitamin A and its derivatives (retinoids and retinoic acid) in animals, the plant hormone abscisic acid, strigolactones in plants, and aromatic compounds like α-ionone and β-ionone used in fragrances (Berman et al. 2015; Saini et al. 2015). In contrast, C50 carotenoids represent a structurally distinct group, with decaprenoxanthin and its glycosylated derivatives being characteristic examples produced by microorganisms such as *Corynebacterium glutamicum* (Saini and Keum 2017).

12.5 Biosynthesis and Regulatory Mechanisms of Carotenoids in Microorganisms

Carotenoids are derived from a C-40 hydrocarbon skeleton, which can be modified to form both cyclic and acyclic xanthophylls through the introduction of oxygen-containing functional groups (Armstrong 1994). Biosynthesis occurs via two distinct metabolic pathways: the mevalonate (MVA) pathway and the 2-C-methyl-D-erythritol 4-phosphate (MEP) pathway (Namitha and Negi 2010). Both pathways converge to produce the essential isoprenoid precursors isopentenyl diphosphate (IPP) and dimethylallyl diphosphate (DMAPP) (Das et al. 2007). The carotenoid biosynthetic pathway involves several key enzymes, such as geranylgeranyl pyrophosphate synthase, phytoene synthase, carotene desaturase, and lycopene cyclase (Namitha and Negi 2010). These enzymes are classified according to their specific catalytic functions. Subsequent modifications of carotenes to produce various C-40 carotenoids are mediated by β-carotene ketolase and β-carotene hydroxylase. Bacteria exclusively use the MEP pathway, while eukaryotes and archaebacteria rely on the MVA pathway. Plants and *Streptomycetes* employ both pathways for carotenoid biosynthesis (Namitha and Negi 2010; Das et al. 2007).

12.5.1 2-C-Methyl-D-erythritol 4-phosphate (MEP) Pathway

The MEP (2-C-methyl-D-erythritol 4-phosphate) pathway begins with the condensation of pyruvate and glyceraldehyde-3-phosphate (G3P), catalyzed by 1-deoxy-D-xylulose-5-phosphate synthase (DXS). This enzyme mediates the formation of 1-deoxy-D-xylulose 5-phosphate (DXP), which is subsequently converted to MEP by 1-deoxy-D-xylulose 5-phosphate reductoisomerase (IspC) (Namitha and Negi 2010). The pathway continues through a series of enzymatic transformations:

- MEP cytidylyltransferase converts MEP to 4-(cytidine 5′-diphospho)-2-C-methyl-D-erythritol (CDP-ME) through the addition of a cytidine monophosphate (CMP) group.
- CDP-ME kinase (IspE) phosphorylates CDP-ME to form CDP-ME2P.
- MECDP synthase (encoded by gcpE) catalyzes the cyclization to 2-C-methyl-D-erythritol 2,4-cyclodiphosphate (MECDP).
- The same enzyme mediates MECDP reduction to 1-hydroxy-2-methyl-2-(E)-butenyl 4-diphosphate (HMBDP).
- The final step involves HMBDP reductase (lytB/ispH) converting HMBDP to isopentenyl diphosphate (IPP) (Wang et al. 2010). IPP serves as the fundamental building block for various isoprenoids, including carotenoids (Adam et al. 2002).

12.5.2 Mevalonate Pathway

The mevalonate pathway begins with the condensation of acetyl-CoA to form acetoacetyl-CoA, a reaction catalyzed by acetyl-CoA acetyl transferase (Walter and Strack 2011). This is followed by the additional condensation of another acetyl-CoA molecule catalyzed by the same enzyme, leading to the formation of 3-hydroxy-3-methyl-glutaryl (HMG). Next, 3-hydroxy-3-methyl-glutaryl-CoA (HMG-CoA) is synthesized from HMG through the action of the enzyme 3-hydroxy-3-methyl-glutaryl-CoA synthase (HMG-CoA synthase). The reduction of HMG-CoA by HMG-CoA reductase then produces mevalonate. Mevalonate is sequentially phosphorylated in two steps: first catalyzed by mevalonate kinase, followed by phosphomevalonate kinase (Walter and Strack 2011; Namitha and Negi 2010). The decarboxylation of phosphorylated mevalonate leads to the synthesis of isopentenyl pyrophosphate (IPP), which can be further transformed into its isomer, dimethylallyl pyrophosphate (DMAPP), catalyzed by the enzyme IPP isomerase. Both IPP and DMAPP serve as essential precursors for the biosynthesis of various carotenoids and play a pivotal role in many cellular processes (Kuzuyama 2002).

The formation of IPP and DMAPP through these pathways leads to the condensation of both compounds to form geranyl diphosphate (GPP). This condensation reaction is facilitated by the enzyme GPP synthase. The addition of IPP to GPP results in the synthesis of farnesyl diphosphate (FPP), followed by the subsequent addition of another IPP molecule, yielding geranylgeranyl diphosphate (GGPP) through the action of the enzyme GGPP synthase. GGPP acts as a key precursor for various groups of plastid isoprenoids (Namitha and Negi 2010). The head-to-head condensation of two GGPP molecules leads to the generation of the C-40 compound phytoene, catalyzed by phytoene synthase (PSY) encoded by *CrtB*. The subsequent reactions involve four consecutive steps in which phytoene is converted to phytofluene, ζ-carotene, neurosporene, and lycopene through sequential desaturation. These enzymatic transformations contribute to the biosynthesis of several carotenoid compounds (Walter and Strack 2011).

Additional desaturation reactions lead to an increase in conjugated double bonds, transforming the colorless phytoene into the brilliant pink pigment lycopene. Lycopene cyclases catalyze the cyclization of lycopene, resulting in the formation of γ-carotene (gamma-carotene) as an intermediate that further transforms to β-carotene (beta-carotene). The addition of an epsilon ring to lycopene, mediated by *CrtL*-e, produces δ-carotene (delta-carotene). Subsequently, a beta ring is added to the opposite end, leading to the formation of α-carotene (alpha-carotene). Conversely, the addition of a beta-ring to lycopene yields γ-carotene (gamma-carotene). The hydroxylation at the C3 site of each ring in α-carotene results in the synthesis of lutein, catalyzed by carotene hydroxylase (*CrtR*) (Namitha and Negi 2010; Walter and Strack 2011).

In a similar way, the addition of a keto group at the C4 site by β-carotene-4-oxygenase/β-carotene ketolase (*CrtO*) results in the formation of echinenone. The addition of another keto group at the C4' site, mediated by *CrtW*, leads to the production of canthaxanthin. The hydroxylation at the C3 position of β-carotene, catalyzed by *CrtR* or *CrtZ*, yields β-cryptoxanthin. Similarly, the addition of a hydroxyl group at the C3' site by carotenoid 3,3′-hydroxylase (*CrtR* or *CrtZ*) produces zeaxanthin (Namitha and Negi 2010; Kushwaha et al. 2014). Further modifications at the C4 and C4' sites of the β-ionone ring through sequential keto group additions lead to the synthesis of astaxanthin, mediated by *CrtO/CrtW* (Walter and Strack 2011). This pathway, involving the transformation of β-carotene into keto-carotenoids such as echinenone, astaxanthin, and canthaxanthin, has been extensively studied in algae, yeast, and non-photosynthetic bacteria (Bhosale and Berstein 2005).

12.6 Role of Carotenoids in Microbial Survival and Environmental Adaptation

Since carotenoids play a universal role in all photosynthetic organisms, such as algae, plants, and non-photosynthetic bacteria, their unique structure provides an additional advantage to these organisms, enabling effective quenching of cellular oxidants such as singlet oxygen or chlorophyll triplets (Frank and Cogdell 1993). This antioxidant capability helps microorganisms support vital pathways, including photosynthesis, photoprotection, cellular membrane stabilization, and ROS detoxification (Goss and Lepetit 2015; Kumar and Morrell-Falvey 2018). Additionally, carotenoids serve as precursors for the biosynthesis of signaling compounds such as strigolactones, trisporic acids, and abscisic acid, which are involved in root-mycorrhizal interactions, reproduction, and stress adaptations (Sahadevan et al. 2013; Sun et al. 2012). Their enhanced production has also been linked to biofilm formation, which helps combat oxidative stress within biofilms (Sahadevan et al. 2013; Zheng et al. 2013; Boles and Singh 2008).

One of the best-known photoprotective mechanisms in photosynthetic organisms involves the dissipation of excess radiant energy, a process known as non-photochemical fluorescence quenching (NPQ) (Goss and Lepetit 2015). This NPQ mechanism has been extensively studied in plants, green algae, and brown algae (Wierzchos et al. 2015; Vílchez et al. 2011). It includes a pH-, light-, and transmembrane proton-dependent reversible conversion of violaxanthin (V), which contains two epoxide groups, to antheraxanthin (A), containing one epoxide group, and then to zeaxanthin (Z), which has no epoxide group (Goss and Lepetit 2015). This cyclic reaction is collectively referred to as the VAZ cycle. Similarly, in diatoms, the Dd-Dt cycle provides photoprotection during intense light conditions (Vílchez et al. 2011). This reaction involves the conversion of mono-epoxy xanthophyll diadinoxanthin (Dd) to diatoxanthin (Dt), which lacks an epoxide group. In cyanobacteria, the NPQ mechanisms involve functional interactions between the orange carotenoid protein (OCP), phycobilisome, and keto-carotenoids such as echinenone and canthaxanthin (Goss and Lepetit 2015). The presence of keto-carotenoids further provides increased adaptation to challenging environmental conditions such as excessive light, UV radiation, dehydration, and desiccation (Vítek et al. 2017; Wierzchos et al. 2015).

In zygomycete fungi, carotenoids such as β-carotene-derived apocarotenoids play an important role in regulating early sexual morphogenesis and mycelial development (Sahadevan et al. 2013). These β-carotene-derived signaling compounds, particularly trisporoids, act as pheromones and aid in the recognition of compatible mating partners (Sahadevan et al. 2013; Sun et al. 2012). Under dynamic light and temperature conditions, polyunsaturated fatty acids (PUFAs), along with xanthophylls (polar carotenoids), help maintain proper structure, fluidity, and functionality of cellular membranes in microorganisms. Xanthophylls such as myxoxanthophyll, zeaxanthin, and deoxymyxoxanthophyll, in combination with PUFAs, restrict lipid movement during the liquid crystalline state, thereby increasing membrane rigidity (Zakar et al. 2017).

12.7 Carotenoid Biosynthesis by the Microorganisms

Currently, commercial carotenoid production primarily relies on extraction from natural sources (fruits, vegetables, and flowers) or chemical synthesis (Soong et al. 2019). However, plant-based carotenoid production faces challenges of seasonal and geographic variability, while chemical synthesis generates hazardous waste, making it environmentally unsustainable (Saini et al. 2015). In contrast, microbial carotenoid production offers a safer, more efficient, and environmentally friendly alternative. This approach provides several advantages over traditional methods, including the utilization of low-cost substrates, reduced production costs, easily controlled and optimized production processes, potential for strain selection to enhance yield, and improved scalability through biotechnological optimization (Saini and Keum 2017). These benefits have driven growing interest in microbial fermentation for natural carotenoid production in recent years (Sanchez et al. 2013).

Numerous studies have documented carotenoid production from microbial sources, including bacteria, fungi, yeasts, and microalgae (Olson and Krinsky 1995; Kushwaha et al. 2014). Carotenogenic microorganisms represent a diverse group comprising halotolerant freshwater algae, photosynthetic and phototrophic bacteria, asporogenous yeasts, and fungi (Table 12.1). Notably, certain species such as *Dunaliella salina*, *Xanthophyllomyces dendrorhous, Haematococcus pluvialis,* and *Blakeslea trispora* have been widely adopted for industrial-scale carotenoid production (Mehta et al. 2003; Raja et al. 2007). These microbial systems demonstrate remarkable versatility, enabling production of various carotenoids from the widely distributed β-carotene and lutein to specialty pigments like astaxanthin and canthaxanthin (Table 12.2).

The advancements in genetic and metabolic engineering have significantly enhanced microbial carotenoid biosynthesis (Sanchez et al. 2013). Since all carotenoids derive from a common precursor, their biosynthetic pathways can be efficiently manipulated using recombinant DNA technology (Namitha and Negi 2010; Sanchez et al. 2013). This approach enables not only increased production but also diversification of carotenoid profiles through combinatorial expression of biosynthetic genes (Dannert et al. 2000). Notably, even non-carotenogenic microorganisms—including *Escherichia coli*, *Saccharomyces cerevisiae*, *Candida utilis*, and *Zymomonas mobilis*—have been successfully engineered to produce novel carotenoids. These strains were transformed with carotenogenic genes from organisms such as *Xanthophyllomyces dendrorhous*, *Rhodotorula glutinis*, and *Erwinia herbicola*, yielding customized cyclic, acyclic, and oxo-carotenoids for specific applications (Misawa and Shimada 1998; Das et al. 2007).

Among these, *E. coli* has emerged as a particularly versatile host, extensively studied for producing zeaxanthin, astaxanthin, and β-carotene. Similarly, *S. cerevisiae* has been investigated for astaxanthin synthesis (Sanchez et al. 2013).

Table 12.1 Microbial sources of carotenoids with functional properties

Microorganism	Carotenoid produced	Functional role(s)	References
Bacteria			
Mycobacterium brevicale	Canthaxanthin	Antioxidant, immune modulation, pigmentation	Schroeder and Johnson (1995)
Mycobacterium lacticola	Astaxanthin	Potent antioxidant (6000× vitamin C), anti-inflammatory	Schroeder and Johnson (1995)
Streptomyces chrestomyceticus	Canthaxanthin	Antioxidant, aquaculture pigmentation	Schroeder and Johnson (1995)
Cyanobacteria			
Anabaena variabilis	Canthaxanthin	Photoprotection, radical scavenging	Schroeder and Johnson (1995)
Aphanizomenon flos-aquae	Canthaxanthin	Oxidative stress protection, light harvesting	Schroeder and Johnson (1995)
Algae			
Chlorella pyrenoidosa	Lutein	Macular pigment, blue light filtration (eye health)	Schroeder and Johnson (1995)
Dictycoccus cinnabarinus	Canthaxanthin	Antioxidant, microbial photoprotection	Schroeder and Johnson (1995)
Dunaliella salina	β-Carotene	Provitamin A precursor (major), antioxidant	Aguilar et al. (2004)
Dunaliella tertiolecta	β-Carotene	Provitamin A, membrane stabilizer in high salinity	Fazeli et al. (2006)
Haematococcus pluvialis	Astaxanthin	World's strongest natural antioxidant, stress response	Orosa et al. (2005)
Spongiococcum excetricum	Lutein	Photoprotective xanthophyll, visual health	Schroeder and Johnson (1995)
Microalgae			
Chlorella vulgaris	Lutein	Macular degeneration prevention, anti-aging	Luengo et al. (2015)
Chlorella saccharophila	Zeaxanthin	Central macular pigment, UV protection	Singh et al. (2015a, b)
Chlorella vulgaris	Astaxanthin	Anti-inflammatory, UV skin protection	Kim et al. (2016)
Haematococcus pluvialis	Astaxanthin	Crosses blood–brain barrier, neuroprotective	Kim et al. (2016)
Fungi/yeast			
Fusarium spp.	Neurosporaxanthin, Torulene	Photoprotection, membrane integrity	Avalos et al. (2015) and Rivera et al. (2014)
Xanthophyllomyces dendrorhous	Astaxanthin, β-carotene	Aquatic feed pigment (astaxanthin), provitamin A (β-carotene)	Domínguez-Bocanegra et al. (2007)
Rhodotorula glutinis	Torulene, Torularhodin	Antioxidant cascade, microbial stress response	Kot et al. (2019)

(continued)

Table 12.1 (continued)

Microorganism	Carotenoid produced	Functional role(s)	References
Cantharellus spp.	Canthaxanthin	Natural food coloring, antioxidant	Haxo (1950)
Blakeslea trispora	Lycopene, β-carotene	Prostate health (lycopene), vitamin A precursor	Pegklidou et al. (2008)
Mucor circinelloides	β-Carotene	Dietary supplement, natural colorant	Naz et al. (2020)
Phaffia rhodozyma	Astaxanthin, β-carotene	Salmonid aquaculture pigmentation, immune booster	Liu et al. (2006)
Rhodotorula mucilaginosa	β-carotene, Torularhodin	Vitamin A precursor, lipid-soluble antioxidant	Dias Rodrigues et al. (2019)

Table 12.2 Biotechnological production of carotenoids for industrial applications

Company	Compound	Microbial source	Applications
Angel Yeast Company	β-carotene	Red yeasts (*Monascus* spp.)	Pharmaceuticals, nutraceuticals, and cosmetic applications
Kemin Industries Inc	Cryptoxanthin	Yeast (*Rhodotorula taiwanensis*)	Food, nutraceuticals, pharmaceuticals, and cosmetics
Allied Biotech Corporation	β-carotene, β-opo-8-carotenal, Lycopene, Canthaxanthin, Lutein	Algae (*Dunaliella salina*)	Natural β-carotene powder (20% extract); enhances survival in harsh conditions
Hetchyn Biotechnology Co., Ltd	Astaxanthin	Algae (*Haematococcus pluvialis*)	Strongest natural antioxidant; superior radical scavenger
Biolifescience	Astaxanthin	Algae (*Haematococcus pluvialis*)	Natural antioxidant for nutraceutical and cosmetic applications
Lycored	Carotenoids	Fungi (*Blakeslea trispora*)	Food, nutraceutical, pharmaceutical, and cosmetics
Dr. Heilbronn	Carotenoids	Algae (*Haematococcus pluvialis*)	Food, nutraceutical, pharmaceutical, and cosmetics

12.8 Industrial Production of Carotenoids

Recently, the rising interest in natural carotenoids has led to their industrial production through fermentation processes. The biotechnology field particularly focuses on enhanced productivity yields of microbial carotenoids through the optimization of fermentation conditions, cell engineering, and synthetic biology applications (Sanchez et al. 2013). Since microorganisms accumulate carotenoids in response to stress conditions, the microorganisms extensively investigated for the industrial production of β-carotene include the fungus *Blakeslea trispora* and the microalga

Xanthophyllomyces dendrorhous, while *Haematococcus* sp. and *X. dendrorhous* are considered prominent candidates for astaxanthin production on a commercial scale. Since the 1970s, numerous studies have been carried out to explore different processes for the production of these carotenoids (Valduga et al. 2009b).

The optimization of the fermentation process encompasses several parameters, including the nutritional composition of the culture medium, pH, temperature, luminosity, aeration rate, and agitation (Sanchez et al. 2013). These factors not only influence microbial cell growth but also the biosynthesis of specific carotenoids (Mezzomo and Ferreira 2016). To maintain efficient production yields and to conduct a specific biosynthesis pathway, meticulous control of the nutritional and processing parameters is essential (Mussagy et al. 2019; Sanchez et al. 2013). Moreover, independent of the upstream bioprocessing, after fermentation, carotenoids remain inside the cells (in the biomass content). Therefore, proper integration of downstream processing is required for efficient extraction and recovery of microbial carotenoids (Mussagy et al. 2019).

12.8.1 Fermentation Conditions

Typically, all microorganisms tend to accumulate these pigments in response to stress conditions that slow their cellular growth, which is maintained during the fermentation process (Das et al. 2007). The known mechanism for carotenoid synthesis involves increased reactive oxygen species (ROS) generation, oxidative stress, limited protein synthesis, unfavorable growth conditions, and nutrient deprivation, which cause either slow growth or growth cessation, leading to carotenoid biosynthesis (Cardoso et al. 2017a). These conditions are controlled using different methods, including controlled oxygen levels, intense aeration, limitations in nutrients such as nitrogen or phosphate, intense light conditions, temperature, salt stress, and the addition of chemical substances like leucine, valine, hydrogen peroxide, duroquinone, ethanol, yeast extract, acetate, or other intermediates related to the mevalonate or TCA cycle (Sanchez et al. 2013).

The addition of such chemical compounds results in: (1) induction of enzymes involved in carotenoid synthesis, (2) induction of superoxide generation, (3) inhibition of growth by blocking the electron transport chain or the TCA cycle function, (4) increased availability of amino acids for other biosynthetic processes, and (5) some compounds also acting as carotenoid precursors (Sanchez et al. 2013).

For instance, a semi-industrial method was utilized for lycopene production through fermentation of the fungus *B. trispora*. The cultures of *B. trispora* were inoculated into a medium containing substances such as imidazole or pyridine, which were added to inhibit the enzymatic activity of lycopene cyclase. This resulted in enhanced accumulation of lycopene. Additionally, other raw materials and physical parameters were investigated to achieve maximal lycopene production (Bhosale 2004).

Similarly, mutant strains of *X. dendrorhous* were developed to achieve profitable industrial production of astaxanthin (De La Fuente et al. 2010). Further improvements were made through exposure to white light and UV light, along with a fed-batch process using glucose feeding (De La Fuente et al. 2010). The microalga *Dunaliella salina* was cultivated in nitrogen-limited media coupled with high salinity and exposed to intense light for higher β-carotene production (Ye et al. 2008). Astaxanthin accumulation was notably elevated under specific light and stress conditions, such as nutrient (N or P) deficiency or salt stress, in *H. pluvialis* (Kim et al. 2006).

Microalgal production systems previously used for carotenoid synthesis were either open pond or raceway systems, which were successful in culturing microbes such as *Dunaliella*, *Chlorella*, and *Spirulina* under adverse growing conditions like high light intensity or salinity (Henríquez et al. 2016). However, due to drawbacks such as low productivity, costly product recovery, large water requirements, and temperature control difficulties, closed culture systems like photobioreactors have replaced conventional open culture systems (Sanchez et al. 2013). Tubular photobioreactors allow high cell density cultures, with three times the biomass compared to open ponds. They also offer other advantages, including easier biomass cultivation and harvesting, lower risk of contamination, better control of growth conditions, and reduced capital investment. These systems can also operate at high light intensities (Ye et al. 2008).

Algae can be cultivated in either mixotrophic (utilizing both organic carbon sources and light energy as electron sources) or heterotrophic (using organic carbon sources in darkness) conditions (Henríquez et al. 2016). Ethanol, acetate, and glucose can serve as organic carbon sources for algal growth. *H. pluvialis* requires mixotrophic or heterotrophic conditions with sodium acetate as a carbon source for astaxanthin synthesis (Orosa et al. 2005). Furthermore, exposing cultures to high light intensity under nitrogen-limited conditions, coupled with bicarbonate supplementation, enhanced astaxanthin accumulation (Hata et al. 2001). Such culture conditions have proven effective for other algae such as *Scenedesmus* sp. and *Chlorella protothecoides*, resulting in increased production of lutein and other carotenoids (Yen et al. 2011; Shi et al. 2002).

12.8.2 Genetic Approaches to Boost Carotenoid Production in Microorganisms

One of the approaches to reduce production costs and increase yields involves the development of hyper-producing strains through bioengineering. Metabolic engineering plays an important role in enhancing cellular properties through the modification of specific biochemical reactions using recombinant DNA technology (Park et al. 2007). Moreover, genetic engineering is particularly valuable in non-carotenoid-producing microorganisms, as it enables the manipulation of well-known microorganisms with well-defined metabolic pathways to improve carotenoid productivity and yields (Ye and Bhatia 2012).

Metabolic engineering tools have been widely studied by various researchers in yeasts such as *Saccharomyces cerevisiae* and *Candida utilis*. Yeast cells have been successfully modified through the insertion of carotenogenic genes from organisms such as *Erwinia uredovora*, *Agrobacterium aurantiacum*, and *Xanthophyllomyces dendrorhous* to produce carotenoids, including β-carotene, lycopene, and astaxanthin (Misawa and Shimada 1998; Miura et al. 1998; Bhataya et al. 2009; Ungureanu et al. 2013).

Another successful instance includes the production of lycopene by the natural non-carotenogenic yeast *Yarrowia lipolytica*, achieved simply through the introduction of two genes, namely *CrtB* (phytoene synthase) and *CrtI* (phytoene desaturase). The appearance of an orange color in transformants after the integration of the heterologous genes *CrtB* and *CrtI* confirmed lycopene formation and significantly increased lycopene productivity (Matthäus et al. 2014).

Similarly, another non-carotenogenic yeast, *Pichia pastoris*, was engineered through the integration of two plasmids, pGAPZA-EBI* and pGAPZA-EBI*L, containing genes encoding enzymes for lycopene and β-carotene biosynthesis. This led to the formation of a recombinant strain capable of producing both lycopene and β-carotene (Araya-Garay et al. 2012).

These approaches have proven effective for increasing the output of specific microbial carotenoids to meet the growing global demand across multiple industries, including animal feed, cosmetics, food and beverages, and pharmaceuticals.

12.8.3 Extraction Techniques for Intracellular Carotenoid Recovery

Carotenoids are intracellularly synthesized compounds in microorganisms and require effective downstream processing for recovery post-fermentation. The recovery process begins with clarification—typically via centrifugation or filtration—to separate biomass from the fermentation broth. The next and most critical phase is cell disruption, where physical, chemical, or biological methods are employed to release carotenoids from microbial cells, followed by separation from cell debris. This disruption–extraction step can be executed either as a single unit operation or in separate stages (Sanchez et al. 2013). The choice of recovery strategy—complete cell disruption or selective membrane permeabilization—is influenced by the structural features of the microbial cells (Mussagy et al. 2019).

Structurally, bacterial cells are generally easier to disrupt than yeast and microalgae due to their relatively fragile cell envelopes. Gram-negative bacteria possess an inner and outer membrane with a thinner peptidoglycan layer, whereas Gram-positive bacteria have a thick cell wall with a single membrane (Saini and Keum 2018). In contrast, yeast and microalgae possess rigid, multilayered cell walls, making them more resistant to disruption (Park et al. 2007; Valduga et al. 2009a). Carotenoids within microbial cells are often tightly bound to intracellular proteins and lipids, necessitating efficient extraction protocols (Mussagy et al. 2019).

Extraction techniques are broadly classified into physical and chemical methods. Physical techniques utilize mechanical forces or temperature fluctuations to lyse cells and release carotenoids. These include ultrasonication, cryogenic grinding, and freeze–thaw cycles (Sanchez et al. 2013). Such methods minimize chemical contamination and are preferred when preserving product integrity is essential.

Chemical extraction methods rely on agents such as acids, bases, surfactants, or volatile organic solvents (VOCs) to solubilize or disrupt cell structures (Saini and Keum 2018). While VOCs like hexane, acetone, ethanol, and tetrahydrofuran are commonly used due to their high extraction efficiency, they pose environmental and health concerns due to volatility, toxicity, and low biodegradability (Saini and Keum 2018; Yara-Varón et al. 2016). Solvent selection is polarity-dependent: nonpolar solvents (e.g., hexane, petroleum ether) are suitable for nonpolar carotenoids, while polar solvents (e.g., DMSO, acetone, ethyl acetate) target polar carotenoids (Mussagy et al. 2019).

Despite the risks, VOCs remain widely used in both traditional (e.g., atmospheric liquid extraction, Soxhlet) and modern (e.g., ultrasound- or enzyme-assisted extraction) techniques because of their membrane permeabilization and solubilizing capabilities (Yara-Varón et al. 2016). Advanced hybrid strategies have been developed to improve extraction efficiency. For example, Soxhlet extraction using DMSO or acetone in conjunction with ultrasonication significantly enhanced carotenoid yield from *Rhodotorula mucilaginosa* (Cheng and Yang 2016). Likewise, food-grade carotenoid extraction now commonly combines enzymatic treatments with solvents like hexane and ethyl acetate (Lavecchia and Zuorro 2008).

Recent developments have shifted focus toward sustainable, green extraction technologies. Green solvents—including supercritical fluids, ionic liquids (ILs), deep eutectic solvents (DESs), and biosolvents—are increasingly replacing hazardous VOCs. These eco-friendly alternatives are characterized by reduced toxicity, biodegradability, and energy efficiency (Anastas and Warner 1998).

- Supercritical fluid extraction (SFE), primarily using CO_2 and ethanol, is a leading nonconventional method, especially effective for nonpolar carotenoids. SFE operates under near-critical conditions, eliminating solvent residues, minimizing thermal degradation, and reducing toxic waste (Mezzomo and Ferreira 2016).
- Ionic liquids (ILs) are salt-based solvents with low vapor pressures and high chemical stability. Their tunable properties allow for optimized carotenoid extraction, especially with imidazolium-, pyridinium-, and ammonium-based ILs. ILs have shown high efficiency in recovering carotenoids from microalgae like *Haematococcus pluvialis* and offer promise for scalable, sustainable production (Saini and Keum 2018).
- Deep eutectic solvents (DESs), typically formed from choline chloride and hydrogen bond donors (e.g., 1,2-propanediol), are biodegradable and have demonstrated potential for intracellular biomolecule extraction. Cicci et al. (2017)

reported successful recovery of carotenoids from *Scenedesmus dimorphus*, achieving yields up to 0.11% of total biomolecules. DESs have also been applied to recover astaxanthin from marine sources such as shrimp and crab.

Another sustainable approach involves biosolvents derived from renewable resources like starch, wood, and plant oils (Sanchez et al. 2013). These novel solvents contribute to reducing the ecological footprint of industrial carotenoid extraction (Anastas and Warner 1998; Yara-Varón et al. 2016).

Process intensification strategies such as ultrasound-assisted extraction (UAE) and microwave-assisted extraction further improve efficiency. UAE utilizes cavitation to enhance solvent penetration and disrupt cellular structures. Gu et al. (2008) demonstrated high carotenoid recovery from *Rhodobacter sphaeroides* using ultrasonication (500 W) and acetone, while Singh et al. (2015a, b) optimized UAE parameters for *Chlorella saccharophila* to recover high yields of β-carotene and zeaxanthin. These techniques not only enhance yield but also minimize energy and solvent consumption. Effective cell disruption is a prerequisite for optimal carotenoid recovery.

Although VOC-based methods deliver high yields, their drawbacks—such as environmental toxicity, safety concerns, and bioactive compound degradation—necessitate greener alternatives. Current innovations combine selective membrane permeabilization, solvent reduction, and physical intensification to develop eco-conscious, industrially viable extraction systems (Saini and Keum 2018).

12.8.4 Co-extracted Metabolites in Carotenoid Recovery

During carotenoid extraction (such as astaxanthin or β-carotene) from microalgae using VOCs, various intracellular microbial metabolites are co-extracted, including fatty acids, lipids, proteins, and carbohydrates (Saini and Keum 2017; Mussagy et al. 2019). This co-extraction complicates downstream processing by requiring additional separation and purification steps (Sanchez et al. 2013). When extracting carotenoids from microalgae, essential fatty acids are simultaneously recovered. For instance, studies on microalgae like *Porphyridium cruentum* and *Isochrysis galbana* have shown that fatty acid extraction results in the recovery of compounds such as eicosapentaenoic acid (EPA), docosahexaenoic acid (DHA), and arachidonic acid (AA) (Saini and Keum 2018; Sanchez et al. 2013). Protein-carotenoid complexes require special handling, such as using wet biomass (Román et al. 2002). Powls and Britton (1977) extracted a violaxanthin-binding protein from photosynthetic routes using hot methanol obtained from the biomass of microalgae *Scenedesmus obliquus*.

Cicci et al. (2017) demonstrated that DES can effectively serve as biocompatible solvents for extracting carotenoids along with other intracellular components, including carbohydrates, proteins, lipids, and pigments such as chlorophyll from *Scenedesmus dimorphus*. This co-extraction of multiple metabolites necessitates a thorough characterization of the extract, which should identify both the target carotenoids and any co-extracted compounds.

The downstream processing plant must then be carefully designed to accommodate not only carotenoid purification and polishing, but also potential recovery of valuable byproducts. When essential fatty acids are among the co-extracted metabolites, their recovery can provide added economic value to the overall process (Mussagy et al. 2019). The specific design of the purification system will depend on the complete extract composition, requiring optimization for both primary product isolation and byproduct valorization (Saini and Keum 2018).

12.8.5 Carotenoids Characterization and Quantification Techniques

After the initial extraction of carotenoids from a microbial source, further analytical procedures are employed to characterize and quantify these pigments. The most frequently applied technique for carotenoid characterization is mass spectrometry (MS) (Rivera et al. 2014). However, MS does not differentiate between stereoisomers. In such cases, nuclear magnetic resonance (NMR) spectroscopy is employed, as it provides detailed information about the molecular structure and can differentiate between different forms of carotenoids (Cardoso et al. 2017b). Another method used for carotenoid characterization is near-infrared Fourier-transform Raman (NIR-FT-Raman) spectroscopy, which relies on the unique vibrational bands of these biomolecules (Sanchez et al. 2013). This technique is generally considered to be fast, nondestructive, and simple, allowing for the identification of predominant carotenoids without sample damage (Rivera et al. 2014; Sanchez et al. 2013).

Chromatographic methods are preferred for the quantification of specific carotenoid molecules, as they allow the separation of individual compounds (Barreiro and Barredo 2018). High-performance liquid chromatography (HPLC) is frequently used in combination with spectrophotometric techniques, providing detailed separation and identification of carotenoids (Arvayo-Enríquez et al. 2013).

HPLC is considered the most accurate technique for the identification and quantification of microbial carotenoids, but its success depends on several process parameters, including the type of detector, stationary phase, mobile phase, and other variables (Arvayo-Enríquez et al. 2013; Barreiro and Barredo 2018). The choice of detector significantly impacts the results, with UV-Vis (ultraviolet-visible) detectors being the most common.

Additional detector options include mass spectrometry and photodiode array detectors. The selection of an appropriate chromatographic column is essential and depends on particle size (smaller particles improve resolution), particle shape (affects separation efficiency), and porosity (influences interactions with carotenoids). C_{18}- and C_{30}-based columns are particularly effective for separating hydrophobic microbial carotenoids. Equally important is the mobile phase selection.

Common solvents include water, acetone, methanol, dichloromethane, isopropanol, and acetonitrile (Saini and Keum 2018). These can be used individually or in mixed solutions. For example, methanol/isopropanol mixtures separate most carotenoids (except lycopene), and acetone effectively resolves β-carotene, torularhodin, γ-carotene, and torulene (Lin et al. 2017; Jin et al. 2017; Al-Yafeai et al. 2018).

12.9 Commercial Applications

Carotenoids, which are vibrant pigments found in fruits, vegetables, and other natural sources, have gained significant commercial value for industrial applications worldwide due to their beneficial properties. They have a wide range of uses, from animal feed and food additives to nutraceutical and pharmaceutical applications (Jaswir et al. 2011). These biomolecules are commonly used as natural food colorants to enhance product appeal. In animal husbandry, carotenoids are added to livestock and poultry feed, where they improve animal health and enhance coloration (Valduga et al. 2009b). Moreover, these compounds possess valuable bioactive properties, including antioxidant effects, antitumor potential, and provitamin A activity (Cardoso et al. 2017b).

Certain carotenoids demonstrate particularly high commercial value, including lutein, β-carotene, lycopene, canthaxanthin, and astaxanthin (Yaqoob et al. 2021). In traditional animal feed applications, these pigments are typically obtained through direct consumption of natural sources such as fruits and plants, eliminating the need for specialized extraction or purification processes (Saini et al. 2015). However, microbial carotenoid production—derived from bacteria, fungi, or microalgae—requires dedicated extraction and purification methods to recover the pigments (Amaya et al. 2014).

Unlike synthetic alternatives, these pigments provide health benefits and have gained prominence due to the growing health consciousness and regulatory restrictions. Microbial-derived carotenoids demonstrate particularly valuable applications (Olson 1999). For example, microbial ß-carotene extracted from *Blakeslea trispora* has applications in the manufacturing of butter, margarine, cakes, milk products, and soft drinks (Martins et al. 2016). Astaxanthin, obtained from *Scenedesmus* sp. and *Xanthophyllomyces dendrorhous*, is used in the feed and nutraceutical industries. It enhances egg yolk pigmentation and contributes to the production of salmonids such as salmon and rainbow trout (Grewe et al. 2007). *Brevibacterium linens* can grow on the surface of Vieux-Pane cheese, producing pigments that effectively color the cheese under specific conditions (Galaup et al. 2015).

Carotenoids serve as coloring agents in tablets and other pharmaceutical formulations, as well as bioactive ingredients in cosmetics, providing protection against oxidation and UV radiation (Masaki 2010; Martins et al. 2016; Sathasivam and Ki 2018). In addition to being used as food additives and nutritional supplements, carotenoids are incorporated into nutraceuticals for disease prevention and improving human life quality (Jaswir et al. 2011; Kushwaha et al. 2014). Their unique functional and biological properties, derived from their chemical structure, make them powerful antioxidants that protect cells from oxidative damage, help reduce inflammation, and play a crucial role in preventing or treating tumors (Park et al. 2009). Some carotenoids can be converted into vitamin A, which is essential for vision and immune function. Due to their significant biological properties, the pharmaceutical industry has rapidly adopted natural carotenoids to enhance product quality and consumer health (Cardoso et al. 2017a; Shahidi and Ambigaipalan 2015).

However, their use is strictly regulated to ensure safety and efficacy, as is the case with other chemical and biological compounds (Martins et al. 2016). Before industrial application, microbial carotenoids must obtain approval from regulatory agencies such as the FDA (United States) and the EFSA (European Union). Therefore, the use of microbial carotenoids is governed by legislation that specifies purity standards, sources, and permissible quantities for each formulation or product (Jaswir et al. 2011).

The industrial production of carotenoids through biotechnological routes has been adopted by several global companies. These manufacturers utilize various microbial sources to produce different carotenoid types, including Astaxanthin from *Haematococcus pluvialis* (Cyanotech) and β-Carotene from *Dunaliella salina* (Parry Nutraceuticals). The global carotenoid market is dominated by producers from the United States and European nations. Additional significant contributions come from manufacturers in Italy, Spain, France, India, Japan, and Australia (Foong et al. 2021).

12.10 Conclusion

Carotenoids are vital pigments with diverse biological and industrial significance. Their production, whether extracted from plants or obtained through microbial fermentation, requires precise control of processes to ensure efficiency and yield, with innovations like metabolic engineering enhancing scalability. Carotenoids are diverse pigments present in plants and microorganisms, crucial for photosynthesis, antioxidant activity, and prevention of oxidative stress. These pigments are widely applied in the food industry, feed supplements, cosmetics, pharmaceuticals, and aquaculture. Industrial interest in microbial carotenoids like astaxanthin, β-carotene, lutein, and lycopene has surged due to sustainable biotechnological methods, which ensure higher yields, cost efficiency, and independence from climatic or soil conditions. The carotenoid market has witnessed substantial growth in recent years as a result of increasing awareness of carotenoids' health benefits, rising demand in food and beverages, the expanding nutraceutical sector, and their application in animal feed. Supported by government regulations favoring natural ingredients, carotenoids are particularly valued in health supplements for their antioxidant properties, aiding in disease prevention and overall wellness.

References

Adam P, Hecht S, Eisenreich W, Kaiser J, Gräwert T, Arigoni D, Bacher A, Rohdlich F (2002) Biosynthesis of terpenes: studies on 1-hydroxy-2-methyl-2-(E)-butenyl 4-diphosphate reductase. Proc Natl Acad Sci USA 99(19):12108–12113. https://doi.org/10.1073/pnas.182412499

Aguilar CP, González M, Cifuentes AS, Silva M (2004) Growth and accumulation of total carotenoids in two strains of *Dunaliella salina* Teod. (*Chlorophyceae*) from the northern and central coast of Perú. J Chil Chem Soc 49(1):69–74. https://doi.org/10.4067/S0717-97072004000100012

Al-Yafeai A, Malarski A, Böhm V (2018) Characterization of carotenoids and vitamin E in *R. rugosa* and *R. canina*: comparative analysis. Food Chem 242:435–442. https://doi.org/10.1016/j.foodchem.2017.09.004

Amaya E, Becquet P, Carné S, Peris S, Miralles P (2014) Carotenoids in animal nutrition. FEFANA Publication. Retrieved from FEFANA

Anastas PT, Warner JC (1998) Green chemistry: theory and practice. Oxford University Press, New York, p 30. https://doi.org/10.1093/oso/9780198506980.001.0001

Araya-Garay JM, Feijoo-Siota L, Rosa-Dos-Santos F, Veiga-Crespo P, Villa TG (2012) Construction of new *Pichia pastoris* X-33 strains for production of lycopene and β-carotene. Appl Microbiol Biotechnol 93(6):2483–2492. https://doi.org/10.1007/s00253-011-3764-7

Armstrong GA (1994) Eubacteria show their true colors: genetics of carotenoid pigment biosynthesis from microbes to plants. J Bacteriol 176(16):4795–4802. https://doi.org/10.1128/jb.176.16.4795-4802.1994

Arvayo-Enríquez H, Mondaca-Fernández I, Gortárez-Moroyoqui P, López-Cervantes J, Rodríguez-Ramírez R (2013) Carotenoids extraction and quantification: a review. Anal Methods 5(12):2916–2924. https://doi.org/10.1039/c3ay26295b

Avalos J, Pardo-Medina J, Parra-Rivero O, Ruger-Herreros M, Rodríguez-Ortiz R, Hornero-Méndez D, Limón MC (2015) Carotenoid biosynthesis in *Fusarium*. J Fungi 3(3):39. https://doi.org/10.3390/jof3030039

Barreiro C, Barredo J-L (2018) Microbial carotenoids: methods and protocols, vol 1852. Humana Press. https://doi.org/10.1007/978-1-4939-8742-9

BCC Research (2018) The global market for carotenoids (FOD025C). BCC Research. https://www.bccresearch.com/market-research/food-and-beverage/carotenoids-market-fod025c.html

Berman J, Zorrilla-López U, Farré G (2015) Nutritionally important carotenoids as consumer products. Phytochem Rev 14(6):727–743. https://doi.org/10.1007/s11101-014-9373-1

Bhataya A, Schmidt-Dannert C, Lee PC (2009) Metabolic engineering of *Pichia pastoris* X-33 for lycopene production. Process Biochem 44(10):1095–1102. https://doi.org/10.1016/j.procbio.2009.05.012

Bhosale P (2004) Environmental and cultural stimulants in the production of carotenoids from microorganisms. Appl Microbiol Biotechnol 63(4):351–361. https://doi.org/10.1007/s00253-003-1441-1

Bhosale P, Berstein PS (2005) Microbial xanthophylls. Appl Microbiol Biotechnol 68(4):445–455. https://doi.org/10.1007/s00253-005-0032-8

Boles BR, Singh PK (2008) Endogenous oxidative stress produces diversity and adaptability in biofilm communities. Proc Natl Acad Sci 105(34):12503–12508. https://doi.org/10.1073/pnas.0801499105

Britton G (1995) Structure and properties of carotenoids in relation to function. FASEB J 9(15):1551–1558. https://doi.org/10.1096/fasebj.9.15.8529834

Cardoso LAC, Kanno KYF, Karp SG (2017a) Microbial production of carotenoids: a review. Afr J Biotechnol 16(4):139–146. https://doi.org/10.5897/AJB2016.15763

Cardoso LAC, Karp SG, Vendruscolo F, Kanno KY, Zoz LI, Carvalho JC (2017b) Biotechnological production of carotenoids and their applications in food and pharmaceutical products. In: Carotenoids. InTech Open, London. https://doi.org/10.5772/67725

Carrillo-Lopez A, Yahia EM, Ramirez-Padilla GK (2010) Bioconversion of carotenoids in five fruits and vegetables to vitamin a measured by retinol accumulation in rat livers. Am J Agric Biol Sci 5(2):215–221. https://doi.org/10.3844/ajabssp.2010.215.221

Cheng YT, Yang CF (2016) Using strain *Rhodotorula mucilaginosa* to produce carotenoids using food wastes. J Taiwan Inst Chem Eng 61:270–275. https://doi.org/10.1016/j.jtice.2016.01.003

Cicci A, Sed G, Bravi M (2017) Potential of choline chloride-based natural deep eutectic solvents (NaDES) in the extraction of microalgal metabolites. Chem Eng Trans 57:61–66. https://doi.org/10.3303/CET1757011

Dannert CS, Umeno D, Arnold FH (2000) Molecular breeding of carotenoid biosynthetic pathways. Nat Biotechnol 18(7):750–753. https://doi.org/10.1038/77319

Das A, Yoon SH, Lee SH, Kim JY, Oh DK, Kim SW (2007) An update on microbial carotenoid production: application of recent metabolic engineering tools. Appl Microbiol Biotechnol 77(3):505–512. https://doi.org/10.1007/s00253-007-1206-x

De La Fuente JL, Rodriguez-Saiz M, Schleissner C, Diez B, Peiro E, Barredo JL (2010) High-titer production of astaxanthin by the semi-industrial fermentation of *Xanthophyllomyces dendrorhous*. J Biotechnol 148(2–3):144–146. https://doi.org/10.1016/j.jbiotec.2010.05.004

Dias Rodrigues TV, Amore TD, Teixeira EC, De Medeiros Burkert JF (2019) Carotenoid production by *Rhodotorula mucilaginosa* in batch and fed-batch fermentation using agroindustrial byproducts. Food Technol Biotechnol 57(3):388–398. https://doi.org/10.17113/ftb.57.03.19.6068

Domínguez-Bocanegra AR, Ponce-Noyola T, Torres-Muñoz JA (2007) Astaxanthin production by *Phaffia rhodozyma* and *Haematococcus pluvialis*: a comparative study. Appl Microbiol Biotechnol 75(4):783–791. https://doi.org/10.1007/s00253-007-0889-9

Fazeli MR, Tofighi H, Samadi N, Jamalifar H (2006) Effects of salinity on β-carotene production by *Dunaliella tertiolecta* DCCBC26 isolated from the Urmia salt lake, north of Iran. Bioresour Technol 97(18):2453–2456. https://doi.org/10.1016/j.biortech.2005.10.016

Fiedor J, Burda K (2014) Potential role of carotenoids as antioxidants in human health and disease. Nutrients 6(2):466–488. https://doi.org/10.3390/nu6020466

Foong LC, Loh CWL, Ng HS, Lan JCW (2021) Recent development in the production strategies of microbial carotenoids. World J Microbiol Biotechnol 37(1):12. https://doi.org/10.1007/s11274-020-02967-3

Frank HA, Cogdell RJ (1993) The photochemistry and functions of carotenoids in photosynthesis. In: Young A, Britton G (eds) Carotenoids in photosynthesis. Springer, London, pp 252–326. https://doi.org/10.1007/978-94-011-1884-1_10

Galaup P, Suthiwong N, Leclercq-Perlat MN, Valla A, Caro Y, Fouillaud M, Guérard F, Dufossé L (2015) First isolation of *Brevibacterium* sp. pigments in the rind of an industrial red-smear-ripened soft cheese. Int J Dairy Technol 68(1):144–147. https://doi.org/10.1111/1471-0307.12174

Goodwin TW (1980) The biochemistry of the carotenoids. Springer, Berlin. https://doi.org/10.1007/978-94-011-6489-3

Goss R, Lepetit B (2015) Biodiversity of NPQ. J Plant Physiol 172:13–32. https://doi.org/10.1016/j.jplph.2014.03.004

Granado F, Olmedilla B, Blanco I (2003) Nutritional and clinical relevance of lutein in human health. Br J Nutr 90(3):487–502. https://doi.org/10.1079/BJN2003927

Grewe C, Menge S, Griehl C (2007) Enantioselective separation of all-E-astaxanthin and its determination in microbial sources. J Chromatogr A 1166(1–2):97–100. https://doi.org/10.1016/j.chroma.2007.08.051

Gu Z, Deming C, Yongbin H, Zhigang C, Feirong G (2008) Optimization of carotenoids extraction from *Rhodobacter sphaeroides*. LWT Food Sci Technol 41(6):1082–1088. https://doi.org/10.1016/j.lwt.2007.07.017

Hata N, Ogbonna JC, Hasegawa Y, Taroda H, Tanaka H (2001) Production of astaxanthin by *Haematococcus pluvialis* in a sequential heterotrophic-photoautotrophic culture. J Appl Phycol 13(4):395–402. https://doi.org/10.1023/A:1011921329568

Haxo F (1950) Carotenoids of the mushroom *Cantharellus cinnabarinus*. Bot Gaz 112(2):228–232. https://doi.org/10.1086/335653

Heba M, Kanzy NF, Hoida AM (2015) Optimization of carotenoids production by yeast strains of *Rhodotorula* using salted cheese whey. Int J Curr Microbiol App Sci 4(1):456–469. https://doi.org/10.20546/ijcmas.2015.401.046

Henríquez V, Escobar C, Galarza J, Gimpel J (2016) Carotenoids in microalgae. In: Stange C (ed) Carotenoids in nature, Subcellular biochemistry, 79, pp 219–236. https://doi.org/10.1007/978-3-319-39126-7_8

Imamura T, Bando N, Yamanishi R (2006) β-carotene modulates the immunological function of RAW264, a murine macrophage cell line, by enhancing the level of intracellular glutathione. Biosci Biotechnol Biochem 70(9):2112–2120. https://doi.org/10.1271/bbb.60092

Jaswir I, Noviendri D, Hasrini RF, Octavianti F (2011) Carotenoids: sources, medicinal properties, and their application in food and nutraceutical industry. J Med Plants Res 5(33):7119–7131. https://doi.org/10.5897/JMPRx11.011

Jayaraj J, Devlin R, Punja Z (2008) Metabolic engineering of novel ketocarotenoid production in carrot plants. Transgenic Res 17(3):489–501. https://doi.org/10.1007/s11248-007-9120-0

Jin H, Lao YM, Zhou J, Zhang HJ, Cai ZH (2017) Simultaneous determination of 13 carotenoids by a simple C18 column-based ultra-high-pressure liquid chromatography method for carotenoid profiling in the astaxanthin-accumulating *Haematococcus pluvialis*. J Chromatogr A 1488:93–103. https://doi.org/10.1016/j.chroma.2017.01.079

Kim ZH, Kim SH, Lee HS, CG (2006) Enhanced production of astaxanthin by flashing light using *Haematococcus pluvialis*. Enzym Microb Technol 39:414–419. https://doi.org/10.1016/j.enzmictec.2005.11.041

Kim DY, Vijayan D, Praveen Kumar R, Han JI, Lee K, Park JY, Chang WS, Lee JS, Oh YK (2016) Cell-wall disruption and lipid/astaxanthin extraction from microalgae: *Chlorella* and *Haematococcus*. Bioresour Technol 199:300–310. https://doi.org/10.1016/j.biortech.2015.08.107

Kot AM, Błazejak S, Kieliszek M, Gientka I, Bryś J (2019) Simultaneous production of lipids and carotenoids by the red yeast *Rhodotorula* from waste glycerol fraction and potato wastewater. Appl Biochem Biotechnol 189:589–607. https://doi.org/10.1007/s12010-018-0293-7

Kumar SV, Morrell-Falvey J (2018) Loss of carotenoids impacts membrane protein and lipid distribution in *Pantoea sp. YR343*. Biophys J 114:98a. https://doi.org/10.1016/j.bpj.2017.11.553

Kushwaha K, Saini A, Saraswat P, Agarwal MK, Saxena J (2014) Colorful world of microbes: carotenoids and their applications. Adv Biol 2014:Article ID 837891. https://doi.org/10.1155/2014/837891

Kuzuyama T (2002) Mevalonate and nonmevalonate pathways for the biosynthesis of isoprene units. Biosci Biotechnol Biochem 66:1619–1627. https://doi.org/10.1271/bbb.66.1619

Landrum JT (2010) Carotenoids: physical, chemical, and biological functions and properties. CRC Press

Lavecchia R, Zuorro A (2008) Improved lycopene extraction from tomato peels using cell-wall degrading enzymes. Eur Food Res Technol 228:153–158. https://doi.org/10.1007/s00217-008-0897-8

Lin X, Gao N, Liu S, Zhang S, Song S, Ji C, Dong X, Su Y, Zhao ZK, Zhu B (2017) Characterization of carotenoid productions and profiles of three *Rhodosporidium toruloides* mutants from *Agrobacterium tumefaciens*-mediated transformation. Yeast 34:335–342. https://doi.org/10.1002/yea.3239

Liu YS, Wu JY, Ho KP (2006) Characterization of oxygen transfer conditions and their effects on *Phaffia rhodozyma* growth and carotenoid production in shake-flask cultures. Biochem Eng J 27(3):331–335. https://doi.org/10.1016/j.bej.2005.08.013

Luengo E, Martínez JM, Bordetas A, Álvarez I, Raso J (2015) Influence of the treatment medium temperature on lutein extraction assisted by pulsed electric fields from *Chlorella vulgaris*. Innov Food Sci Emerg Technol 29:15–22. https://doi.org/10.1016/j.ifset.2015.01.013

Maldonade R, Adilma R, Scamparini P, Rodriguez Amaya B (2007) Selection and characterization of carotenoid-producing yeasts from the Campinas region, Brazil. Braz J Microbiol 38(1):65–70. https://doi.org/10.1590/S1517-83822007000100012

Martins N, Roriz CL, Morales P, Barros L, Ferreira I (2016) Food colorants: challenges, opportunities, and current desires of agro-industries to ensure consumer expectations and regulatory practices. Trends Food Sci Technol 52:1–15. https://doi.org/10.1016/j.tifs.2016.03.009

Masaki H (2010) Role of antioxidants in the skin: anti-aging effects. J Dermatol Sci 58(2):85–90. https://doi.org/10.1016/j.jdermsci.2010.03.003

Mata-Gómez LC, Montañez JC, Méndez-Zavala A, Aguilar CN (2014) Biotechnological production of carotenoids by yeasts: an overview. Microb Cell Fact 13:1–12. https://doi.org/10.1186/1475-2859-13-12

Matthäus F, Ketelhot M, Gatter M, Barth G (2014) Production of lycopene in the non-carotenoid-producing yeast *Yarrowia lipolytica*. Appl Environ Microbiol 80(5):1660–1669. https://doi.org/10.1128/AEM.03434-13

Mehta BJ, Obraztsova IN, Cerda-Olmedo E (2003) Mutants and intersexual heterokaryons of *Blakeslea trispora* for production of β-carotene and lycopene. Appl Environ Microbiol 69(7):4043–4048. https://doi.org/10.1128/AEM.69.7.4043-4048.2003

Mezzomo N, Ferreira SRS (2016) Carotenoids: functionality, sources, and processing by supercritical technology: a review. J Chem 2016:1–16. https://doi.org/10.1155/2016/3164312

Misawa N, Shimada H (1998) Metabolic engineering for the production of carotenoids in non-carotenogenic bacteria and yeasts. J Biotechnol 59(3):169–181. https://doi.org/10.1016/S0168-1656(97)00171-4

Miura Y, Kondo K, Saito T, Shimada H, Fraser PD, Misawa N (1998) Production of the carotenoids lycopene, β-carotene, and astaxanthin in the food yeast *Candida utilis*. Appl Environ Microbiol 64(4):1226–1231. https://doi.org/10.1128/aem.64.4.1226-1231.1998

Mussagy CU, Winterburn J, Santos-Ebinuma VC, Pereira JFB (2019) Production and extraction of carotenoids produced by microorganisms. Appl Microbiol Biotechnol 103(3):1095–1114. https://doi.org/10.1007/s00253-018-9557-5

Namitha KK, Negi PS (2010) Chemistry and biotechnology of carotenoids. Crit Rev Food Sci Nutr 50(8):728–760. https://doi.org/10.1080/10408398.2010.499821

Naz T, Nosheen S, Li S, Nazir Y, Mustafa K, Liu Q, Garre V, Song Y (2020) Comparative analysis of β-carotene production by *Mucor circinelloides* strains CBS 277.49 and WJ11 under light and dark conditions. Meta 10(2):38. https://doi.org/10.3390/metabo10020038

Olson JA (1999) Carotenoids and human health. Arch Latinoam Nutr 49(1):7–11. https://doi.org/10.1080/07315724.1999.10718870

Olson JA, Krinsky NI (1995) Introduction: the colorful, fascinating world of the carotenoids: important physiologic modulators. FASEB J 15(1):1547–1550. https://doi.org/10.1096/fasebj.9.15.8529831

Orosa M, Franqueira D, Cid A, Abalde JJ (2005) Analysis and enhancement of astaxanthin accumulation in *Haematococcus pluvialis*. Bioresour Technol 96(3):373–378. https://doi.org/10.1016/j.biortech.2004.05.003

Park PK, Kim EY, Chu KH (2007) Chemical disruption of yeast cells for the isolation of carotenoid pigments. Sep Purif Technol 53(2):148–152. https://doi.org/10.1016/j.seppur.2006.06.020

Park SY, Nomura SMY, Murphy SP, Wilkens LR, Henderson BE, Kolonel LN (2009) Carotenoid intake and colorectal cancer risk: the multiethnic cohort study. J Epidemiol 19(2):63–71. https://doi.org/10.2188/jea.JE20081010

Pegklidou K, Mantzouridou F, Tsimidou MZ (2008) Lycopene production using *Blakeslea trispora* in the presence of 2-methyl imidazole: yield, selectivity, and safety aspects. J Agric Food Chem 56(12):4482–4490. https://doi.org/10.1021/jf800272k

Powls R, Britton G (1977) A series of mutant strains of *Scenedesmus obliquus* with abnormal carotenoid compositions. Arch Microbiol 113(3):275–280. https://doi.org/10.1007/BF00492035

Raja R, Hemaiswarya S, Rengasamy R (2007) Exploitation of *Dunaliella* for β-carotene production. Appl Microbiol Biotechnol 74(3):517–523. https://doi.org/10.1007/s00253-006-0777-8

Research and Markets (2024) Global carotenoids market report 2024: Booming dietary supplements and food & beverages markets drives carotenoids market growth – A $3.4 billion market by 2029. Retrieved from https://www.globenewswire.com/news release/2024/06/05/2893694/28124/en/Global-Carotenoids-Market-Report-2024-Booming-Dietary-Supplements-and-Food-Beverages-Markets-Drives-Carotenoids-Market-Growth-A-3-4-Billion-Market-by-2029.html

Rivera SM, Christou P, Canela-Garayoa R (2014) Identification of carotenoids using mass spectrometry. Mass Spectrom Rev 33(4):353–372. https://doi.org/10.1002/mas.21390

Román RB, Alvárez-Pez JM, Fernández FGA, Grima EM (2002) Recovery of pure b-phycoerythrin from the microalga *Porphyridium cruentum*. J Biotechnol 93(1):73–85. https://doi.org/10.1016/S0168-1656(01)00384-7

Sahadevan Y, Richter-Fecken M, Kaerger K, Voigt K, Boland W (2013) Early and late trisporoids differentially regulate β-carotene production and gene transcript levels in the Mucoralean fungi *Blakeslea trispora* and *Mucor mucedo*. Appl Environ Microbiol 79(23):7466–7475. https://doi.org/10.1128/AEM.02438-13

Saini RK, Keum YS (2017) Progress in microbial carotenoids production. Indian J Microbiol 57(2):129–130. https://doi.org/10.1007/s12088-017-0653-5

Saini RK, Keum YS (2018) Carotenoid extraction methods: a review of recent developments. Food Chem 240:90–103. https://doi.org/10.1016/j.foodchem.2017.07.099

Saini RK, Nile SH, Park SW (2015) Carotenoids from fruits and vegetables: chemistry, analysis, occurrence, bioavailability, and biological activities. Food Res Int 76(3):735–750. https://doi.org/10.1016/j.foodres.2015.07.047

Saini RK, Sivanesan I, Keum Y-S (2019) Emerging roles of carotenoids in the survival and adaptations of microbes. Indian J Microbiol 59(1):125–127. https://doi.org/10.1007/s12088-018-0772-7

Sanchez S, Ruiz B, Rodríguez-Sanoja R, Flores-Cotera LB (2013) Microbial production of carotenoids. In: Microbial production of food ingredients, enzymes and nutraceuticals. Woodhead Publishing, pp 194–233. https://doi.org/10.1533/9780857093547.2.194

Sathasivam R, Ki JS (2018) A review of the biological activities of microalgal carotenoids and their potential use in healthcare and cosmetic industries. Mar Drugs 16(1):26. https://doi.org/10.3390/md16010026

Schroeder WA, Johnson EA (1995) Singlet oxygen and peroxyl radicals regulate carotenoid biosynthesis in *Phaffia rhodozyma*. J Biol Chem 270(31):18374–18379. https://doi.org/10.1074/jbc.270.31.18374

Shahidi F, Ambigaipalan P (2015) Novel functional food ingredients from marine sources. Curr Opin Food Sci 2:123–129. https://doi.org/10.1016/j.cofs.2015.03.012

Shi XM, Jiang Y, Chen F (2002) High-yield production of lutein by the green microalga *Chlorella protothecoides* in heterotrophic fed-batch culture. Biotechnol Prog 18(4):723–727. https://doi.org/10.1021/bp020057h

Singh D, Barrow CJ, Mathur AS, Tuli DK, Puri M (2015a) Optimization of zeaxanthin and β-carotene extraction from *Chlorella saccharophila* isolated from New Zealand marine waters. Biocatal Agric Biotechnol 4:166–173. https://doi.org/10.1016/j.bcab.2015.09.006

Singh D, Gupta A, Wilkens SL, Mathur AS, Tuli DK, Barrow CJ, Puri M (2015b) Understanding response surface optimization to the modeling of astaxanthin extraction from a novel strain *Thraustochytrium* sp. Algal Res 11:113–120. https://doi.org/10.1016/j.algal.2015.05.020

Soong YHV, Liu N, Yoon S, Lawton C, Xie D (2019) Cellular and metabolic engineering of oleaginous yeast *Yarrowia lipolytica* for bioconversion of hydrophobic substrates into high-value products. Eng Life Sci 19(6):423–443. https://doi.org/10.1002/elsc.201800236

Sun J, Li H, Sun X, Yuan Q (2012) Trisporic acid stimulates gene transcription of terpenoid biosynthesis in *Blakeslea trispora*. Process Biochem 47(12):1889–1893. https://doi.org/10.1016/j.procbio.2012.07.015

Ungureanu C, Marchal L, Chirvase AA, Foucault A (2013) Centrifugal partition extraction: a new method for direct metabolites recovery from culture broth—case study of torularhodin recovery from *Rhodotorula rubra*. Bioresour Technol 132:406–409. https://doi.org/10.1016/j.biortech.2013.01.033

Valduga E, Valério A, Tatsch PO (2009a) Assessment of cell disruption and carotenoids extraction from *Sporidiobolus salmonicolor* (CBS 2636). Food Bioprocess Technol 2(3):234–238. https://doi.org/10.1007/s11947-008-0062-y

Valduga E, Tatsch PO, Tiggemann LT (2009b) Carotenoids production: microorganisms as a source of natural dyes. Quim Nova 32(9):2429–2436. https://doi.org/10.1590/S0100-40422009000900036

Vershinin A (1999) Biological functions of carotenoids: diversity and evolution. Biofactors 10(2–3):99–104. https://doi.org/10.1002/biof.5520100202

Vílchez C, Forján E, Cuaresma M (2011) Marine carotenoids: biological functions and commercial applications. Mar Drugs 9(3):319–333. https://doi.org/10.3390/md9030319

Vítek P, Ascaso C, Artieda O, Casero MC, Wierzchos J (2017) Discovery of carotenoid red-shift in endolithic cyanobacteria from the Atacama Desert. Sci Rep 7:11116. https://doi.org/10.1038/s41598-017-11460-y

Wackenroder H (1831) Geiger's pharmacy magazine 33

Walter MH, Strack D (2011) Carotenoids and their cleavage products: biosynthesis and functions. Nat Prod Rep 28(4):663–692. https://doi.org/10.1039/C0NP00036A

Wang W, Li J, Wang K, Huang C, Zhang Y, Olfi Eld E (2010) Organometallic mechanism of action and inhibition of the 4Fe-4S isoprenoid biosynthesis protein GcpE (IspG). Proc Natl Acad Sci U S A 107(25):11189–11193. https://doi.org/10.1073/pnas.1000869107

Wierzchos J, DiRuggiero J, Vítek P, Artieda O, Souza-Egipsy V, Škaloud P, Tisza M, Dávila AF, Vílchez C, Garbayo I, Ascaso C (2015) Adaptation strategies of endolithic chlorophototrophs to survive the hyperarid and extreme solar radiation environment of the Atacama Desert. Front Microbiol 6:Article 934. https://doi.org/10.3389/fmicb.2015.00934

Yabuzaki J (2017) Carotenoids database: structures, chemical fingerprints, and distribution among organisms. Database (Oxford) 2017:bax004

Yaqoob S, Riaz M, Shabbir A, Zia-ul-Haq M, Alwakeel SS, Bin-Jumah M (2021) Commercialization and marketing potential of carotenoids. In: Zia-ul-Haq M, Dewanjee S, Riaz M (eds) Carotenoids: structure and function in the human body. Springer, pp 799–826. https://doi.org/10.1007/978-3-030-46459-2_24

Yara-Varón E, Fabiano-Tixier AS, Balcells M, Canela-Garayoa R, Bily A, Chemat F (2016) Is it possible to substitute hexane with green solvents for extraction of carotenoids? A theoretical versus experimental solubility study. RSC Adv 6:27750–27759. https://doi.org/10.1039/C6RA01339B

Ye VM, Bhatia SK (2012) Pathway engineering strategies for production of beneficial carotenoids in microbial hosts. Biotechnol Lett 34(8):1405–1414. https://doi.org/10.1007/s10529-012-0919-4

Ye ZW, Jiang JG, Wu GH (2008) Biosynthesis and regulation of carotenoids in *Dunaliella*: progresses and prospects. Biotechnol Adv 26(4):352–360. https://doi.org/10.1016/j.biotechadv.2008.03.002

Yen HW, Sun CH, Ma TW (2011) The comparison of lutein production by *Scenedesmus* sp. in the autotrophic and the mixotrophic cultivation. Appl Biochem Biotechnol 164(3):353–361. https://doi.org/10.1007/s12010-010-9148-1

Young AJ (1991) The photoprotective role of carotenoids in higher plants. Physiol Plant 83(4):702–708. https://doi.org/10.1111/j.1399-3054.1991.tb02490.x

Zaccarim BR, De Oliveira F, Passarini MRZ, Duarte AWF, Sette LD, Jozala AF, De Carvalho SEV (2018) Sequencing and phylogenetic analyses of *Talaromyces amestolkiae* from the Amazon, a producer of natural colorants. Biotechnol Prog 35(1):e2684. https://doi.org/10.1002/btpr.2684

Zakar T, Herman E, Vajravel S, Kovacs L, Knoppova J, Komenda J, Domonkos I, Kis M, Gombos Z, Laczko-Dobos H (2017) Lipid and carotenoid cooperation-driven adaptation to light and temperature stress in *Synechocystis sp. PCC6803*. Biochimica et Biophysica Acta (BBA) – Bioenergetics 1858(5):337–350. https://doi.org/10.1016/j.bbabio.2017.03.003

Zheng YT, Toyofuku M, Nomura N, Shigeto S (2013) Correlation of carotenoid accumulation with aggregation and biofilm development in *Rhodococcus sp. SD-74*. Anal Chem 85(15):7295–7301. https://doi.org/10.1021/ac401188f

MIX
Papier aus verantwortungsvollen Quellen
Paper from responsible sources
FSC® C105338

If you have any concerns about our products,
you can contact us on
ProductSafety@springernature.com

In case Publisher is established outside the EU,
the EU authorized representative is:
Springer Nature Customer Service Center GmbH
Europaplatz 3, 69115 Heidelberg, Germany

Printed by Libri Plureos GmbH
in Hamburg, Germany